Vibration Analysis, Instruments, and Signal Processing

Jyoti Kumar Sinha

CRC Press is an imprint of the
Taylor & Francis Group, an **informa** business

CRC Press
Taylor & Francis Group
6000 Broken Sound Parkway NW, Suite 300
Boca Raton, FL 33487-2742

First issued in paperback 2020

ISBN-13: 978-1-4822-3144-1 (hbk)
ISBN-13: 978-0-367-73865-5 (pbk)

Library of Congress Cataloging-in-Publication Data

Sinha, Jyoti Kumar.
Vibration analysis, instruments, and signal processing / author, Jyoti Kumar Sinha.
pages cm
Includes bibliographical references and index.
ISBN 978-1-4822-3144-1 (hardback)
1. Vibration--Measurement. 2. Damping (Mechanics) I. Title.

TA355.S525 2014
620.3--dc23 2014032230

Visit the Taylor & Francis Web site at
http://www.taylorandfrancis.com

and the CRC Press Web site at
http://www.crcpress.com

Contents

Preface

Over the past four decades, the technology in vibration instrumentation and measurements, signal processing, and analytical simulation using finite element (FE) methods has advanced significantly. There are several dedicated books that have recorded these advancements. However, it has been consistently observed that several persons (students, researchers, designers, and maintenance personnel in industry) involved in, say, vibration-related works or research, do not fully comprehend the interrelation between theory and experiments. These individuals can be grouped as (1) good in vibration data collection but may not be aware of the applicable basic theory, and vice versa, (2) good in signal processing but may not know the basics of either the theory or vibration data collection and measurement procedures, and (3) involved in dynamic qualifications (FE analysis and modal testing) using standard commercially available software without knowing much about the basic principles and methods. It is imperative that persons involved in vibration-based analysis have at least a basic understanding of the different processes so that they can more effectively solve vibration-related problems.

This book aims to communicate the fundamental principles of all three facets of vibration-based analysis (i.e., instruments and measurement, signal processing, and theoretical analysis) in a simplified tutorial manner, which is not readily available in literature. The unique content of this book will therefore be very useful for a diverse audience who are interested in vibration analysis. The target audience includes students (all levels), researchers, and engineers (involved in vibration-based condition monitoring). There are several chapters related to day-to-day requirements in vibration measurements and analysis so that the reader may become aware of the basics. He or she can then consult the many dedicated books in an area of interest if needed. A chapter is also added for real-life case studies relating the basic theory, types of measurements needed, and requisite signal processing that ultimately lead to effective fault diagnosis.

The author acknowledges his teachers and colleagues, particularly Prof. M.I. Friswell, Prof. Arthur Lees, Mr. K.K. Meher, Prof. R.I.K. Moorthy, Prof. P.M. Mujumdar, Mr. A. Rama Rao, Mr. R.K. Sinha, and Mr. B.C.B.N. Suryam. I am really thankful to Prof. R.I.K. Moorthy for his guidance at the beginning of my professional career and in research in the area of structural dynamics and vibration. I also acknowledge a number of the postgraduate students with industrial experience in the "Maintenance (Reliability) Engineering and Asset Management (MEAM/REAM)" MSc course and my diligent PhD students for their useful suggestions during the book's preparation.

Finally, I dedicate this book to my parents, Mr. Jagdish Prasad Sinha and Mrs. Chinta Sinha, my wife, Sarita Sinha, and my son, Aarambh Sinha. My parents stay miles away from me but give unconditional love to me every moment of my life. My wife and son are my real strength, with whom I share all the moments of my life.

Jyoti K. Sinha

About the Author

Dr. Jyoti K. Sinha joined the School of MACE, University of Manchester, in January 2007. He previously worked as a senior research scientist in the Bhabha Atomic Research Centre, India, for nearly 16 years. He also worked as a research fellow in Cranfield University, UK, for 18 months. He earned his bachelor's degree (Mech. Eng.) from BIT, Sindri (India), and master's degree (Aerospace Eng.) from IIT Bombay, Mumbai (India). He completed his PhD from the University of Wales Swansea (now Swansea University), UK.

His research area is in vibration and structural dynamics, using both experiments and analytical methods, including finite element (FE) model updating, vibration control, and health monitoring techniques. He is the recipient of the prestigious "Young Scientist (Boyscast) Fellowship" from the Department of Science and Technology, India, for his outstanding work in vibration diagnosis and structural dynamics related to the number of rotating machines and structural components of nuclear power plants (NPPs). Dr. Sinha has been extensively involved in a number of projects (nearly £2.5M) related to vibration and dynamics. He has also been instrumental in the development of a number of innovative techniques in his research area. He is the author of more than 60 technical reports and 135 technical papers in both international journals and conferences combined. He has delivered a number of invited lectures. Dr. Sinha is also the associate editor of two international journals, *Structural Health Monitoring: An International Journal* and *Advances in Vibration Engineering* (now *Journal of Vibration Engineering and Technology*), editorial board member of the journal *Structural Monitoring and Maintenance*, coauthor of two books, and a member of the Technical Committee on Rotor Dynamics, IFTOMM (2011 onward).

In academia, nine PhD and EngD students have completed their programs since January 2007 under Dr. Sinha's supervision in the different areas of vibration, rotordynamics, and signal processing. He has completed supervision of nearly 75 MSc dissertations, and he currently supervises a number of PhD students and MSc dissertations. Presently, Dr. Sinha is the principal investigator of different industrial projects related to steam and wind turbines in the areas of vibration-based condition monitoring, fault diagnosis, and life prediction.

1

Introduction

1.1 Introduction

Vibration measurements, tests, and analyses are becoming popular tools for several applications, a few of which are listed below:

- Design qualification and optimization
- Machine installation
- Health monitoring
  - Mechanical and civil structures
  - Machines
  - Human body, electrical panels, etc.

Over the decades, a significant advancement has been made in the experimental facilities and capabilities in terms of vibration instruments, sensors, and the computation power to analyze the experimental data. However, to achieve the requirements listed above, one needs a better understanding of dynamics and vibration behavior by both the experiments and theory. The correlation of the experimental observations with the theory is essential to suggesting a simple but effective and reliable solution. It has been observed that the following four aspects are very important to tackle any problem:

1. Understand the object(s) (e.g., structures, machines) and the objectives.
2. What kind of vibration instruments and experiments are required for measurements? This depends on the history of previous failures, malfunctioning in case of machinery, etc.
3. What kind of data processing is required for the measured vibration signals?

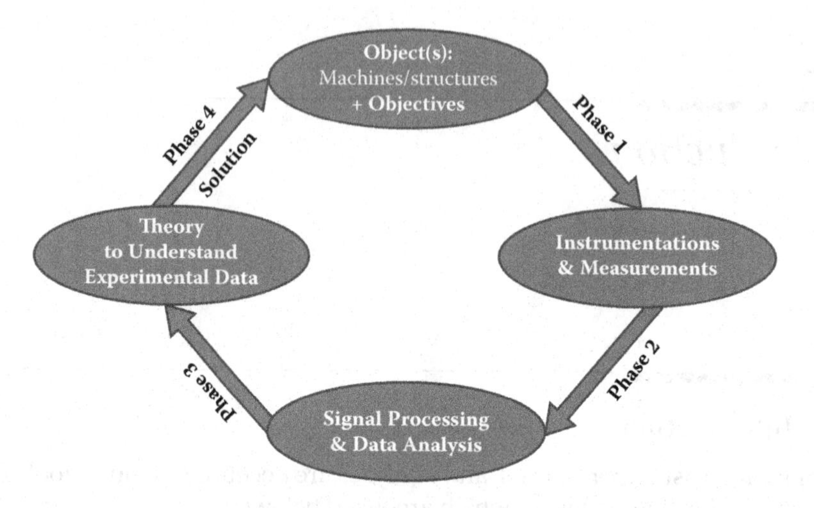

FIGURE 1.1
Typical abstract representations of different steps for analyzing any dynamic system.

4. Meet the expectation of an easy remedial suggestion (solution). In fact, it is always observed and experienced that a kind of "magic solution" is expected. It is only possible if the measurements, data processing, and their correlation to the theory are fully understood.

This book aims to communicate the above facets in a simplified tutorial manner, which is not readily available in a book. An abstract form of representation of the above facets is also illustrated in Figure 1.1 for better understanding.

1.2 Layout of the Chapters

The complete book is presented in 10 chapters (including Chapter 1) and aims to address a number of topics often useful for many day-to-day analyses and experiments in an illustrative and tutorial style, so the basic concept can be understood in a much better way. Although the book is divided into 10 chapters, it can be grossly classified into the following four categories.

1. Basic theories and analysis methods
2. Instrumentations and signal processing methods
3. Combined analysis and experimental approaches
4. Case studies relating 1 to 3

1.2.1 Basic Theories and Analysis Methods

Chapters 2 to 4 provide the basic understanding and concept of the vibration theory, mathematical modeling of structures and machines using the finite element (FE) method, and the vibration response computation using the FE model for the load applied.

Chapter 2 discusses a simplified vibration theory through a single degree of freedom (SDOF) system of a mass and a spring. A better understanding of vibration theory on a simplified system like a spring-mass system is often useful to understand several complex problems related to structures and machines.

Chapter 3 is on finite element (FE) modeling. This chapter introduces the concept of FE modeling at a very basic level through a few simple examples. This may provide a better concept and understanding even when developing an FE model using the commercial FE codes. The concept of the theoretical modal analysis using the FE model and eigenvalues and eigenvectors analysis is also discussed to introduce the concept of the mode shape at a natural frequency.

Chapter 4 is related to the use of the FE model discussed in Chapter 3. It discusses how the equation of motion in matrix form for any system can be integrated to solve for the responses (displacement, velocity, and acceleration) at all DOFs due to the time-varying external loadings. Broadly, two different approaches, the direct integration (DI) method and the mode superposition (MS) method, are used in practice and discussed in this chapter. The step-by-step concepts through a few examples are used to explain both methods. The Newmark-β method is used as the integration method for solving the dynamic equation of motion in matrix form for both the DI and MS methods to compute the responses.

1.2.2 Instrumentations, Signal Processing, and Experimental Methods

This book contains three exclusive chapters related to instrumentation, signal processing, and modal experiments, respectively. These are important elements for any vibration experiment.

Chapter 5 introduces the basic concept and working principles of different vibration sensors that include the proximity probes for displacement measurement, seismometers, accelerometers, and force sensors with their specifications. The concept of the experimental setup for the vibration measurements, experiments, and data collection is explained systematically for any experimental work. The basic understanding of the analogue-to-digital conversion (ADC) and the sampling frequency are discussed through the aid of a number of examples. The data acquisition (DAQ) device used to collect the experimental data in digital form from the analogue signals from the different sensors is also discussed so that appropriate DAQ and setting selection is possible for the experimental data collection.

Chapter 6 focuses on signal processing. It is a very important step to get the meaningful outputs from the measured vibration signals. Signal processing generally used in vibration, in both time and frequency domains, is explained through a number of vibration signals. A simplified concept for the fast Fourier transformation (FFT) and its use in the spectrum, cross-spectrum, frequency response function (FRF), and coherence are discussed. The use of the window function and averaging process while estimating the spectrum, FRF, etc., is explained through the aid of a number of signals. The usefulness of filters and the aliasing effect during data collection is also explained in this chapter.

Chapter 7 deals with experimental modal testing and analysis. Conducting modal experiments on any structure, machine, or equipment may be a straightforward procedure, but the data analysis needed to extract the natural frequencies, mode shapes, and modal damping accurately is a complex process. A number of modal analysis software codes are already available commercially to do such data analysis. In this chapter, a very simplified test procedure and the data analysis based on the theory of a SDOF system are outlined to extract the modal properties. It is generally observed to be an effective practical approach for many applications, including industrial structures and machines.

1.2.3 Combined Analysis and Experimental Methods

FE model updating and vibration-based condition monitoring and diagnosis require the knowledge of both analysis (Section 1.2.1) and experiments (Section 1.2.2). These are presented in the following chapters. The basic theories are explained through examples.

Chapter 8 describes the FE model updating method. The development of an FE model (Chapter 3) for any structure or machine is generally based on the material properties, physical dimensions, and ideally assumed boundary conditions. Such an a priori FE model may not be a true reflection of in situ dynamic behavior, and hence the FE model needs to be updated to get the model close to in situ behavior. The FE model updating approach uses the experimental modal parameters to update the model using updating parameters. The step-by-step procedure of this updating method is explained through a few simple laboratory examples and a case study.

Chapter 9 is dedicated to vibration-based condition monitoring, which is one of the most popular topics for monitoring the health of any structure or machine. A simplified concept about the requisite instrumentation, data collection, data analysis, diagnosis features, and the related International Organization for Standardization (ISO) codes is discussed in the chapter, mainly related to rotating machines.

1.2.4 Case Studies

Chapter 10 is exclusively on a few case studies, from laboratory examples to industrial examples, in order to provide a clear concept for analyzing different problems and the usefulness and relation of Sections 1.2.1 to 1.2.3 in each case study, and then how to achieve the magic solution to any problem.

2

Single Degree of Freedom (SDOF) System

2.1 A Single Degree of Freedom (SDOF) System

A *single degree of freedom (SDOF) system* is explained in this chapter. An abstract representation of a SDOF system is shown in Figure 2.1. It consists of a spring (k) and a mass (m) with the assumption that the mass has a SDOF to oscillate in the y-direction only as per the figure. While at rest, the system is said to be in static equilibrium under gravity. However, when the system is disturbed from rest, it oscillates about the static equilibrium position along the y-direction. In dynamics, only the effect of the external disturbance other than gravity on the system is studied. Hence, the dynamic behavior of a structure, in general, remains unaffected due to its orientation in the space. For example, the behavior of a cantilever beam will always be the same, whether its orientation is vertical or horizontal or in an inclined plane. Although a SDOF system is the simplest form of any structure or machine, it is a key element for understanding the dynamics and vibration behavior of even a complex system. Hence, it is important to understand the vibration theory involved in such a simple SDOF system.

2.2 Equation of Motion

To analyze a static problem, the usual practice is to balance the forces and moments acting on the static system. Similarly, for a dynamic system where the system status is changing with time, the vibrating system (or object) is assumed to be in *dynamic equilibrium* at a point of time by applying a fictitious force to bring the system to rest. This fictitious force is called an *inertia force*. The complete concept is known as the D'Alembert principle, which is based on Newton's second law of motion. The equation of motion at a time, t, for the SDOF system shown in Figure 2.1 is written as

$$F_i(t) + F_s(t) = 0 \tag{2.1}$$

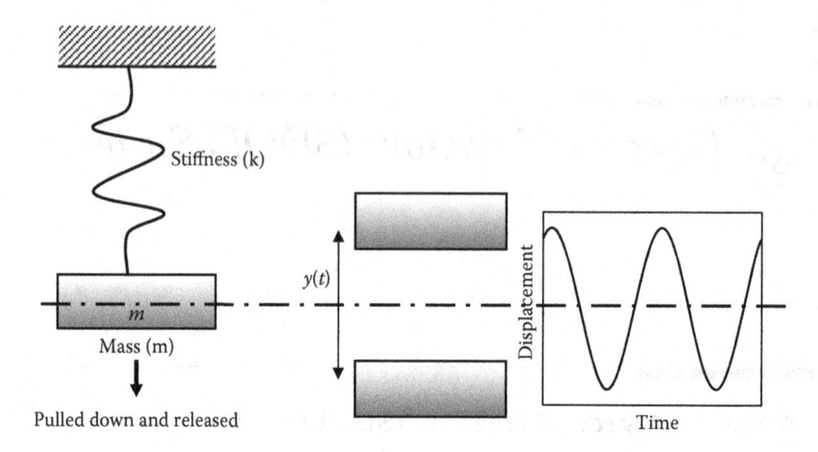

FIGURE 2.1
A SDOF system and its behavior during free vibration.

where $F_i(t)$ and $F_s(t)$ are the inertia force and the stiffness force at a time t, respectively. So Equation (2.1) becomes

$$m\ddot{y}(t) + ky(t) = 0 \tag{2.2}$$

It is known that the system undergoes cyclic (harmonic) motion; therefore, Equation (2.2) may be written as

$$\ddot{y}(t) + \omega_n^2 y(t) = 0 \tag{2.3}$$

where $\omega_n = \sqrt{\dfrac{k}{m}}$ is known as the circular (or angular) frequency. So, the period of oscillation is

$$T = \frac{2\pi}{\omega_n} \tag{2.4}$$

The frequency related to time period, T, is

$$f_n = \frac{1}{T} = \frac{\omega_n}{2\pi} = \frac{1}{2\pi}\sqrt{\frac{k}{m}} \tag{2.5}$$

where f_n is called the *natural frequency* of the system, as it depends on the properties (k, m) of the system. The unit of frequency is cycle/s or Hz (Hertz).

The general solution of Equation (2.3), which is a homogeneous second-order linear differential equation, is given by

$$y(t) = A \sin \omega_n t + B \cos \omega_n t \tag{2.6}$$

where the constants A and B are evaluated from initial conditions; $y(t=0)=y(0)$ and $\dot{y}(t=0)=\dot{y}(0)$. Therefore, Equation (2.6) becomes

$$y(t)=\frac{\dot{y}(0)}{\omega_n}\sin\omega_n t+y(0)\cos\omega_n t \tag{2.7}$$

2.2.1 Example 2.1: SDOF System

A SDOF system consists of a spring of stiffness, k = 3947.8 N/m, attached to a mass; m = 1 kg is considered here. The natural frequency (f_n) therefore becomes equal to 10 Hz using Equation (2.5). It is assumed that the mass is now displaced by a distance $y(0)$ = 1, with the initial velocity $\dot{y}(0)=0$; then the oscillatory response of the mass in the SDOF system can be computed using Equation (2.7) (see Figure 2.2). This indicates the system will continue to oscillate with the time period of the oscillation, $T=1/10=0.1$ s for an infinite duration.

2.2.2 Example 2.2: A Massless Bar with a Tip Mass

The system shown in Figure 2.3(a) consists of a massless steel bar of length 1 m and diameter 10 mm. A mass of 1 kg is attached to the tip of the bar.

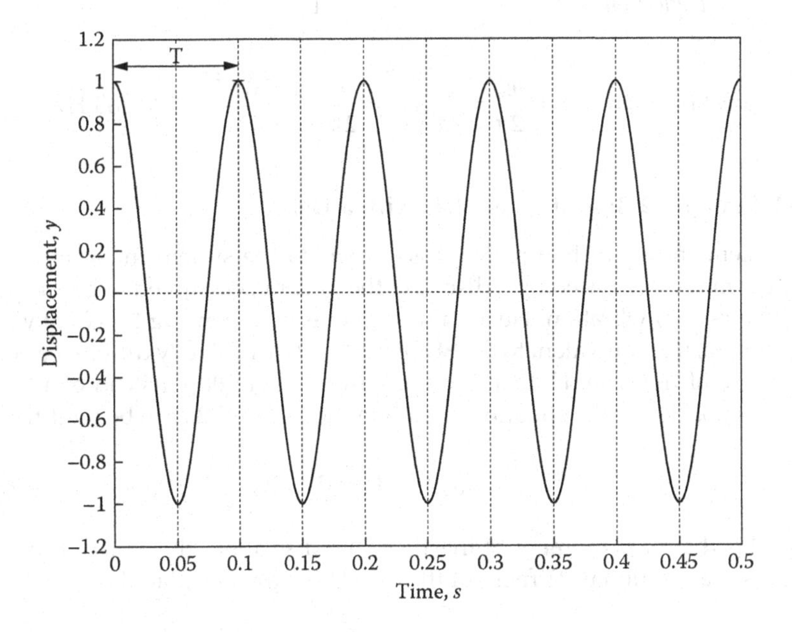

FIGURE 2.2
Response of a SDOF system of Example 2.1.

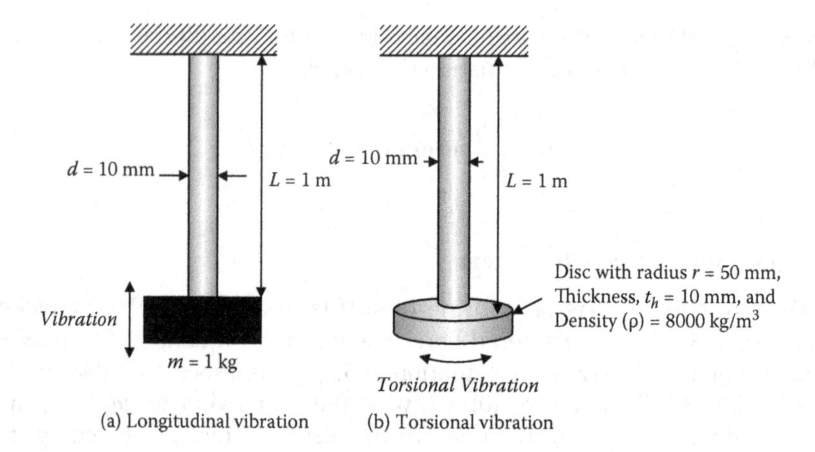

FIGURE 2.3
Two typical SDOF systems: (a) Example 2.2: Longitudinal vibration. (b) Example 2.3: Torsional vibration. System dynamic behavior at resonance.

This system is also equivalent to a SDOF system. The bar is acting as a spring with stiffness (k) that can be estimated from Hooke's law of elasticity as

$$k = \frac{Force}{Deflection} = \frac{EA}{L} = \frac{E\left(\frac{\pi}{4}d^2\right)}{L} = \frac{2e11\left(\frac{\pi}{4}0.01^2\right)}{1} = 1.5708e7 \text{ N/m} \tag{2.8}$$

The natural frequency $f_n = \frac{\omega_n}{2\pi} = \frac{1}{2\pi}\sqrt{\frac{k}{m}} = \frac{1}{2\pi}\sqrt{\frac{1.5708e7}{1}} = 630.783$ Hz.

2.2.3 Example 2.3: A Massless Bar with a Disc

The system shown in Figure 2.3(b) is similar to the system in Example 2.2. Here the mass is replaced by a disc, and the system is assumed to be a typical case of torsional vibration. Hence, the system is experiencing torsional vibration only. So, it is equivalent to the SDOF system with the only difference being the degree of freedom. Here, it is an angular motion ($\theta(t)$) instead of a linear motion ($y(t)$). Hence, the equation of motion (Equation (2.2)) can be modified as

$$I\ddot{\theta}(t) + k_\theta\theta(t) = 0 \tag{2.9}$$

where I is the mass moment of inertia of the disc about the axis of rotation, and k_θ is the rotational stiffness of the bar. These are calculated as

$$\text{Mass moment of inertia of the disc, } I = \frac{1}{2}mr^2 \tag{2.10}$$

Rotational stiffness, $k_\theta = \frac{Torsion}{Rotation} = \frac{GJ}{L}$ (basis similar to Equation (2.8)) (2.11)

where $J = \frac{\pi}{32}d^4$ is the area moment of inertia of the cross section of the bar about the axis of twist (rotation), and m is the mass of the disc. G is the shear modulus, which is equal to 75 GPa. The solution of Equation (2.9) is similar to that of Equation (2.2), and hence the torsional natural frequency of the bar with a disc is estimated as

$$f_{n,\theta} = \frac{1}{2\pi}\sqrt{\frac{k_\theta}{I}} \tag{2.12}$$

The example is just trying to highlight that torsion study is also similar to linear motion study.

Given data, disc density $\rho = 8000\,\text{kg/m}^3$, $I = \frac{1}{2}mr^2 = \frac{1}{2}(\rho\pi r^2 t_h)r^2 = 7.854e{-}4\,\text{kg-m}^2$,

$J = \frac{\pi}{32}d^4 = \frac{\pi}{32}0.01^4 = 9.817e{-}10\,\text{m}^4$, $k_\theta = \frac{GJ}{L} = \frac{(75e9)(9.817e-10)}{1} = 73.631\,\text{Nm/rad}$.

Hence, the torsional natural frequency $f_{n,\theta} = \frac{1}{2\pi}\sqrt{\frac{k_\theta}{I}} = \frac{1}{2\pi}\sqrt{\frac{73.631}{7.854e-4}}$ = 48.731 Hz.

2.3 Damped SDOF System

In practice, none of the real systems can freely oscillate with the same amplitude of vibration for an infinite time without the aid of any external source of excitation (see Example 2.1). The natural tendency of any system is to decay down to the equilibrium position after a few oscillations when the system is disturbed from its natural position. This phenomenon simply indicates that there is some kind of energy dissipation during oscillation. This inherent tendency is classified as damping of the system in the domain of structural dynamics. Hence, the equivalent dynamic system of a spring and mass SDOF system is often represented as shown in Figure 2.4(a), where the dashpot is used to represent the damping in the system. Figure 2.4(b) typically shows the decay motion during the free vibration of the system.

2.3.1 Equation of Motion for Free Vibration

The equation of motion (Equation (2.1)) for the damped system undergoing free vibration when disturbed from an initial equilibrium position at time t is now written as

$$F_i(t) + F_c(t) + F_s(t) = 0 \tag{2.13}$$

$$\Rightarrow \text{Inertia force} + \text{Damping force} + \text{Stiffness force} = 0$$

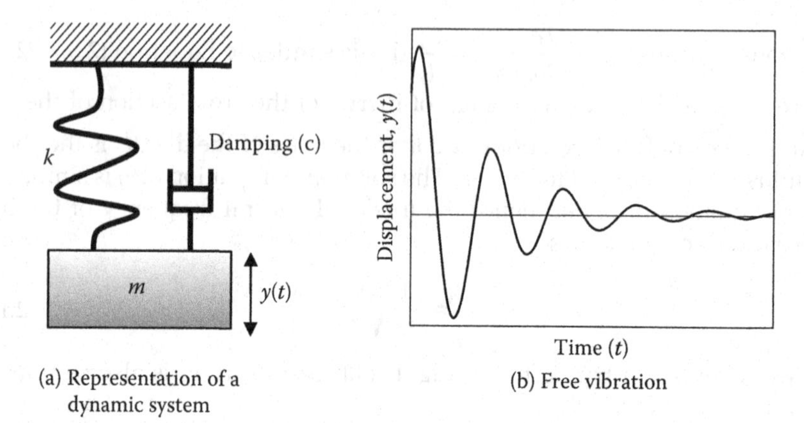

FIGURE 2.4
A damped SDOF system and its behavior. (a) Representation of a dynamic system. (b) Free vibration.

where the additional term, $F_c(t)$, is called the damping force, which is assumed to be dependent on the velocity of the vibration, and is defined as $F_c(t) = c\dot{y}(t)$, where c is the damping constant of the system. Hence, Equation (2.13) can be written as

$$m\ddot{y}(t) + c\dot{y}(t) + ky(t) = 0 \tag{2.14}$$

Let us assume the solution of Equation (2.14) in the form

$$y(t) = y_0 e^{st} \tag{2.15}$$

where y_0 and s are constants. Upon substitution, Equation (2.14) becomes

$$(ms^2 + cs + k)\, y_0 e^{st} = 0 \tag{2.16}$$

Equation (2.16) will be satisfied for all values of t when

$$ms^2 + cs + k = 0$$

$$\Rightarrow s^2 + \frac{c}{m}s + \frac{k}{m} = 0 \tag{2.17}$$

Equation (2.17) has two roots:

$$s_{1,2} = -\frac{c}{2m} \pm \sqrt{\left(\frac{c}{2m}\right)^2 - \frac{k}{m}} \tag{2.18}$$

Hence, the general solution of Equation (2.15) is given by

$$y(t) = Ae^{s_1 t} + Be^{s_2 t} \tag{2.19}$$

where A and B are constants that are estimated from the initial conditions, $y(t = 0) = y(0)$ and $\dot{y}(t = 0) = \dot{y}(0)$, of the system. However, the system behavior is completely dependent on the discriminant factor (DF) $= \sqrt{\left(\frac{c}{2m}\right)^2 - \frac{k}{m}}$ of the roots s_1 and s_2. All three possibilities are discussed here.

2.3.2 Critically Damped System: Case 1: Limiting Case When *DF* = 0

The expression $\sqrt{\left(\frac{c}{2m}\right)^2 - \frac{k}{m}} = 0$ gives $c = 2m\sqrt{\frac{k}{m}} = 2m\omega_n = 2\sqrt{km}$. This damping is called *critical damping*, c_c. For this case, both roots are equal. Hence, solution (2.19) becomes

$$y(t) = (A + Bt)e^{st} \tag{2.20}$$

where $s = s_1 = s_2 = -\frac{c}{2m}$. In practice, the damping (c) of the system is expressed in terms of the critical damping (c_c) by a nondimensional number (ζ), called the *damping ratio*. The root, s, is now written as

$$s = -\frac{c}{2m} = -\frac{c}{c_c}\frac{c_c}{2m} = -\zeta\frac{2m\omega_n}{2m} = -\zeta\omega_n \tag{2.21}$$

Here, the damping, c, is equal to the critical damping, c_c, i.e., $\zeta = 1$, and the root, $s = -\omega_n$. Upon substitution, Equation (2.20) is written as

$$y(t) = (A + Bt)e^{-\omega_n t} \tag{2.22}$$

Substitution of initial conditions ($y(0)$ and $\dot{y}(0)$) in Equation (2.22) gives

$$y(t) = (y(0) + (\dot{y}(0) + \omega_n y(0))t)e^{-\omega_n t} \tag{2.23}$$

For Example 2.1 of Section 2.2.1 with the damping ratio $\zeta = 1$, the response, $y(t)$, of the system is shown in Figure 2.5. There is no oscillatory motion in the system response if the system is disturbed from the equilibrium position. Hence, this condition is known as a *critically damped system* and is the limiting case between the oscillatory and nonoscillatory motion.

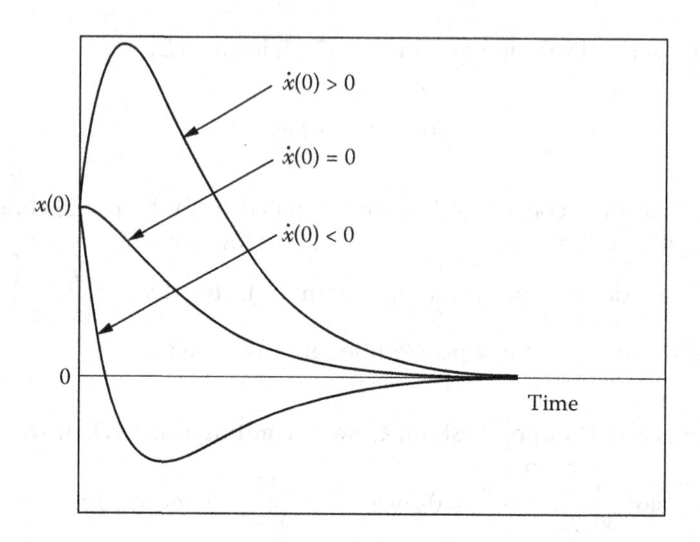

FIGURE 2.5
Response of the critically damped system.

2.3.3 Overdamped System: Case 2: When $DF \geq 0$

The expression $\sqrt{\left(\frac{c}{2m}\right)^2 - \frac{k}{m}} > 0$ gives $c > c_c$ and $\zeta > 1$. This system is called the *overdamped system,* where the two roots, s_1 and s_2, remain real. Since the damping ratio is more than 1, the motion will be nonoscillatory. The general solution in Equation (2.19) now becomes

$$y(t) = Ae^{\left(-\zeta+\sqrt{\zeta^2-1}\right)\omega_n t} + Be^{\left(-\zeta-\sqrt{\zeta^2-1}\right)\omega_n t} \tag{2.24}$$

where

$$A = \frac{\dot{y}(0) + \left(\zeta + \sqrt{\zeta^2 - 1}\right)\omega_n y(0)}{2\omega_n\sqrt{\zeta^2 - 1}} \quad \text{and} \quad B = \frac{-\dot{y}(0) - \left(\zeta - \sqrt{\zeta^2 - 1}\right)\omega_n y(0)}{2\omega_n\sqrt{\zeta^2 - 1}} \tag{2.25}$$

To illustrate the behavior of an overdamped system compared to a critically damped system, Example 2.1 of Section 2.1.1 is considered again with damping ratio $\zeta = 1$ and 2, but the same initial conditions are used for both cases. It is obvious from the responses shown in Figure 2.6 that the overdamped system requires a larger time ($t_{os} > t_{cs}$) to return to the equilibrium position than the critically damped system.

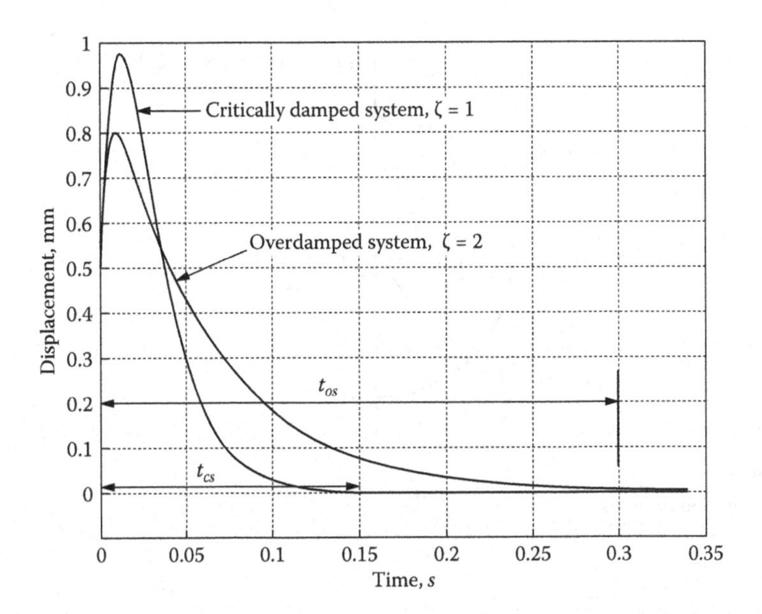

FIGURE 2.6
Comparison of dynamic behavior between an overdamped and a critically damped system.

2.3.4 Underdamped System: Case 3: When $DF \leq 0$

For this case, the expression $\sqrt{\left(\frac{c}{2m}\right)^2 - \frac{k}{m}} < 0$ gives $c < c_c$ and $\zeta < 1$. This is the case of an *underdamped system*, and the system will have an oscillatory motion. Hence, this case represents most of the real-life systems, but both roots do not remain real. The roots (Equation (2.18)) are written as

$$s_{1,2} = -\frac{c}{2m} \pm j\sqrt{\frac{k}{m} - \left(\frac{c}{2m}\right)^2} \tag{2.26}$$

$$\Rightarrow s_{1,2} = -\zeta\omega_n \pm j\sqrt{1-\zeta^2}\,\omega_n \tag{2.27}$$

where $j = \sqrt{-1}$ is an imaginary number. The roots are now complex conjugates. They are known as *poles* of the system in structural dynamics and are often represented in the real and imaginary axes as shown in Figure 2.7.

The roots (poles) are also expressed as

$$s_{1,2} = -\zeta\omega_n \pm j\omega_d \tag{2.28}$$

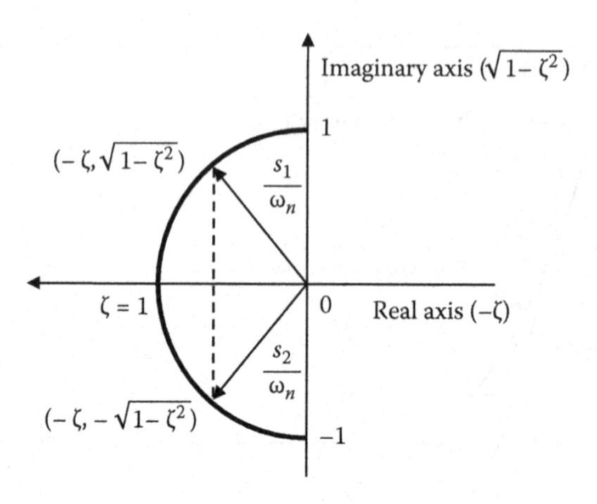

FIGURE 2.7
Representation of poles.

where $\omega_d = \omega_n\sqrt{1-\zeta^2}$ is the *damped natural frequency* of the SDOF system. The general solution (Equation (2.19)) becomes

$$y(t) = e^{-\zeta\omega_n t}(Ae^{j\sqrt{1-\zeta^2}\omega_n t} + Be^{-j\sqrt{1-\zeta^2}\omega_n t}) \tag{2.29}$$

$$\Rightarrow y(t) = e^{-\zeta\omega_n t}(Ae^{j\omega_d t} + Be^{-j\omega_d t}) \tag{2.30}$$

Equation (2.30) can further be written as

$$y(t) = e^{-\zeta\omega_n t}(C_1 \cos\omega_d t + C_2 \sin\omega_d t) \tag{2.31}$$

where $C_1 = y(0)$, $C_2 = \dfrac{\dot{y}(0) + \zeta\omega_n y(0)}{\omega_d}$. Equation (2.31) can also be written as

$$y(t) = Ye^{-\zeta\omega_n t}\sin(\omega_d t + \theta) \tag{2.32}$$

where $Y = \sqrt{C_1^2 + C_2^2}$, and the angle $\theta = \tan^{-1}\left(\dfrac{C_1}{C_2}\right)$.

Example 2.1 of Section 2.2.1 is again used here to illustrate the dynamic behavior of the *underdamped system*. Here, the damping ratio (ζ) is assumed to be 8%. The motion of the system when disturbed from the equilibrium position is shown in Figure 2.8.

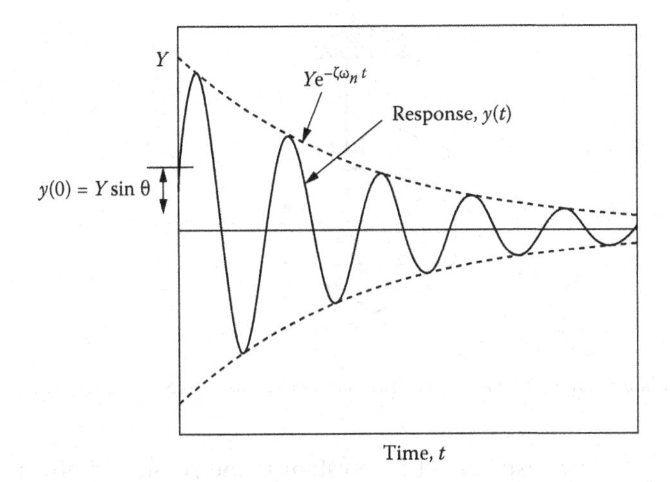

FIGURE 2.8
Response of the underdamped SDOF system.

2.4 Forced Vibration

When a system vibrates under the influence of an external force, this is called forced vibration. It is schematically shown in Figure 2.9. The equation of motion (2.13) is now modified as

$$F_i(t) + F_c(t) + F_s(t) = F(t) \tag{2.33}$$

$$\Rightarrow m\ddot{y}(t) + c\dot{y}(t) + ky(t) = F(t) \tag{2.34}$$

Hence, Equation (2.34) will have two solutions:

1. *Complimentary solution, $y_c(t)$,* when

$$m\ddot{y}(t) + c\dot{y}(t) + ky(t) = 0$$

   It is merely the response of the system under free vibration, and the solution, $y_c(t)$, is Equation (2.31) or (2.32).

2. *Particular integration due to the applied force, $y_p(t)$.* So the complete solution is given by

$$y(t) = y_c(t) + y_p(t) \tag{2.35}$$

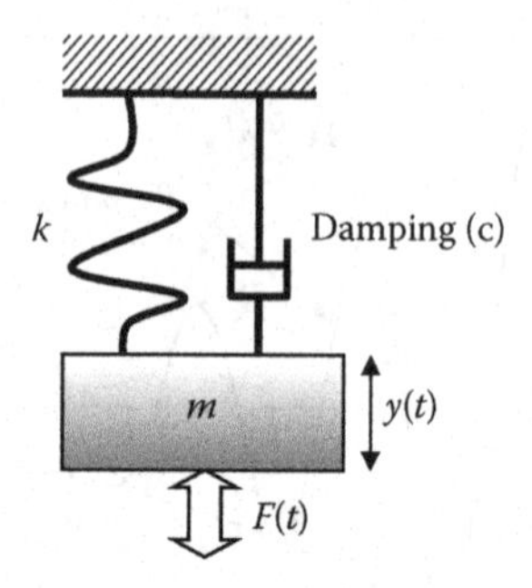

FIGURE 2.9
An underdamped SDOF system vibrating under an external force.

> The complete response of the system is the combined effect of the free vibration and the response due to the externally applied force. The response ($y_c(t)$) is called the *transient response*, whereas the response ($y_p(t)$) is the *steady-state response*.

For particular integration, the general solution is assumed to be $y_p(t) = Ye^{j\omega t}$ for the applied force, $F(t) = F_0 e^{j\omega t}$, where ω is the forcing frequency in rad/s. Upon substitution, Equation (2.34) becomes

$$(-m\omega^2 + jc\omega + k)Ye^{j\omega t} = F_0 e^{j\omega t} \tag{2.36}$$

Hence, the displacement, $y_p(t)$, is given by

$$y_p(t) = \frac{F_0 e^{j\omega t}}{(k - m\omega^2) + jc\omega} \tag{2.37}$$

$$\Rightarrow y_p(t) = \frac{(F_0/k)}{\left(1 - \left(\frac{\omega}{\omega_n}\right)^2\right) + j\left(2\zeta\frac{\omega}{\omega_n}\right)} e^{j\omega t} \tag{2.38}$$

Equation (2.38) is a complex quantity. Any complex quantity ($a + jb$) can be expressed as $Ae^{j\theta}$, where, $A = \sqrt{a^2 + b^2}$, and $\theta = \tan^{-1}\left(\frac{b}{a}\right)$. Hence, Equation (2.38) is expressed as

$$y_p(t) = \frac{(F_0/k)e^{j\omega t}}{\sqrt{\left(1 - \left(\frac{\omega}{\omega_n}\right)^2\right)^2 + \left(2\zeta\frac{\omega}{\omega_n}\right)^2} e^{j\varphi}} \tag{2.39}$$

$$\Rightarrow y_p(t) = \frac{(F_0/k)}{\sqrt{(1-r^2)^2 + (2\zeta r)^2}} e^{j(\omega t - \varphi)} \tag{2.40}$$

$$\Rightarrow y_p(t) = Ye^{j(\omega t - \varphi)} \tag{2.41}$$

where $r = \frac{\omega}{\omega_n}$, Y is the steady-state response (displacement), and φ is the phase of the response with respect to the exciting force. The phase angle is given by

$$\varphi = \tan^{-1} \frac{2\zeta r}{(1-r^2)} \tag{2.42}$$

The vector relationship among all the four components of Equation (2.33) or (2.36) for the steady-state condition can now be represented as shown in Figure 2.10.

Equation (2.40) can further be written as

$$y_p(t) = \left(\frac{F_0}{k}\right) De^{j(\omega t - \varphi)} \tag{2.43}$$

where $\left(\frac{F_0}{k}\right)$ is equivalent to the static deflection of the system, and $D = \frac{1}{\sqrt{(1-r^2)^2 + (2\zeta r)^2}}$ is called an *amplification* or *magnification factor.* When the forcing frequency ($\omega = 2\pi f$) becomes equal to the system natural frequency ($\omega_n = 2\pi f_n$), the phenomenon is called *resonance,* the amplification in the response that takes place during each cycle of the oscillations of the system. The factor for the maximum possible amplification in the system response at resonance is given by

$$D = \frac{1}{2\zeta}, \text{ when } r = 1 \tag{2.44}$$

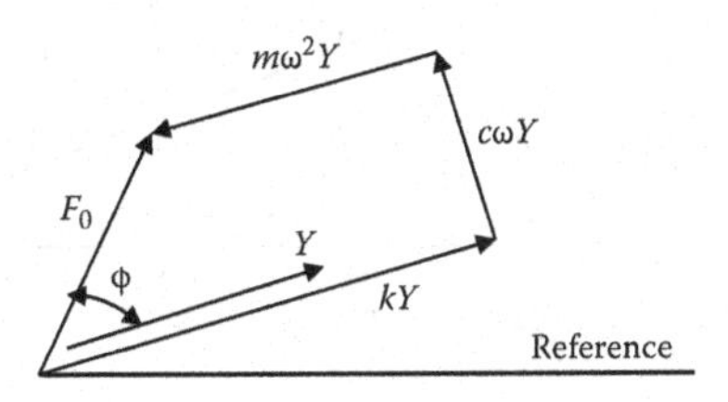

FIGURE 2.10
Graphical representation of the system dynamic behavior.

Hence, if a system with damping $\zeta = 0$, then $D \to \infty$, $y_p(t) \to \infty$, and $y(t) \to \infty$ over a period of oscillation irrespective of the amplitude of the exciting force and the system natural frequency. The buildup of such a vibration response at resonance for Example 2.1 of Section 2.1.1, when the damping $\zeta = 0$ and applied force $F(t) = \sin \omega_n t$, is shown in Figure 2.11. It can be seen from Figure 2.11 that the response is linearly increasing with each cycle of oscillation and will reach to infinity over a period of time. If the amplitude of the applied force is larger, then the system will reach an infinite displacement in a shorter time. It is important to note that the substitution of Equations (2.32) and (2.43) in Equation (2.35) may not lead to the vibration response shown in Figure 2.11, because the term D becomes equal to infinity for the damping $\zeta = 0$. It is better to use the method suggested in Chapter 4 for this solution of a SDOF system or any other suggested methods in the literature for the solving of this equation when damping $\zeta = 0$.

The buildup of the vibration response of the above system with the same applied force but damping $\zeta = 10\%$ is shown in Figure 2.12 for comparison. The displacement response is calculated using Equation (2.36). The transient and steady-state responses are clearly shown in Figure 2.12. Comparing Equations (2.41) and (2.43), the maximum steady-state response (displacement) is written as

$$Y = \frac{F_0}{k} D \tag{2.45}$$

For Example 2.1 with damping $\zeta = 10\%$, $Y = \dfrac{1}{3947.8}\left(\dfrac{1}{2\zeta}\right) = 1.266$ mm.

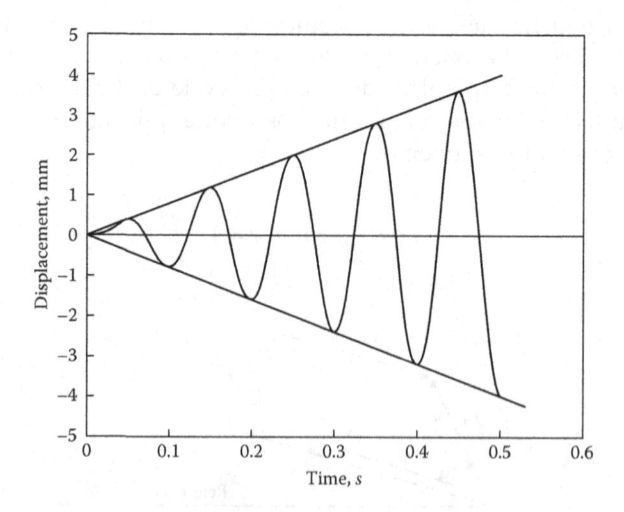

FIGURE 2.11
Response buildup of a system at the resonance when $\zeta = 0$.

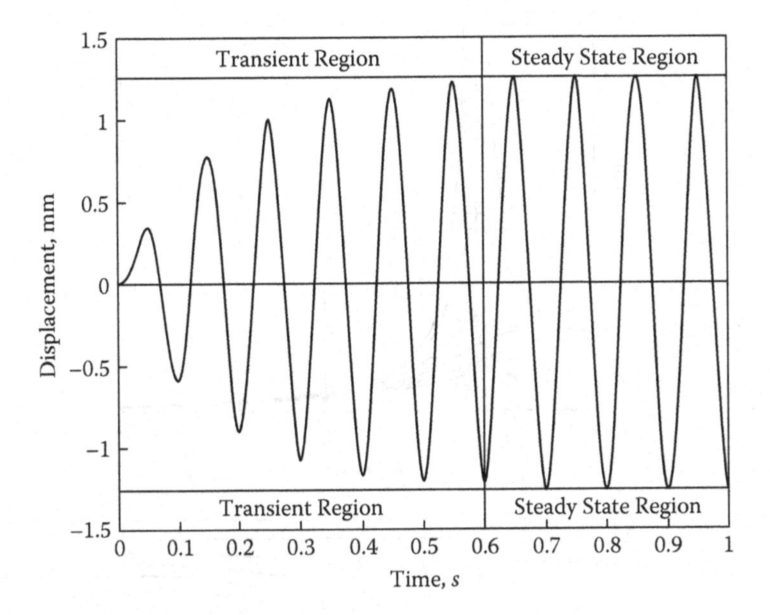

FIGURE 2.12
Response buildup of a system at the resonance when z = 8%.

2.4.1 The System Vibration Behavior

A plot of the maximum amplification and change in phase at different exciting (forcing) frequencies and the damping ratios for a SDOF system, computed from Equations (2.45) and (2.42), is shown in Figure 2.13. It is obvious from Figure 2.13 that the maximum amplification in the vibration amplitude is at resonance; i.e., the exciting frequency becomes equal to the natural frequency of the system, $r = 1$, but the amplitude at resonance decreases with increase in the damping. The vibration amplitude is generally increasing from the static deflection, $\frac{F_0}{k}$ (amplification, $D = 1$), to a maximum vibration amplitude, $\frac{F_0}{k}\left(\frac{1}{2\zeta}\right)$, at zero exciting frequency ($r = 0$) to the resonance case ($r = 1$), respectively. However, the amplitude starts decreasing after resonance, with an increase in the exciting frequency, i.e., when $r > 1$ and it goes below the static deflection, $\frac{F_0}{k}$ (amplification, $D < 1$), at the higher exciting frequency. It is also important to note that the phase angle of the vibration response with respect to the exciting frequency is observed to be changing from 0° to 180° from the zero exciting frequency ($r = 0$) to the exciting frequency much higher than the resonance frequency ($r > 1$), but always passes through 90° at resonance ($r = 1$) irrespective of the system damping (see Figure 2.13(b)).

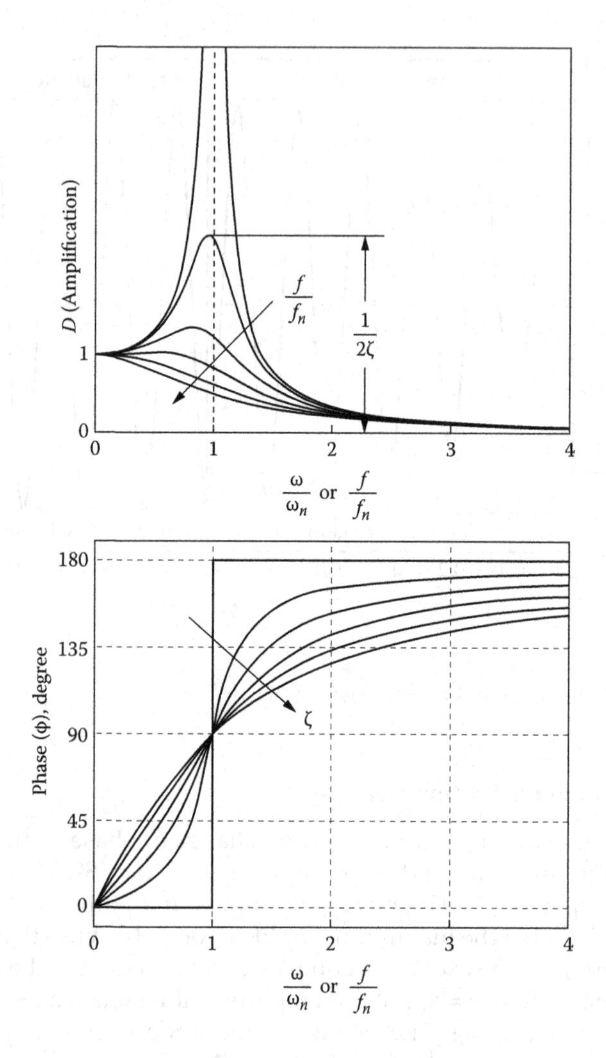

FIGURE 2.13
Steady-state amplification and phase relation between displacement and force at different exciting frequencies.

Figure 2.14 is the modified form of Figure 2.10, which represents the dynamic behavior of the system at resonance. Figure 2.14 clearly indicates that the stiffness force (related to displacement response) canceled out the inertia force (related to acceleration response) at resonance, and ideally the system should not have any strength at resonance to resist the increase in the vibration deflection. However, as seen from Figure 2.14, the applied force is totally balanced by the damping force only, and hence the amplification of the vibration amplitude at resonance only depends on the system damping ratio (see Equation (2.44)).

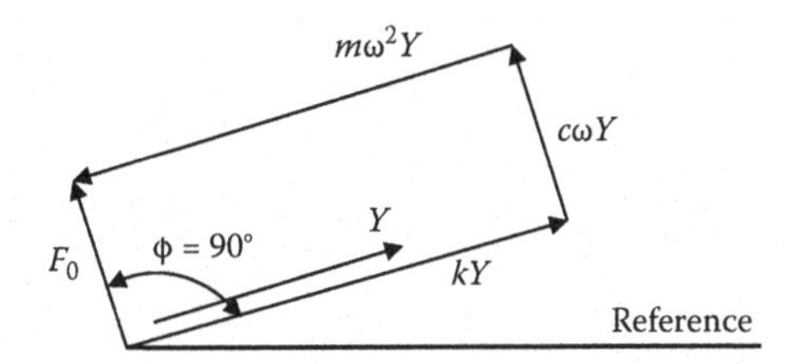

FIGURE 2.14
Graphical representation of the system dynamic behavior at the resonance.

2.5 Summary

The concept of the vibration theory is explained through a simple SDOF example. The significance of the system mass and stiffness in estimating the system natural frequency is explained. The existence of the damping in a system is also introduced. The vibration theory discussion in this chapter is useful to understand the dynamic behavior for large and complex systems as well.

3

Introduction to Finite Element Modeling

3.1 Basic Concept

Real-life systems and structures, such as civil structures, bridges, piping systems, etc., are continuous systems. The mathematical modeling using the standard differential equations based on theory may be complicated for such systems, and their solution would be even more complex. However, the mathematical models are important for several requirements, which include design, impact of design modification, vibration behavior, etc. Any small change in the system may require complete new formulation all over again using the standard differential equations based on the elasticity theories (Fenner, 1986). Considering these difficulties, nowadays the finite element (FE) modeling method is most popular, well accepted, and useful to meet the requirements perhaps in a much simpler than the earlier approach. The simplified concept of the FE method is presented in this chapter through a few simple examples to aid understanding and visualizing of FE modeling and analysis either through in-house developed computational code or through any commercially available FE codes. The well-known textbooks by Zienkiewicz and Taylor (1994) and Cook et al. (1989) can be referred to for further details of the FE method and modeling if needed.

3.1.1 Degree of Freedom (DOF)

It is always assumed that a simple helical coil spring can take the load and related deflection in a direction only; hence, it has just one degree of freedom at any point of the spring. It is illustrated in Figure 3.1. However, other structures may have a number of DOFs at a location, which is illustrated through a beam, as shown in Figure 3.2, subjected to the following loadings:

1. **Axial load in the x-direction:** It produces the deflection of the beam in the x-direction only at any point on the beam. This deflection DOF is represented here by x_e, where e denotes the location on the beam.
2. **Torsion about the x-direction:** It produces the twist angle in the beam about the x-axis, and this degree of freedom is represented by θ_{x_e}.

FIGURE 3.1
A spring undergoes displacement in only one direction (y-direction).

3. **Transverse loads and bending moments in the x-y-axes plane:** These loadings generate bending displacement in the y-direction, i.e., y_e and bending rotation of the beam about the z-axis, which is represented by θ_{z_e} at point e on the beam.
4. **Transverse loads and bending moment in the x-z-axes plane:** Similar to number 3, the related deflection and bending rotation by these loads are z_e and θ_{y_e}, respectively.

Hence, there are the following degrees of freedom (DOFs) at the beam location, e:

$$\mathbf{d}_e = \begin{bmatrix} x_e & y_e & z_e & \theta_{x_e} & \theta_{y_e} & \theta_{z_e} \end{bmatrix}^T \tag{3.1}$$

It is generally considered that the six DOFs at a point on any structure system can represent the complete state of deformation of the point in the x-y-z-coordinate system. These DOFs at location are pictorially represented in Figure 3.3.

3.1.2 Concept of Node, Element, and Meshing in the FE Model

In the FE model, the continuous system is divided into a number of small parts; each part is generally referred to as an element. The number of points on each element for connecting the element to other elements of the complete system is referred to as node numbers. A few illustrative examples are shown in Figures 3.4 to 3.6. The beam in Figure 3.4 is divided into four elements, and each element has two nodes. Figure 3.5 shows a thin plate that is divided into four elements with four nodes on each element. A solid 3D element with eight nodes is shown in Figure 3.6. Each node on an element can have DOFs up to six, but depending upon the type of element, analysis requirements, etc., the DOFs can be reduced; for example, the spring element has just a single DOF per node.

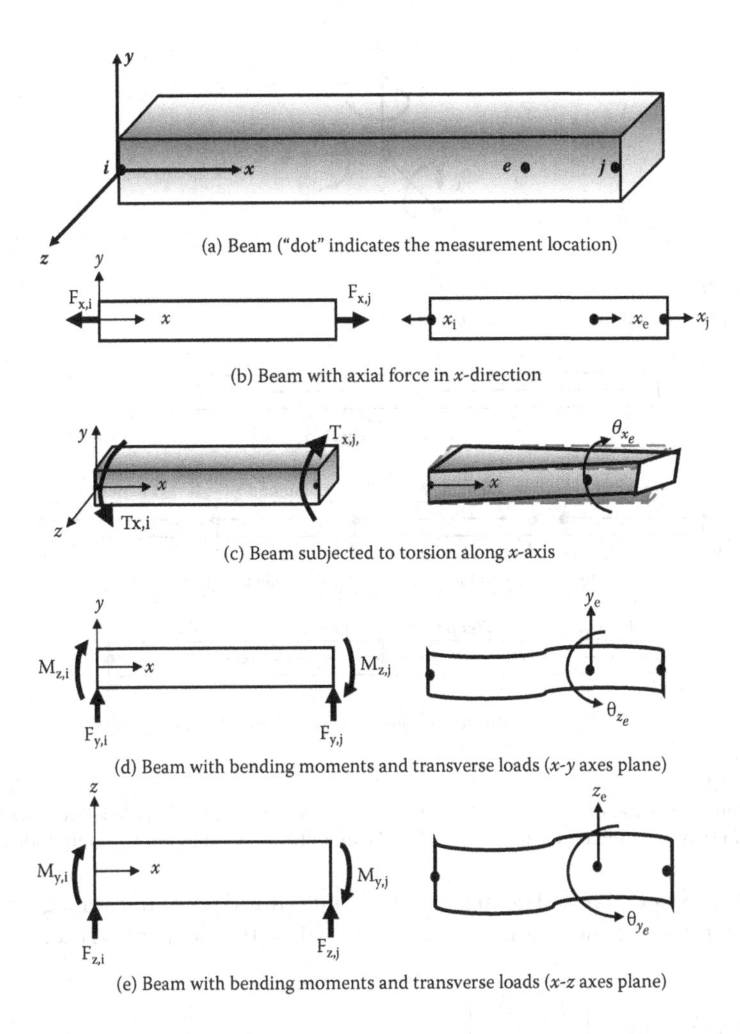

FIGURE 3.2
Beam with different loadings and associated deflection and rotation. (a) Beam (dot indicates the measurement location). (b) Beam with axial force in the x-direction. (c) Beam subjected to torsion along the x-axis. (d) Beam with bending moments and transverse loads (x-y-axes plane). (e) Beam with bending moments and transverse loads (x-z-axes plane).

The assembly of all elements by connecting them at their nodes is known as meshing. Since this mathematical approach is just an approximate method for modeling any continuous system, using fine meshing (i.e., use of small-size elements) often represents the real model more accurately.

3.1.3 Element Mass and Stiffness Matrices for a Spring

Consider the massless spring element of stiffness, k, in Figure 3.7, which has nodes i and j. It is assumed that the nodes are associated with the masses

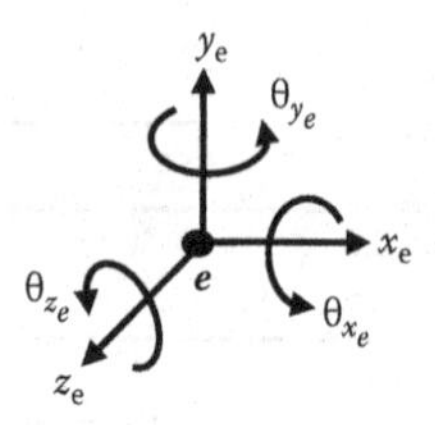

FIGURE 3.3
Possible DOFs at a point in *x*-*y*-*z*-coordinate axes.

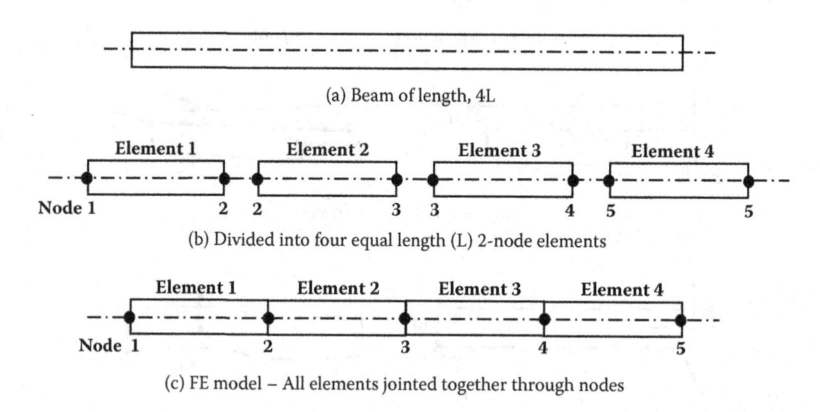

FIGURE 3.4
FE modeling process of beam structure. (a) Beam of length 4*L*. (b) Divided into four equal-length (*L*) two-node elements. (c) FE model: All elements joined together through nodes.

m_i and m_j, respectively. The mass and stiffness matrices of the spring element corresponding to the displacement vectors, $\mathbf{d}_{el} = [y_i \quad y_j]^T$, are written as

$$\text{Mass matrix, } \mathbf{m}_{el} = \begin{bmatrix} m_i & 0 \\ 0 & m_j \end{bmatrix}$$

$$\text{Stiffness matrix, } \mathbf{k}_{el} = \begin{bmatrix} k & -k \\ -k & k \end{bmatrix}$$

The size of these element matrices is (2 × 2) because the spring element has two nodes and each node has just one DOF. However, if an element has n number of nodes and each node six DOFs, then the size of the mass and stiffness matrices corresponding to the DOFs, $\mathbf{d}_{el} = \begin{bmatrix} x_1 & y_1 & z_1 & \theta_{x1} & \theta_{y1} & \theta_{z1} & x_2 & y_2 & z_2 & \theta_{x2} & \theta_{y2} & \theta_{z3} & \cdots & x_n & y_n & z_n & \theta_{x_n} & \theta_{y_n} & \theta_{zn} \end{bmatrix}^T$, is $(6n \times 6n)$. This concept is further explained through a few simple examples and the assembly of all element matrices into the global system mass

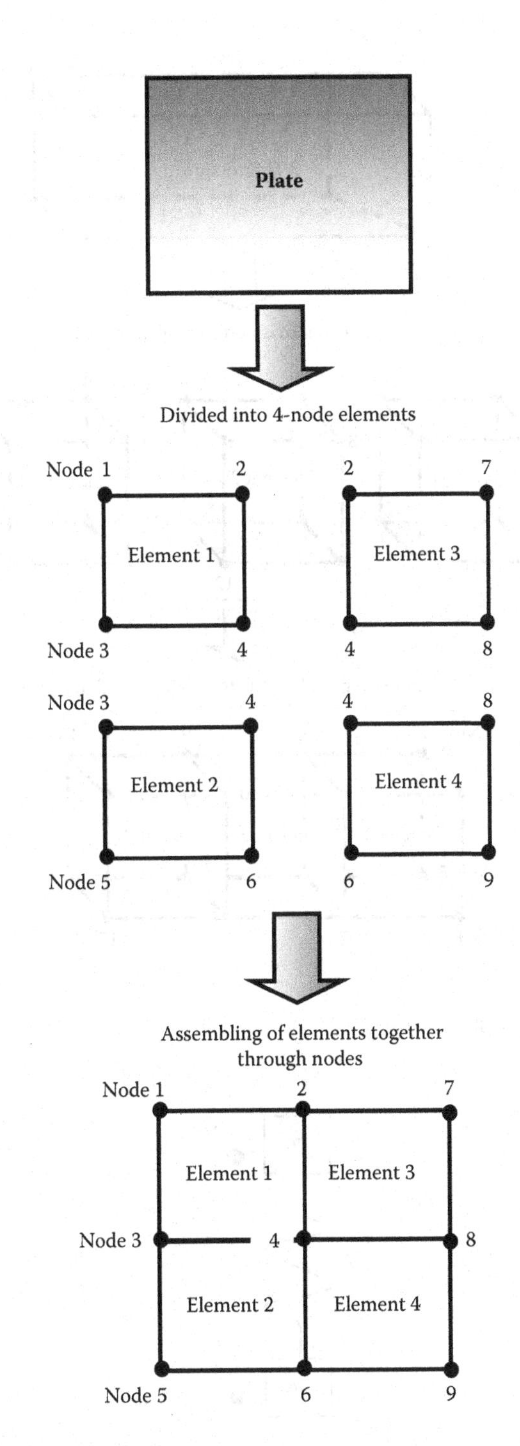

FIGURE 3.5
FE modeling process of plate-type structure.

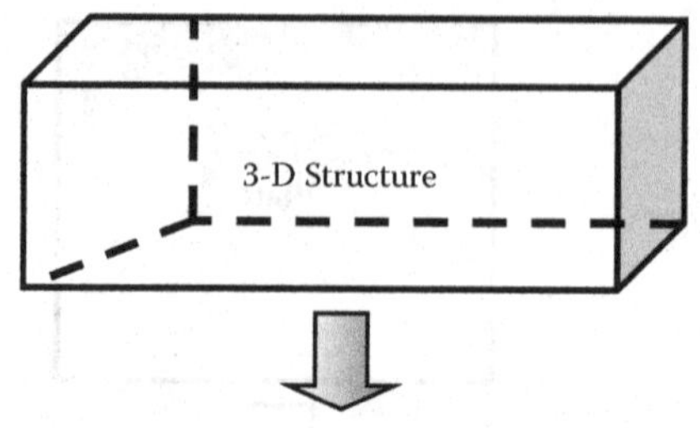

Divided into 8-node 3-D elements

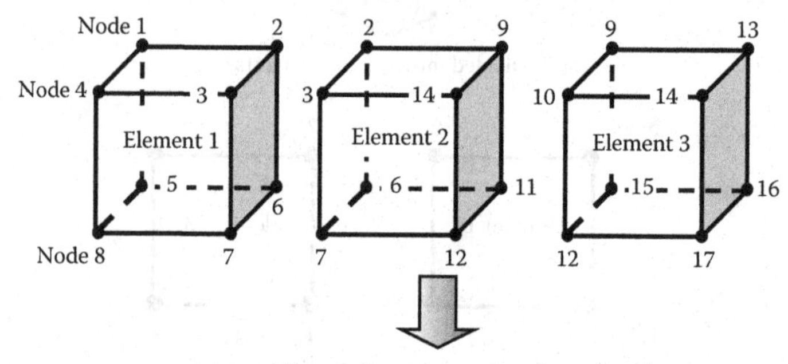

Assembling of elements together through nodes

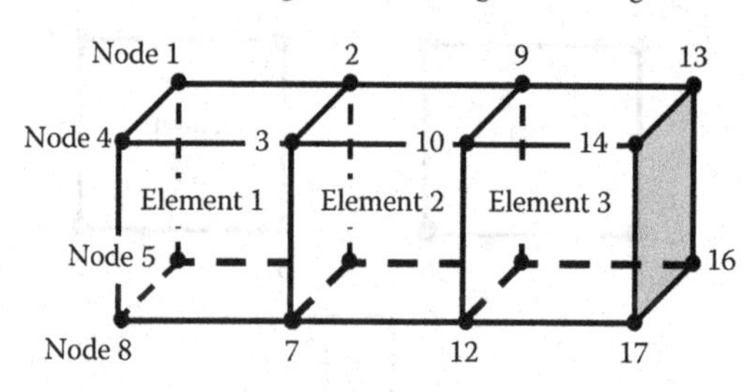

FIGURE 3.6
FE modeling process of 3D structure.

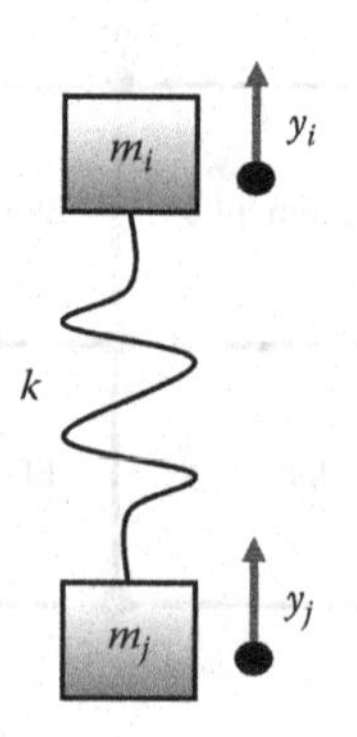

FIGURE 3.7
A spring with lumped masses at nodes i and j.

and stiffness matrices. The construction of these matrices is based on the physical dimensions and material properties. Mechanics and elasticity theories (Fenner, 1986) are used to estimate the parameters of these matrices. Similarly, the damping matrix can be constructed to simulate the dynamics and vibration behavior of the system that is discussed in Section 3.8.

3.2 Modeling Procedure for Discrete Systems

Multi-degrees of freedom (MDOFs) or discrete systems are made of a number of spring-mass-damper systems that are discussed here to explain the mathematical modeling.

3.2.1 Example 3.1: Three-DOF System

In Chapter 2 only a SDOF system is discussed, and its governing equation of the motion is given by

$$m\ddot{y}(t) + c\dot{y}(t) + ky(t) = F(t) \tag{3.2}$$

Now consider a MDOF system consisting of three SDOF systems shown in Figure 3.8. This system has four nodes, nodes 0, 1, 2, and 3, and the displacement vectors corresponding to the DOFs for the dynamic study are given by $\mathbf{y}(t) = [y_0(t)\ \ y_1(t)\ \ y_2(t)\ \ y_3(t)]^T$. For this MDOF system, Equation (3.2) can be modified as

$$\mathbf{M}\ddot{\mathbf{y}}(t) + \mathbf{C}\dot{\mathbf{y}}(t) + \mathbf{K}\mathbf{y}(t) = \mathbf{F}(t) \tag{3.3}$$

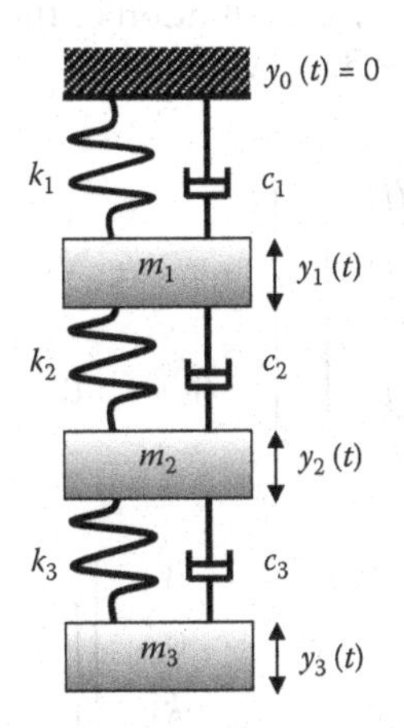

FIGURE 3.8
Three-DOF system of Example 3.1.

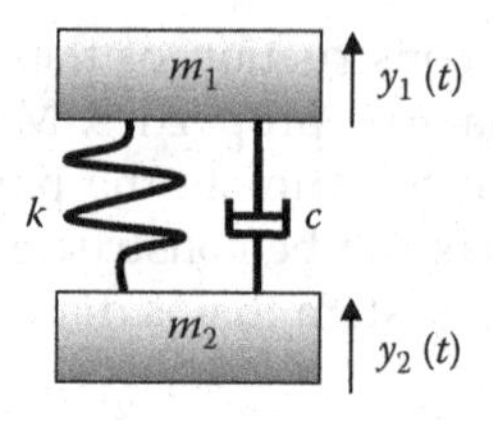

FIGURE 3.9
Element 2 of Figure 3.8.

where **M**, **C**, and **K** are the system mass, damping, and stiffness matrices, respectively. The element mass, damping, and stiffness matrices (as per Section 3.1.3) for the first element of the spring (Nodes 1 and 2) in Figure 3.9 of Example 3.1 are written as

$$\text{Mass matrix, } \mathbf{m}_{el2} = \begin{bmatrix} m_1 & 0 \\ 0 & m_2 \end{bmatrix}$$

$$\text{Damping matrix, } \mathbf{c}_{el2} = \begin{bmatrix} c & -c \\ -c & c \end{bmatrix}$$

$$\text{Stiffness matrix, } \mathbf{k}_{el2} = \begin{bmatrix} k & -k \\ -k & k \end{bmatrix}$$

These matrices correspond to the displacement DOF vectors, $\begin{Bmatrix} y_1(t) \\ y_2(t) \end{Bmatrix}$.

By combining all the matrices, the equation of the motion (Equation (3.3)) for the example shown in Figure 3.8 is written as

$$\begin{bmatrix} 0 & 0 & 0 & 0 \\ 0 & m_1 & 0 & 0 \\ 0 & 0 & m_2 & 0 \\ 0 & 0 & 0 & m_3 \end{bmatrix} \begin{Bmatrix} \ddot{y}_0(t) \\ \ddot{y}_1(t) \\ \ddot{y}_2(t) \\ \ddot{y}_3(t) \end{Bmatrix} + \begin{bmatrix} c_1 & -c_1 & 0 & 0 \\ -c_1 & c_1+c_2 & -c_2 & 0 \\ 0 & -c_2 & c_2+c_3 & -c_3 \\ 0 & 0 & -c_3 & c_3 \end{bmatrix} \begin{Bmatrix} \dot{y}_0(t) \\ \dot{y}_1(t) \\ \dot{y}_2(t) \\ \dot{y}_3(t) \end{Bmatrix}$$

$$+ \begin{bmatrix} k_1 & -k_1 & 0 & 0 \\ -k_1 & k_1+k_2 & -k_2 & 0 \\ 0 & -k_2 & k_2+k_3 & -k_3 \\ 0 & 0 & -k_3 & k_3 \end{bmatrix} \begin{Bmatrix} y_0(t) \\ y_1(t) \\ y_2(t) \\ y_3(t) \end{Bmatrix} = \begin{Bmatrix} 0 \\ 0 \\ F_2(t) \\ 0 \end{Bmatrix} \tag{3.4}$$

Equation (3.4) indicates that the time-dependent force is acting only on the mass, m_2. The advantage of this modeling is that the construction of mass and stiffness matrices is independent of the applied force and boundary conditions (BCs). Once the matrices are constructed, then BCs can be applied. Here the top end (node 0) is fixed, i.e., the displacement $y_0(t) = 0$. Therefore, the first column and the first row of the matrices (the column and the row corresponding to zero DOF) in Equation (3.4) are removed to satisfy the BCs. Hence, the final dynamic equation with BCs is written as

$$\begin{bmatrix} m_1 & 0 & 0 \\ 0 & m_2 & 0 \\ 0 & 0 & m_3 \end{bmatrix} \begin{Bmatrix} \ddot{y}_1(t) \\ \ddot{y}_2(t) \\ \ddot{y}_3(t) \end{Bmatrix} + \begin{bmatrix} c_1 + c_2 & -c_2 & 0 \\ -c_2 & c_2 + c_3 & -c_3 \\ 0 & -c_3 & c_3 \end{bmatrix} \begin{Bmatrix} \dot{y}_1(t) \\ \dot{y}_2(t) \\ \dot{y}_3(t) \end{Bmatrix}$$

$$+ \begin{bmatrix} k_1 + k_2 & -k_2 & 0 \\ -k_2 & k_2 + k_3 & -k_3 \\ 0 & -k_3 & k_3 \end{bmatrix} \begin{Bmatrix} y_1(t) \\ y_2(t) \\ y_3(t) \end{Bmatrix} = \begin{Bmatrix} 0 \\ F_2(t) \\ 0 \end{Bmatrix} \tag{3.5}$$

The solution of Equation (3.5) will have both the complementary solution (transient) and the particular integration (steady-state responses), similar to a SDOF system, but these solutions are not straightforward like the SDOF system. Several numerical integration methods are suggested in the literature to solve such equations of motion; a typical computational method is discussed in Chapter 4.

3.2.2 Example 3.2: Another Three-DOF System

Another example of a three-DOF system (without damping) is shown in Figure 3.10. The equation of the motion for this system can be written as

$$\begin{bmatrix} m_1 & 0 & 0 \\ 0 & m_2 & 0 \\ 0 & 0 & m_3 \end{bmatrix} \begin{Bmatrix} \ddot{y}_1(t) \\ \ddot{y}_2(t) \\ \ddot{y}_3(t) \end{Bmatrix} + \begin{bmatrix} k_1 + k_2 & -k_1 & -k_2 \\ -k_1 & k_1 & 0 \\ -k_2 & 0 & k_2 \end{bmatrix} \begin{Bmatrix} y_1(t) \\ y_2(t) \\ y_3(t) \end{Bmatrix} = \begin{Bmatrix} F_1(t) \\ F_2(t) \\ F_3(t) \end{Bmatrix} \tag{3.6}$$

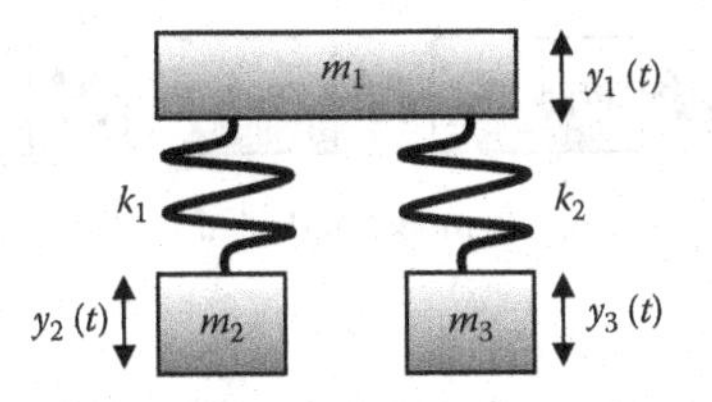

FIGURE 3.10
Another three-DOF system of Example 3.2.

3.3 Extension of FE Modeling Approach to Continuous Systems

As it is known, most of the existing systems are continuous systems that are not generally like a spring, damper, or mass, as in a SDOF system, or a combination of springs, dampers, and masses, as in a MDOF system. However, in FE modeling, any continuous system is divided into a number of small elements and then assembled like a MDOF system. Each element is then converted into the equivalent stiffness and mass depending on the material and geometrical properties of the element.

3.3.1 Example 3.3: A Simple Continuous System of Steel Bar

A 5 m (5L) long bar of diameter 10 mm (cross-section area, A) is a continuous system with material properties—elastic constant E (200 GN/m^2) and density ρ (7800 kg/m^3)—as shown in Figure 3.11, which is undergoing vibration along the bar axis (x-direction).

The bar is discretized into five equal parts for FE modeling. Hence, there are five elements and six nodes, as shown in Figure 3.11(b). Each bar element

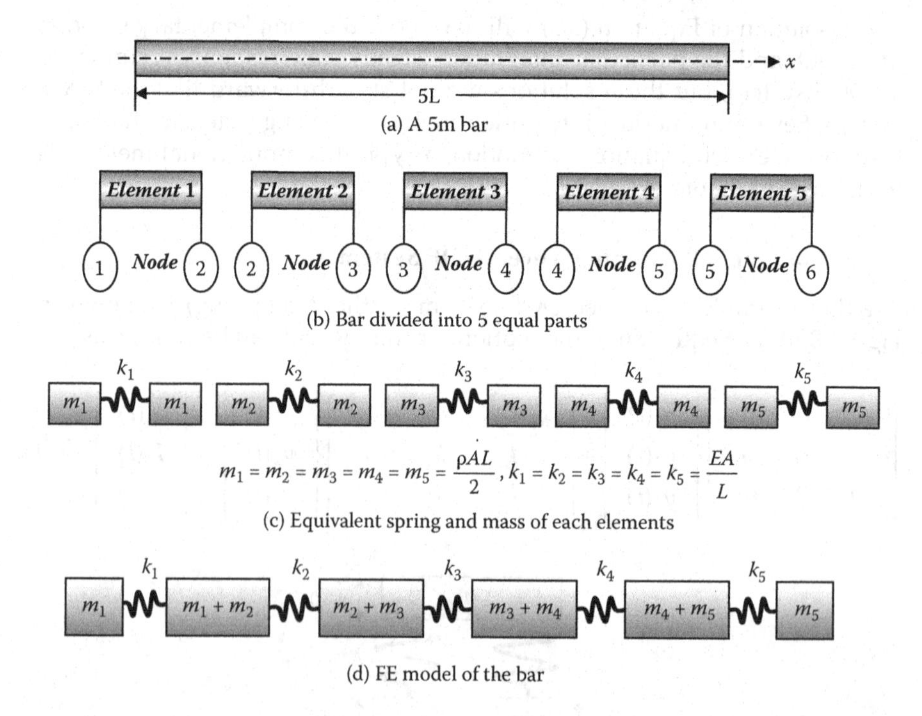

FIGURE 3.11
Simple illustration of FE modeling of a continuous system in Example 3.3. (a) A 5 m bar. (b) Bar divided into five equal parts. (c) Equivalent spring and mass of each element. (d) FE model of the bar.

is then converted to its equivalent massless spring of stiffness, $k = \frac{EA}{L}$, and half of the element mass equal to $\frac{\rho AL}{2}$ is lumped to each node as shown in Figure 3.11(c). The FE model of the bar for axial displacement/vibration is shown in Figure 3.11(d). The global mass and stiffness matrices are then constructed based on the similar concept in Example 3.1, which are written as

Mass matrix,

$$\mathbf{M} = \begin{bmatrix} m_1 & 0 & 0 & 0 & 0 & 0 \\ 0 & m_1 + m_2 & 0 & 0 & 0 & 0 \\ 0 & 0 & m_2 + m_3 & 0 & 0 & 0 \\ 0 & 0 & 0 & m_3 + m_4 & 0 & 0 \\ 0 & 0 & 0 & 0 & m_4 + m_5 & 0 \\ 0 & 0 & 0 & 0 & 0 & m_5 \end{bmatrix} \tag{3.7}$$

$$\text{Stiffness matrix, } \mathbf{K} = \begin{bmatrix} k_1 & -k_1 & 0 & 0 & 0 & 0 \\ -k_1 & k_1 + k_2 & -k_2 & 0 & 0 & 0 \\ 0 & -k_2 & k_2 + k_3 & -k_3 & 0 & 0 \\ 0 & 0 & -k_3 & k_3 + k_4 & -k_4 & 0 \\ 0 & 0 & 0 & -k_4 & k_4 + k_5 & -k_5 \\ 0 & 0 & 0 & 0 & -k_5 & k_5 \end{bmatrix} \tag{3.8}$$

The matrices in Equation (3.7) and (3.8) are corresponding to the displacement vectors $\mathbf{x} = [x_1 \;\; x_2 \;\; x_3 \;\; x_4 \;\; x_5 \;\; x_6]^T$.

3.3.2 Example 3.4: Beam Structure

A steel beam of length $2L = 1.2$ m and cross-section area $A = 20 \times 20$ mm, shown in Figure 3.12(a), is considered now. The beam cross section is shown in Figure 3.12(b). Material properties—elastic constant E and density ρ—are 210 GN/m^2 and 7800 kg/m^3, respectively. The beam is assumed to be deforming in the x-y-plane due to the transverse loading.

The beam is divided into two elements. The deformation in the y-direction will also create bending rotation about the z-axis so that each node is likely to have two DOFs: bending deflection (y) and bending rotation (θ_z). Both elements with their DOFs are shown in Figure 3.12(c). Initially, it is important to understand the construction of the mass and stiffness matrices for each beam element so that the global system matrices can be constructed for this problem.

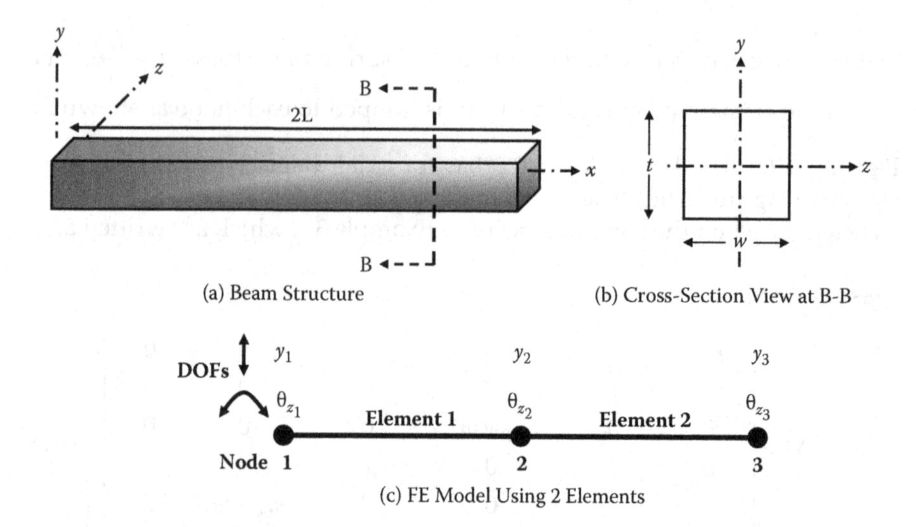

FIGURE 3.12
Beam structure of Example 3.4. (a) Beam structure. (b) Cross-section view at B-B. (c) FE model using two elements.

3.4 Element Mass and Stiffness Matrices

For a better understanding of constructing the mass and stiffness matrices, a beam element is considered, which is shown in Figure 3.13. The element mass ($\mathbf{m}_{el}$) and stiffness ($\mathbf{k}_{el}$) corresponding to the four DOFs are written as

$$\mathbf{d}_{el} = \begin{bmatrix} d_1 & d_2 & d_3 & d_4 \end{bmatrix}^T = \begin{bmatrix} y_i & \theta_{z_i} & y_j & \theta_{z_j} \end{bmatrix}^T$$

$$\mathbf{m}_{el} = \begin{bmatrix} m_{11} & m_{12} & m_{13} & m_{14} \\ m_{21} & m_{22} & m_{23} & m_{24} \\ m_{31} & m_{32} & m_{33} & m_{34} \\ m_{41} & m_{42} & m_{43} & m_{44} \end{bmatrix} \tag{3.9}$$

and

$$\mathbf{k}_{el} = \begin{bmatrix} k_{11} & k_{12} & k_{13} & k_{14} \\ k_{21} & k_{22} & k_{23} & k_{24} \\ k_{31} & k_{32} & k_{33} & k_{34} \\ k_{41} & k_{42} & k_{43} & k_{44} \end{bmatrix} \tag{3.10}$$

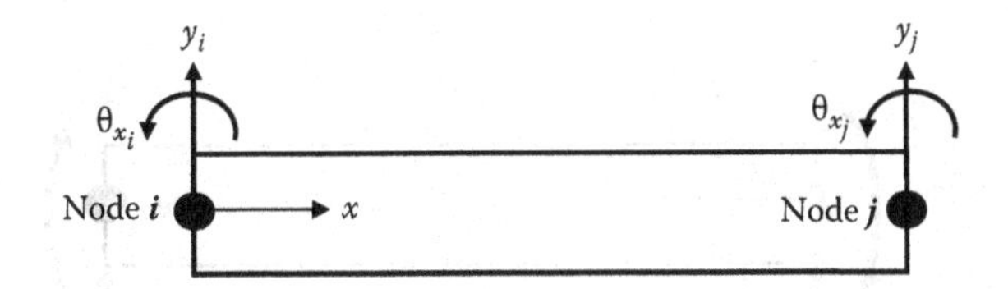

FIGURE 3.13
A two-node beam element with two DOFs (displacement and rotation) at each node.

Similar to the bar element, the lumped mass matrix (**m**) can be written as

$$\mathbf{m}_{el} = \begin{bmatrix} m_{11} & 0 & 0 & 0 \\ 0 & 0 & 0 & 0 \\ 0 & 0 & m_{33} & 0 \\ 0 & 0 & 0 & 0 \end{bmatrix} = \begin{bmatrix} \frac{\rho A \ell}{2} & 0 & 0 & 0 \\ 0 & 0 & 0 & 0 \\ 0 & 0 & \frac{\rho A \ell}{2} & 0 \\ 0 & 0 & 0 & 0 \end{bmatrix} \tag{3.11}$$

where $m_{11} = m_{33} = \frac{\rho A \ell}{2}$ is the beam element length and $m_{13} = m_{31} = 0$ are the masses related to the bending deflection DOFs, $m_{22} = m_{44} = 0$ and $m_{24} = m_{42} = 0$ are masses related to the rotational DOFs, and $m_{12} = m_{21} = 0$, $m_{14} = m_{41} = 0$, $m_{23} = m_{32} = 0$, and $m_{34} = m_{23} = 0$ are the masses related to the combined deflection and rotational DOFs.

However, obtaining the beam element stiffness matrix, **k**, is difficult and not straightforward. The *stiffness influence coefficient* (SIC) method (Thomson and Dahleh, 1998) is used to estimate the elements of the stiffness matrix in Equation (3.10). For a simple spring, the applied force, F, is given by

$$F = ky \tag{3.12}$$

where k is the spring stiffness. It is also called the SIC for the spring, which is a ratio of the applied force, F, to the displacement, y. Similarly, the applied forces and moments at nodes for the beam element in Figure 3.14 and the nodal displacement and rotation vectors are written as

$$\begin{Bmatrix} F_i \\ M_i \\ F_j \\ M_j \end{Bmatrix} = \begin{bmatrix} k_{11} & k_{12} & k_{13} & k_{14} \\ k_{21} & k_{22} & k_{23} & k_{24} \\ k_{31} & k_{32} & k_{33} & k_{34} \\ k_{41} & k_{42} & k_{43} & k_{44} \end{bmatrix} \begin{Bmatrix} y_i \\ \theta_{zi} \\ y_j \\ \theta_{zj} \end{Bmatrix} \tag{3.13}$$

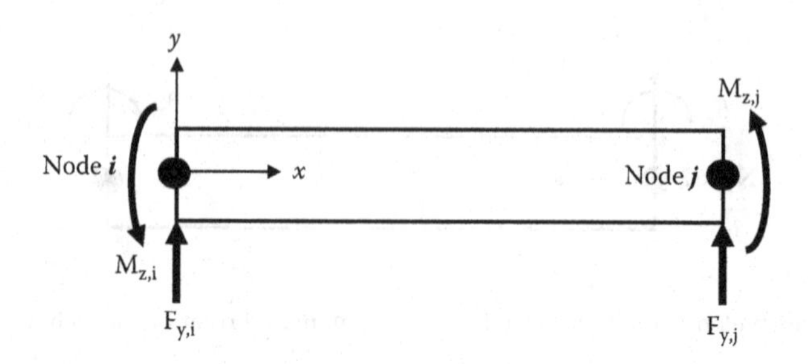

FIGURE 3.14
Typical forces and moments at nodes of the beam element in the *x*-*y*-axes plane.

where the elements of the stiffness matrix in Equation (3.13) are the SICs. These SICs can be estimated using elasticity theory (Fenner, 1986). The forces and moments in Figure 3.14 are assumed as $F_i = F_{y,i}$, $F_j = F_{y,j}$, $m_i = z_{z,i}$ and $M_j = M_{z,j}$ in eq (3.13).

$$EI\frac{d^2y}{dx^2} = -M(x) \tag{3.14}$$

where $M(x)$ is the moment at a distance x. From node i. I is the second moment of the cross-section area. For the example considered here, $I = I_{zz} = \frac{1}{12}wt^3 = \frac{1}{12}(0.020)(0.020)^3 = 1.3333e\text{-}08$ m^4, where I_{zz} is the second moment of the area about the neutral axis along the z-axis. Assuming $y_i = 1$, $y_j = 0$, and $\theta_{zi} = \theta_{zj} = 0$ yields $F_i = k_{11}$, $M_i = k_{21}$, $F_j = k_{31}$, and $M_j = k_{41}$ from Equation (3.13). Hence, the first column SICs, $[k_{11}\ k_{21}\ k_{31}\ k_{41}]^T$, of the stiffness matrix in Equation (3.13) can be estimated using Equation (3.14) as follows:

$$EI\frac{d^2y}{dx^2} = F_i x - M_i$$

where the clockwise moment is assumed to be positive, causing the positive displacement. Twice integration of the above equation yields the following:

$$EI\frac{dy}{dx} = \frac{F_i x^2}{2} - M_i x + C_1 \tag{3.15}$$

$$EIy = \frac{F_i x^3}{6} - \frac{M_i x^2}{2} + C_1 x + C_2 \tag{3.16}$$

Now the boundary conditions are applied at $x = \ell = 0$, where $y = y_j = 0$ and $\frac{dy}{dx} = \theta_{zj} = 0$. Substituting these values into Equations (3.15) and (3.16) yields the integration constants as

$$C_2 = -\frac{M_i\ell^2}{2} + \frac{F_i\ell^3}{3}$$

Substituting $x = 0$ into Equations (3.15) and (3.16) gives

$$F_i = k_{11} = \frac{12EI}{\ell^3} \text{ and } M_i = k_{21} = \frac{6EI}{\ell^2}$$

When $x = \ell$, $F_j = -F_i = k_{31} = -\frac{12EI}{\ell^3}$ and $M_j = -M_i + F_i L = k_{41} = \frac{6EI}{\ell^2}$. Similarly, other SICs in the second, third, and fourth columns of the stiffness matrix can be estimated, and the beam element stiffness matrix in Equation (3.10) corresponding to the DOF vectors, $\mathbf{d}_{el} = [y_i \; \theta_{zi} \; y_j \; \theta_{zj}]^T$, is written as

$$\mathbf{k}_{el} = \begin{bmatrix} k_{11} & k_{12} & k_{13} & k_{14} \\ k_{21} & k_{22} & k_{23} & k_{24} \\ k_{31} & k_{32} & k_{33} & k_{34} \\ k_{41} & k_{42} & k_{43} & k_{44} \end{bmatrix} = \frac{EI}{\ell^3} \begin{bmatrix} 12 & 6\ell & -12 & 6\ell \\ 6\ell & 4\ell^2 & -6\ell & 2\ell^2 \\ -12 & -6\ell & 12 & -6\ell \\ 6\ell & 2\ell^2 & -6\ell & 4\ell^2 \end{bmatrix} \tag{3.17}$$

The element stiffness matrix in Equation (3.17) is derived based on the assumption that data in the beam cross section remain in the plane after deformation; this is known as the Euler-Bernoulli beam theory. The stiffness matrix for any type of elements can also be derived based on elasticity theories in a similar fashion.

3.5 Construction of Global Mass and Stiffness Matrices

Lumped mass and stiffness matrices related to the DOF vectors $\mathbf{d}_{el1} = [y_1 \;\; \theta_{z1} \;\; y_2 \;\; \theta_{z2}]^T$ and $\mathbf{d}_{el2} = [y_2 \;\; \theta_{z2} \;\; y_3 \;\; \theta_{z3}]^T$ for elements 1 and 2 shown in Figure 3.12(c) are written as

Element 1:

$$\mathbf{m}_{el1} = \begin{bmatrix} \frac{\rho AL}{2} & 0 & 0 & 0 \\ 0 & 0 & 0 & 0 \\ 0 & 0 & \frac{\rho AL}{2} & 0 \\ 0 & 0 & 0 & 0 \end{bmatrix} \text{ and } \mathbf{k}_{el1} = \frac{EI}{L^3}\begin{bmatrix} 12 & 6L & -12 & 6L \\ 6L & 4L^2 & -6L & 2L^2 \\ -12 & -6L & 12 & -6L \\ 6L & 2L^2 & -6L & 4L^2 \end{bmatrix} \quad (3.18)$$

where $A = wt = 0.020 \times 0.020 = 4 \times 10^{-4}$ m^2, $\ell = L = 0.6$ m, and $I = \frac{1}{12}wt^3 = \frac{1}{12}(0.020)(0.020)^3 = 1.3333e\text{-}08$ m^4.

Element 2:

$$\mathbf{m}_{el2} = \mathbf{m}_{el1} \text{ and } \mathbf{k}_{el2} = \mathbf{k}_{el1} \quad (3.19)$$

Hence, the global mass, **M**, and stiffness, **K**, matrices corresponding to the displacement DOF vectors, $d = [y_1 \quad \theta_{z1} \quad y_2 \quad \theta_{z2} \quad y_3 \quad \theta_{z3}]^T$, are assembled as

$$\mathbf{M} = \begin{bmatrix} \frac{\rho AL}{2} & 0 & 0 & 0 & 0 & 0 \\ 0 & 0 & 0 & 0 & 0 & 0 \\ 0 & 0 & \frac{\rho AL}{2}+\frac{\rho AL}{2} & 0 & 0 & 0 \\ 0 & 0 & 0 & 0 & 0 & 0 \\ 0 & 0 & 0 & 0 & \frac{\rho AL}{2} & 0 \\ 0 & 0 & 0 & 0 & 0 & 0 \end{bmatrix} \quad (3.20)$$

and

$$\mathbf{K} = \frac{EI}{L^3}\begin{bmatrix} 12 & 6L & -12 & 6L & 0 & 0 \\ 6L & 4L^2 & -6L & 2L^2 & 0 & 0 \\ -12 & -6L & 12+12 & -6L+6L & -12 & 6L \\ 6L & 2L^2 & -6L+6L & 4L^2+4L^2 & -6L & 2L^2 \\ 0 & 0 & -12 & -6L & 12 & -6L \\ 0 & 0 & 6L & 2L^2 & -6L & 4L^2 \end{bmatrix} \quad (3.21)$$

The mass and stiffness matrices after substituting the values of the parameters are

$$\mathbf{M} = \begin{bmatrix} 0.936 & 0 & 0 & 0 & 0 & 0 \\ 0 & 0 & 0 & 0 & 0 & 0 \\ 0 & 0 & 1.872 & 0 & 0 & 0 \\ 0 & 0 & 0 & 0 & 0 & 0 \\ 0 & 0 & 0 & 0 & 0.936 & 0 \\ 0 & 0 & 0 & 0 & 0 & 0 \end{bmatrix} \text{kg}$$

and

$$\mathbf{K} = \begin{bmatrix} 155.56 & 46.667 & 155.56 & 46.667 & 0 & 0 \\ 46.667 & 18.667 & -46.667 & 9.333 & 0 & 0 \\ 155.56 & -46.667 & 311.11 & 0 & -155.56 & 46.667 \\ 46.667 & 9.333 & 0 & 37.333 & -46.667 & 9.333 \\ 0 & 0 & -155.56 & -46.667 & 155.56 & -46.667 \\ 0 & 0 & 46.667 & 9.333 & -46.667 & 18.667 \end{bmatrix} \times 10^3$$

3.6 Concept of the Formal FE Method

As discussed earlier, the FE method treats a continuous system as a discrete system; therefore, the mathematical FE models can compute the responses at the nodes only. The accuracy of the model depends upon the fineness of the element size so that the model approaches close to the actual system. The lumped mass matrix discussed earlier neglects the effect of bending rotation inertia and often needs a large number of elements with small size to get accurate results. A few illustrative examples related to the different element sizes and their impacts on the results are discussed in Section 3.7. The FE method is further improved by introducing the element shape functions (ESFs). The advantages of the ESFs are:

1. The mass matrix can be improved compared to the lumped mass matrix concept discussed earlier. This often enhances the accuracy of the results, even with fewer elements compared to the lumped mass matrix.
2. The FE model may be deemed a continuous system to the extent possible by using the ESFs. It is because the global matrices generally estimate the results only at nodes, but the displacement DOF vectors at any location within an element can then easily be estimated using the ESFs.

3.6.1 Element Shape Functions (ESFs)

The beam element shown in Figure 3.13 is considered again. Let's assume that the functions $\mathbf{N} = [N_1 \; N_2 \; N_3 \; N_4]$ are the ESFs related to the displacement vectors $\mathbf{d}_{el} = [y_i \; \theta_{zi} \; y_j \; \theta_{zj}]^T$, respectively. The shape functions are defined in the nondimensional form such that the displacement at a location e is given by

$$y_e = N_1 y_i + N_2\theta_{zi} + N_3 y_j + N_4\theta_{zj} \tag{3.22}$$

where $N_1 = 1 - 3\left(\frac{x}{\ell}\right)^2 + 2\left(\frac{x}{\ell}\right)^3$, $N_2 = \ell\left(\frac{x}{\ell}\right) - 2\ell\left(\frac{x}{\ell}\right)^2 + \ell\left(\frac{x}{\ell}\right)^3$, $N_3 = 3\left(\frac{x}{\ell}\right)^2 - 2\left(\frac{x}{\ell}\right)^3$, and $N_4 = -\ell\left(\frac{x}{\ell}\right)^2 + \ell\left(\frac{x}{\ell}\right)^3$. Differentiation of Equation (3.22) with respect to x gives

$$\theta_{ze} = \frac{dy_e}{dx} = \frac{dN_1}{dx} y_i + \frac{dN_2}{dx}\theta_{zi} + \frac{dN_3}{dx} y_j + \frac{dN_4}{dx}\theta_{zj} \tag{3.23}$$

where $\frac{dN_1}{dx} = -6\left(\frac{1}{\ell}\right)\left(\frac{x}{\ell}\right) + 6\left(\frac{1}{\ell}\right)\left(\frac{x}{\ell}\right)^2$, $\frac{dN_2}{dx} = 1 - 4\left(\frac{x}{\ell}\right) + 3\left(\frac{x}{\ell}\right)^2$, $\frac{dN_3}{dx} = 6\left(\frac{1}{\ell}\right)\left(\frac{x}{\ell}\right) - 6\left(\frac{1}{\ell}\right)\left(\frac{x}{\ell}\right)^2$, and $\frac{dN_4}{dx} = -2\left(\frac{x}{\ell}\right) + 3\left(\frac{x}{\ell}\right)^2$.

When $x = 0$, then $y_e = y_i$, $\theta_{ze} = \theta_{zi}$, and when $x = \ell$, then $y_e = y_j$, $\theta_{ze} = \theta_{zj}$. Hence, the defined ESFs satisfy the required conditions. Note that the definition of the ESFs is derived from Equations (3.15) and (3.16) in the nondimensional form based on elasticity theories. In a similar fashion, the ESFs for any type and any shape of element can be defined.

3.6.2 Generalized Mass and Stiffness Matrices

The kinetic energy of the beam element shown in Figure 3.13 can be written as

$$T = \frac{1}{2}\mathbf{m}_{el}\dot{\mathbf{d}}_{el}^2 = \frac{1}{2}\int_{x=0}^{x=\ell} \dot{y}_e^2(\rho A)\,dx = \frac{1}{2}\int_{x=0}^{x=\ell} (\mathbf{N}^T\mathbf{N})\dot{\mathbf{d}}_{el}^2(\rho A)\,dx \tag{3.24}$$

$$\text{Mass matrix, } \mathbf{m}_{el} = (\rho A)\int_{x=0}^{x=\ell} (\mathbf{N}^T\mathbf{N})\,dx = \frac{\rho A\ell}{420}\begin{bmatrix} 156 & 22\ell & 54 & -13\ell \\ 22\ell & 4\ell^2 & 13\ell & -3\ell^2 \\ 54 & 13\ell & 156 & -22\ell \\ -13\ell & -3\ell^2 & -22\ell & 4\ell^2 \end{bmatrix} \tag{3.25}$$

The mass matrix in Equation (3.25) is called the *consistent mass matrix* for the beam element in Figure 3.13. The stiffness matrix in Equation (3.10) is estimated as

$$\mathbf{k}_{el} = \begin{bmatrix} k_{11} & k_{12} & k_{13} & k_{14} \\ k_{21} & k_{22} & k_{23} & k_{24} \\ k_{31} & k_{32} & k_{33} & k_{34} \\ k_{41} & k_{42} & k_{43} & k_{44} \end{bmatrix} = EI \int_{x=0}^{x=\ell} \mathbf{B}^T \mathbf{B}\, dx \tag{3.26}$$

where the derivative of the ESF $\mathbf{B} = \frac{d\mathbf{N}}{dx} = \begin{bmatrix} \frac{dN_1}{dx} & \frac{dN_2}{dx} & \frac{dN_3}{dx} & \frac{dN_4}{dx} \end{bmatrix}$.

Each element (SIC) of the stiffness matrix in Equation (3.26) is written as

$$k_{pq} = EI \int_{x=0}^{x=\ell} \frac{dN_p}{dx} \frac{dN_q}{dx}\, dx\,,\ p = 1, 2, 3, 4 \text{ and } q = 1, 2, 3, 4 \tag{3.27}$$

Note that Equations (3.26) and (3.27) yield exactly the same stiffness matrix as in Equation (3.17).

3.7 FE Modeling for the Beam in Example 3.4

Once again, the beam in Example 3.4 is considered here. It is also assumed that the beam is divided into two elements each with length L, as earlier shown in Figure 3.12. The global stiffness $\mathbf{K}$ matrix corresponding to the displacement DOF vectors, $d = [y_1\ \theta_{z1}\ y_2\ \theta_{z2}\ y_3\ \theta_{z3}]^T$, remains the same as in Equation (3.21), but the global mass matrix for the complete beam based on the concept of the consistent mass in Equation (3.25) is now given by

$$\mathbf{M} = \frac{\rho AL}{420} \begin{bmatrix} 156 & 22L & 54 & -13L & 0 & 0 \\ 22L & 4L^2 & 13L & -3L^2 & 0 & 0 \\ 54 & 13L & 156+156 & -22L+22L & 54 & -13L \\ -13L & -3L^2 & -22L+22L & 4L^2+4L^2 & 13L & -3L^2 \\ 0 & 0 & 54 & 13L & 156 & -22L \\ 0 & 0 & -13L & -3L^2 & -22L & 4L^2 \end{bmatrix} \tag{3.28}$$

$$\mathbf{M} = \begin{bmatrix} 695.31 & 58.834 & 240.69 & -34.766 & 0 & 0 \\ 58.834 & 6.418 & 34.766 & -4.814 & 0 & 0 \\ 240.69 & 34.766 & 1390.62 & 0 & 240.69 & -34.766 \\ -34.766 & -4.814 & 0 & 12.837 & 34.766 & -4.814 \\ 0 & 0 & 240.69 & 34.766 & 695.31 & -58.834 \\ 0 & 0 & -34.766 & -4.814 & -58.834 & 6.418 \end{bmatrix} \times 10^{-3}$$

3.7.1 Applying Boundary Conditions (BCs)

Once the global mass and stiffness matrices are constructed, the BCs and forces can be applied to simulate the required conditions. Application of forces and moments, and then response estimation, are discussed in Chapter 4. Therefore, the application of BCs is only discussed here. Different beam conditions are assumed to illustrate how these BCs can be incorporated into the global mass and stiffness matrices.

3.7.1.1 Cantilever Condition at Node 1

This means that the bending deflection and rotation at node 1 are 0, i.e., $y_1 = 0$ and $\theta_{z1} = 0$. The displacement DOF vectors are now $d = [y_2 \quad \theta_{z2} \quad y_3 \quad \theta_{z3}]^T$. This means that the rows and columns corresponding to the DOFs, y_1 and θ_{z1}, must be removed from the global mass and stiffness matrices. In this case, the first two rows and columns need to be removed from Equations (3.21) and (3.26) corresponding to the stiffness and mass matrices, respectively, to simulate cantilever BCs at node 1; hence, the resultant system matrices are given by

$$\mathbf{M}_{Cantilever} = \frac{\rho AL}{420} \begin{bmatrix} 156+156 & -22L+22L & 54 & -13L \\ -22L+22L & 4L^2+4L^2 & 13L & -3L^2 \\ 54 & 13L & 156 & -22L \\ -13L & -3L^2 & -22L & 4L^2 \end{bmatrix} \tag{3.29}$$

$$\mathbf{K}_{Cantilever} = \frac{EI}{L^3} \begin{bmatrix} 12+12 & -6L+6L & -12 & 6L \\ -6L+6L & 4L^2+4L^2 & -6L & 2L^2 \\ -12 & -6L & 12 & -6L \\ 6L & 2L^2 & -6L & 4L^2 \end{bmatrix} \tag{3.30}$$

3.7.1.2 Simply Supported (SS) Beam at Nodes 1 and 3

The bending displacements at nodes 1 and 3 are 0, i.e., $y_1 = 0$ and $y_3 = 0$. The remaining displacement DOF vectors are now $d = [\theta_{z1} \quad y_2 \quad \theta_{z2} \quad \theta_{z3}]^T$.

This means that the rows and columns corresponding to the DOFs, y_1 and y_3, must be removed from the global mass and stiffness matrices. In this case, the first and fifth rows and columns, corresponding to the stiffness and mass matrices, respectively, need to be removed from Equations (3.21) and (3.26) to simulate simply supported BCs at nodes 1 and 3; hence, the resultant system matrices are given by

$$\mathbf{M}_{SS} = \frac{\rho AL}{420}\begin{bmatrix} 4L^2 & 13L & -3L^2 & 0 \\ 13L & 156+156 & -22L+22L & -13L \\ -3L^2 & -22L+22L & 4L^2+4L^2 & -3L^2 \\ 0 & -13L & -3L^2 & 4L^2 \end{bmatrix} \tag{3.31}$$

$$\mathbf{K}_{SS} = \frac{EI}{L^3}\begin{bmatrix} 4L^2 & -6L & 2L^2 & 0 \\ -6L & 12+12 & -6L+6L & 6L \\ 2L^2 & -6L+6L & 4L^2+4L^2 & 2L^2 \\ 0 & 6L & 2L^2 & 4L^2 \end{bmatrix} \tag{3.32}$$

3.7.1.3 Clamped–Clamped (CC) Beam at Nodes 1 and 3

Node 1: $y_1 = 0$ and $\theta_{z1} = 0$; node 2: $y_3 = 0$ and $\theta_{z3} = 0x$. Thus, the mass and stiffness matrices corresponding to the remaining displacement DOF vectors, $d = [y_2 \quad \theta_{z2}]^T$, are written as

$$\mathbf{M}_{CC} = \frac{\rho AL}{420}\begin{bmatrix} 156+156 & -22L+22L \\ -22L+22L & 4L^2+4L^2 \end{bmatrix} \tag{3.33}$$

$$\mathbf{K}_{CC} = \frac{EI}{L^3}\begin{bmatrix} 12+12 & -6L+6L \\ -6L+6L & 4L^2+4L^2 \end{bmatrix} \tag{3.34}$$

3.8 Modal Analysis

Modal analysis is the process of estimating the natural frequencies and their associated shape of deflection, called mode shapes, from the mathematical model of any structural systems.

3.8.1 Natural Frequencies and Mode Shape Estimation

It is obvious from Chapter 2 that the resultant of the inertial and stiffness forces is zero at resonance. This means the equation of motion for any system by neglecting the damping and the external force can be written as

$$\mathbf{M}\ddot{\mathbf{y}}(t) + \mathbf{K}\mathbf{y}(t) = 0 \tag{3.35}$$

where **M** and **K** are the system mass and stiffness matrices of size $(n \times n)$, respectively. Equation (3.35) for a SDOF system can be written as

$$m\ddot{y}(t) + ky(t) = 0 \tag{3.36}$$

If $y(t) = y_0 \sin(\omega t)$, where $\omega = 2\pi f$ is the rotational frequency, then Equation (3.36) becomes $(-\omega^2 m + k)y_0 = 0$. The solution of $(-\omega^2 m + k) = 0$ gives the natural frequency of the SDOF systems, $f_n = \frac{\omega_n}{2\pi} = \frac{\omega}{2\pi} = \frac{1}{2\pi}\sqrt{\frac{k}{m}}$. Similarly, Equation (3.35) can also be written as

$$(-\omega^2 \mathbf{M} + \mathbf{K})\mathbf{y}_0 = 0 \tag{3.37}$$

$$\Rightarrow (\mathbf{K} - \lambda \mathbf{M})\varphi = 0 \tag{3.38}$$

Equation (3.38) is just analogous to Equation (3.37), which represents the equation of the eigenvalues and eigenvectors, where $\lambda = \omega^2$ are a set of the eigenvalues, $\lambda = [\lambda_1 \ \lambda_2 \ \ldots \ \lambda_n]$, and the corresponding eigenvector matrices are $\varphi = [\varphi_1 \ \varphi_2 \ \ldots \ \varphi_n]$, respectively. Following are the steps to compute the natural frequencies and mode shapes:

Step 1: For the determinant of $(\mathbf{K} - \lambda\mathbf{M}) = 0 \Rightarrow |\mathbf{K} - \lambda\mathbf{M}| = 0$, compute the eigenvalues, $\lambda = [\lambda_1 \ \lambda_2 \ \ldots \ \lambda_n]$, where $\lambda_1 \le \lambda_2 \ldots \lambda_{n-1} \le \lambda_n$.

Step 2: The computed eigenvalues are then converted into the frequencies to obtain the natural frequencies of the system. The frequencies $\begin{bmatrix} f_{n1} & f_{n2} & \cdots & f_{n_n} \end{bmatrix}$ are referred as the first, second, ..., nth natural frequencies of the system or modes 1, 2, ..., n, respectively.

Natural frequencies in rad/s,

$$\begin{bmatrix} \omega_{n1} & \omega_{n2} & \cdots & \omega_{n_n} \end{bmatrix} = \begin{bmatrix} \sqrt{\lambda_1} & \sqrt{\lambda_2} & \cdots & \sqrt{\lambda_n} \end{bmatrix} \tag{3.39}$$

Natural frequencies in Hz,

$$\begin{bmatrix} f_{n1} & f_{n2} & \cdots & f_{n_n} \end{bmatrix} = \begin{bmatrix} \frac{\sqrt{\lambda_1}}{2\pi} & \frac{\sqrt{\lambda_2}}{2\pi} & \cdots & \frac{\sqrt{\lambda_n}}{2\pi} \end{bmatrix} \quad (3.40)$$

Step 3: The eigenvectors of each computed eigenvalue are then estimated as $(\mathbf{K}-\lambda_p\mathbf{M})\varphi_p = 0$, where λ_p is the pth eigenvalue (pth natural frequency) and φ_p is the corresponding eigenvector of size $(n \times 1)$ for the pth mode of the system. The eigenvectors φ_p define the shape of deformation of the system at the pth mode. The computed eigenvector is then normalized by the mass matrix such that $\varphi_p^T\mathbf{M}\varphi_p = 1$, $\varphi_p^T\mathbf{M}\varphi_q = 0$, where p or $q = 1, 2, \ldots, n$, except $p \neq q$ and $\varphi_p^T\mathbf{K}\varphi_p = \omega_p^2$.

Note that the condition $\varphi_p^T\mathbf{M}\varphi_q = 0$ indicates the pth mode is orthogonal to the qth mode, which means that when the system is excited at the pth natural frequency, it will not vibrate at the qth natural frequency, and vice versa.

The mass and stiffness matrices of Example 3.1 (Figure 3.8) in Equation (3.5) are considered again to illustrate computation of eigenvalues (natural frequencies) and the corresponding eigenvectors (mode shapes). Assume that the masses $m_1 = m_2 = m_3 = 2$ kg and stiffnesses $k_1 = k_2 = k_3 = 5$ kN/m.

$$\mathbf{M} = \begin{bmatrix} m_1 & 0 & 0 \\ 0 & m_2 & 0 \\ 0 & 0 & m_3 \end{bmatrix} = \begin{bmatrix} 2 & 0 & 0 \\ 0 & 2 & 0 \\ 0 & 0 & 2 \end{bmatrix} \text{kg}$$

$$\mathbf{K} = \begin{bmatrix} k_1 + k_2 & -k_2 & 0 \\ -k_2 & k_2 + k_3 & -k_3 \\ 0 & -k_3 & k_3 \end{bmatrix} = \begin{bmatrix} 10 & -5 & 0 \\ -5 & 10 & -5 \\ 0 & -5 & 5 \end{bmatrix} \times 10^3 \text{N/m}$$

The eigenvalues and eigenvectors are now computed as per Equation (3.38) as

$$|\mathbf{K} - \lambda\mathbf{M}| = 0$$

$$\Rightarrow \left| \begin{bmatrix} 10 & -5 & 0 \\ -5 & 10 & -5 \\ 0 & -5 & 5 \end{bmatrix} 10^3 - \lambda \begin{bmatrix} 2 & 0 & 0 \\ 0 & 2 & 0 \\ 0 & 0 & 2 \end{bmatrix} \right| = 0 \quad (3.41)$$

The solution of the determinant in Equation (3.41) gave three eigenvalues, $[\lambda_1 \ \lambda_2 \ \lambda_3]$, which indicates three natural frequencies for the three-DOF system in Example 3.1. Once the natural frequencies are known, then the mode shapes for each natural frequency (eigenvalue) are computed from Equation (3.38). The computed three eigenvalues and natural frequencies are $\begin{bmatrix} \lambda_1 & \lambda_2 & \lambda_3 \end{bmatrix} = \begin{bmatrix} \omega_{n1}^2 & \omega_{n2}^2 & \omega_{n3}^2 \end{bmatrix} = \begin{bmatrix} 495.2 & 3887.4 & 8117.4 \end{bmatrix}$ (radian/s)^3 and $\begin{bmatrix} f_{n1} & f_{n2} & f_{n3} \end{bmatrix} = \begin{bmatrix} 3.54 & 9.92 & 14.34 \end{bmatrix}$ Hz, respectively. The corresponding mass normalized mode shapes (eigenvectors), $\varphi = [\varphi_1 \; \varphi_2 \; \varphi_3]$, for the three modes are given in Equation (3.42). Note that the element of the mode shapes is represented by $\phi_{yi,j}$, where i and j represent node number and mode number, respectively. Note that y is the displacement DOF in the y-direction.

$$\varphi = \begin{bmatrix} \varphi_1 & \varphi_2 & \varphi_3 \end{bmatrix} = \begin{bmatrix} \phi_{y1,1} & \phi_{y1,2} & \phi_{y1,3} \\ \phi_{y2,1} & \phi_{y2,2} & \phi_{y2,3} \\ \phi_{y3,1} & \phi_{y3,2} & \phi_{y3,3} \end{bmatrix} = \begin{bmatrix} -0.2319 & +0.5211 & -0.4179 \\ -0.4179 & +0.2319 & +0.5211 \\ -0.5211 & -0.4179 & -0.2319 \end{bmatrix} \tag{3.42}$$

The numerical values of the mode shapes shown in Equation (3.42) are generally nondimensional numbers. They only indicate the shape of deformation of the structure when vibrating at a particular natural frequency. Columns 1, 2, and 3 represent mode shapes for modes 1, 2, and 3, respectively, at DOFs related to three nodes.

3.8.2 Concept of Mode Shapes

The first column of Equation (3.42) indicates that all three masses—m_1, m_2, and m_3—at nodes 1 to 3 are moving in phase (i.e., in the same direction) at all times when the system is vibrating at the first natural frequency, 3.54 Hz. The vibration amplitude of the masses, m_1, m_2, and m_3, will follow the ratio: 0.2319:0.4179:0.5211 (i.e., lowest vibration for mass m_1). Similarly, the second column of Equation (3.42) represents the second mode shape, where masses m_1 and m_2 vibrate in phase (in the same direction), but mass m_3 vibrates out of phase (opposite direction) with respect to the other masses at the second natural frequency, 9.92 Hz. At third natural frequency, masses m_1 and m_3, vibrates in phase but out of phase with mass m_2.

Example 3.4 with the cantilever boundary condition is used to further explain the concept of the mode shapes. The left end of the beam is assumed to be clamped to satisfy the condition of the cantilever beam, as shown in Figure 3.15. Now the beam is divided into 10 elements with two DOFs per node, as shown in Figure 3.16. The global mass (consistent) and stiffness matrices are constructed, and then BCs are also applied at node 1

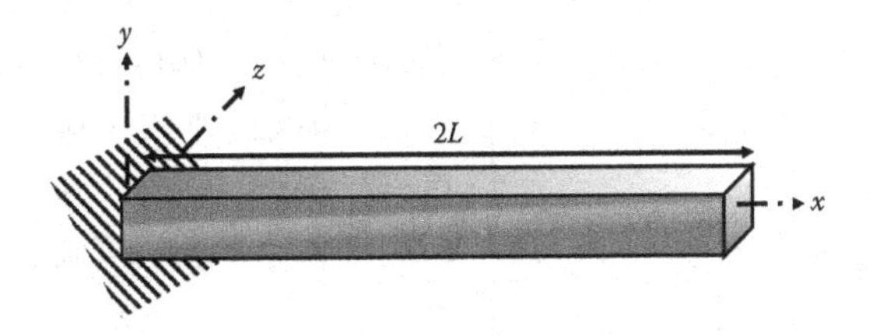

FIGURE 3.15
The cantilever beam of Example 3.4.

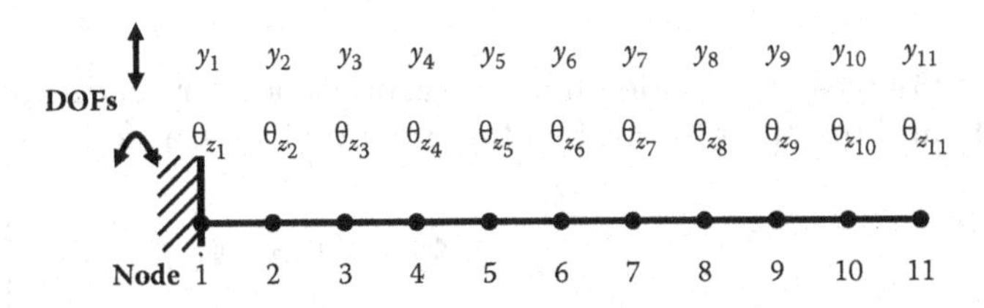

FIGURE 3.16
The FE model of the cantilever beam of Example 3.4 (y_1 to y_{11}, the bending deflection in the y-direction θ_{z1} to θ_{z11}, the bending rotation about the z-axis for nodes 1 to 11, respectively).

TABLE 3.1
Natural Frequencies and Mode Shape-Related Bending Displacement DOFs as in Equation (3.44)

		Mode Shapes			
Node No.	Location, m	$\varphi_{y,1}$	$\varphi_{y,2}$	$\varphi_{y,3}$	$\varphi_{y,4}$
1	0	0	0	0	0
2	0.12	1.7338e–02	9.5750e–02	–2.3586e–01	–3.9871e–01
3	0.24	6.6019e–02	3.1120e–01	–6.2515e–01	–7.8063e–01
4	0.36	1.4107e–01	5.4386e–01	–7.8207e–01	–4.4936e–01
5	0.48	2.3761e–01	7.0650e–01	–5.4389e–01	3.2674e–01
6	0.60	3.5094e–01	7.3771e–01	–2.0370e–02	7.3230e–01
7	0.72	4.7664e–01	6.0934e–01	4.8994e–01	3.3819e–01
8	0.84	6.1074e–01	3.2773e–01	6.7988e–01	–4.1146e–01
9	0.96	7.4987e–01	–7.2395e–02	4.0837e–01	–6.6594e–01
10	1.08	8.9140e–01	–5.4140e–01	–2.3630e–01	–5.3982e–02
11	1.20	1.0336e–00	–1.0337e+00	–1.0341e+00	1.0355e+00
Natural frequencies		f_{n1} = 11.64 Hz	f_{n1} = 72.96 Hz	f_{n1} = 204.33 Hz	f_{n1} = 400.69 Hz, respectively

for the cantilever condition. The natural frequencies and mode shapes are then computed through the eigenvalue and eigenvector analysis using Equation (3.38). The first four computed bending modes are 11.64, 72.96, 204.33, and 400.69 Hz. The corresponding estimated mode shapes as per Step 3 in Section 3.8.1 are listed in Tables 3.1–3.2 as per Eq.(3.41) to (3.43).

$$\varphi = \begin{bmatrix} \varphi_1 & \varphi_2 & \varphi_3 & \varphi_4 \end{bmatrix} = \begin{bmatrix} \phi_{y1,1} & \phi_{y1,2} & \phi_{y1,3} & \phi_{y1,4} \\ \phi_{\theta_{z1},1} & \phi_{\theta_{z1},2} & \phi_{\theta_{z1},3} & \phi_{\theta_{z1},4} \\ \phi_{y2,1} & \phi_{y2,2} & \phi_{y2,3} & \phi_{y2,4} \\ \phi_{\theta_{z2},1} & \phi_{\theta_{z2},2} & \phi_{\theta_{z2},3} & \phi_{\theta_{z2},4} \\ \vdots & \vdots & \vdots & \vdots \\ \phi_{y11,1} & \phi_{y11,2} & \phi_{y11,3} & \phi_{y11,4} \\ \phi_{\theta_{z11},1} & \phi_{\theta_{z11},2} & \phi_{\theta_{z11},3} & \phi_{\theta_{z11},4} \end{bmatrix} \quad (3.43)$$

Equation (3.43) can be divided into two groups, one for bending displacement DOFs and the other for the bending rotation DOFs, as follows:

$$\varphi_y = \begin{bmatrix} \varphi_{y,1} & \varphi_{y,2} & \varphi_{y,3} & \varphi_{y,4} \end{bmatrix} = \begin{bmatrix} \phi_{y1,1} & \phi_{y1,2} & \phi_{y1,3} & \phi_{y1,4} \\ \phi_{y2,1} & \phi_{y2,2} & \phi_{y2,3} & \phi_{y2,4} \\ \phi_{y3,1} & \phi_{y3,2} & \phi_{y3,3} & \phi_{y3,4} \\ \phi_{y4,1} & \phi_{y4,2} & \phi_{y4,3} & \phi_{y4,4} \\ \phi_{y5,1} & \phi_{y5,2} & \phi_{y5,3} & \phi_{y5,4} \\ \phi_{y6,1} & \phi_{y6,2} & \phi_{y6,3} & \phi_{y6,4} \\ \phi_{y7,1} & \phi_{y7,2} & \phi_{y7,3} & \phi_{y7,4} \\ \phi_{y8,1} & \phi_{y8,2} & \phi_{y8,3} & \phi_{y8,4} \\ \phi_{y9,1} & \phi_{y9,2} & \phi_{y9,3} & \phi_{y9,4} \\ \phi_{y10,1} & \phi_{y10,2} & \phi_{y10,3} & \phi_{y10,4} \\ \phi_{y11,1} & \phi_{y11,2} & \phi_{y11,3} & \phi_{y11,4} \end{bmatrix} \quad (3.44)$$

$$\varphi_y = \begin{bmatrix} \varphi_{\theta_z,1} & \varphi_{\theta_z,2} & \varphi_{\theta_z,3} & \varphi_{\theta_z,4} \end{bmatrix} = \begin{bmatrix} \phi_{\theta_{z1},1} & \phi_{\theta_{z1},2} & \phi_{\theta_{z1},3} & \phi_{\theta_{z1},4} \\ \phi_{\theta_{z2},1} & \phi_{\theta_{z2},2} & \phi_{\theta_{z2},3} & \phi_{\theta_{z2},4} \\ \phi_{\theta_{z3},1} & \phi_{\theta_{z3},2} & \phi_{\theta_{z3},3} & \phi_{\theta_{z3},4} \\ \phi_{\theta_{z4},1} & \phi_{\theta_{z4},2} & \phi_{\theta_{z4},3} & \phi_{\theta_{z4},4} \\ \phi_{\theta_{z5},1} & \phi_{\theta_{z5},2} & \phi_{\theta_{z5},3} & \phi_{\theta_{z5},4} \\ \phi_{\theta_{z6},1} & \phi_{\theta_{z6},2} & \phi_{\theta_{z6},3} & \phi_{\theta_{z6},4} \\ \phi_{\theta_{z7},1} & \phi_{\theta_{z7},2} & \phi_{\theta_{z7},3} & \phi_{\theta_{z7},4} \\ \phi_{\theta_{z8},1} & \phi_{\theta_{z8},2} & \phi_{\theta_{z8},3} & \phi_{\theta_{z8},4} \\ \phi_{\theta_{z9},1} & \phi_{\theta_{z9},2} & \phi_{\theta_{z9},3} & \phi_{\theta_{z9},4} \\ \phi_{\theta_{z10},1} & \phi_{\theta_{z10},2} & \phi_{\theta_{z10},3} & \phi_{\theta_{z10},4} \\ \phi_{\theta_{z11},1} & \phi_{\theta_{z11},2} & \phi_{\theta_{z11},3} & \phi_{\theta_{z11},4} \end{bmatrix} \quad (3.45)$$

TABLE 3.2

Natural Frequencies and Mode Shape-Related Bending Rotation DOFs as in Equation (3.45)

		Mode Shapes			
Node No.	**Location, *m***	$\varphi_{\theta_z,1}$	$\varphi_{\theta_z,2}$	$\varphi_{\theta_z,3}$	$\varphi_{\theta_z,4}$
1	0	0	0	0	0
2	0.12	2.8201e−01	1.4451e+00	−3.2451e+00	−4.8011e+00
3	0.24	5.2243e−01	2.0020e+00	−2.6872e+00	−5.2295e−01
4	0.36	7.2167e−01	1.7530e+00	3.0596e−01	5.6056e+00
5	0.48	8.8081e−01	8.7125e−01	3.4987e+00	6.0436e+00
6	0.60	1.0018e+00	−3.9034e−01	4.7847e+00	3.9502e−02
7	0.72	1.0877e+00	−1.7395e+00	3.2673e+00	−5.9133e+00
8	0.84	1.1427e+00	−2.9038e+00	−3.0741e−01	−5.2515e+00
9	0.96	1.1724e+00	−3.6934e+00	−4.0809e+00	1.5756e+00
10	1.08	1.1839e+00	−4.0568e+00	−6.3242e+00	7.9614e+00
11	1.20	1.1857e+00	−4.1182e+00	−6.7640e+00	9.4901e+00
Natural frequencies		f_{n1} = 11.64 Hz	f_{n2} = 72.96 Hz	f_{n3} = 204.33 Hz	f_{n4} = 400.69 Hz

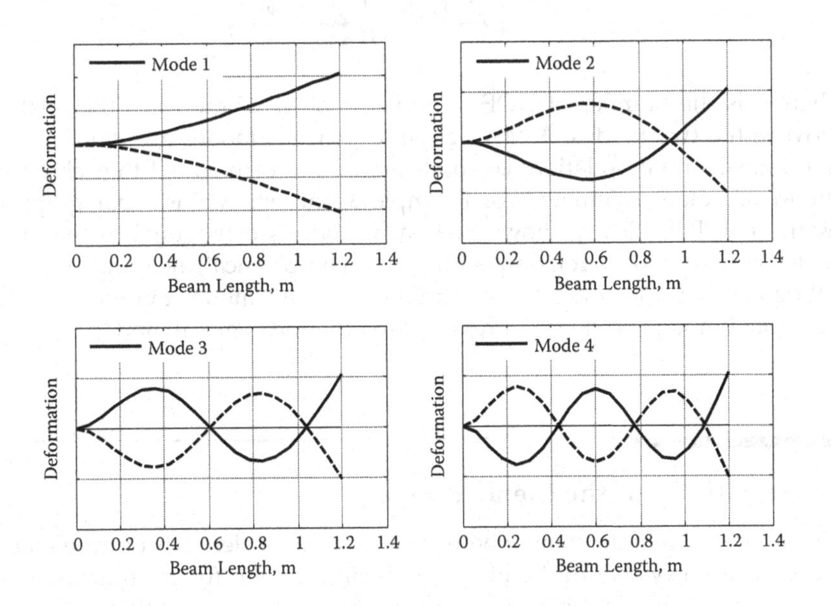

FIGURE 3.17
Mode shapes of the cantilever beam of Example 3.4.

The mode shapes related to the bending deflection in Table 3.1 for modes 1 to 4 are shown in Figure 3.17 and represent the deformation pattern at time *t*. To further explain the concept of the mode shape deformation pattern, the beam is assumed to be vibrating at the first mode, f_{n1} = 11.64 Hz, where the time period $T = {}^{1}\!/_{f_{n1}}$ = 0.068. This means that each point on the beam is undergoing a complete sinusoidal cycle in a time period T as per the first mode shape. It is also explained pictorially in Figure 3.18. Figures on the left-hand side column show the vibration of node 11 (free end) with a circle pointer, whereas the figures on the right-hand side column show the beam deformation corresponding to the instant of the circle pointer during the beam vibration. Arrowheads in Figure 3.18 show the direction of beam motion at a specific instant of vibration of the beam.

Modes are generally orthogonal to one another. It is generally examined by the modal assurance criterion (MAC), which is defined and computed as

$$MAC_{ij} = \frac{\left|\varphi_i^T \varphi_j\right|^2}{\left(\varphi_i^T \varphi_i\right)\left(\varphi_j^T \varphi_j\right)} \tag{3.46}$$

$$MAC_{ij} = \frac{\left|\sum_{k=1}^{n} \phi_{k,i}\phi_{k,j}\right|^2}{\left(\sum_{k=1}^{n} \phi_{k,i}\phi_{k,i}\right)\left(\sum_{k=1}^{n} \phi_{k,j}\phi_{k,j}\right)} \tag{3.47}$$

where n is the number of DOFs. The value of MAC indicates the relation between the two modes. A MAC close to 1 means modes are 100% related and 0 means no correlation; i.e., both modes are orthogonal to each other. The MAC values calculated for Example 3.4 are shown as a bar graph in Figure 3.19. This clearly shows that each mode is orthogonal to the other modes and has 100% relation with itself. The physical meaning of mode orthogonality is that when the system is excited at a mode, it vibrates only at the excited mode and does not respond to other orthogonal modes.

3.9 Sensitivity of the Element Size

It is important to understand whether the number of elements or the element size used for a system in the FE model is fine enough for accurate analysis and results. Example 3.4 of the beam with cantilever condition is considered again. Now the natural frequencies are again computed for the cantilever beam starting from 4 equal-size elements to 60 equal-size elements to observe the impact of the element size on the computed natural frequencies. The variation in the computed natural frequencies with respect to the number of elements for the first four modes is shown in Figure 3.20. Both the lumped and consistent mass matrices are used to observe the usefulness of the consistent mass matrix. It is obvious from Figure 3.20 that the computed

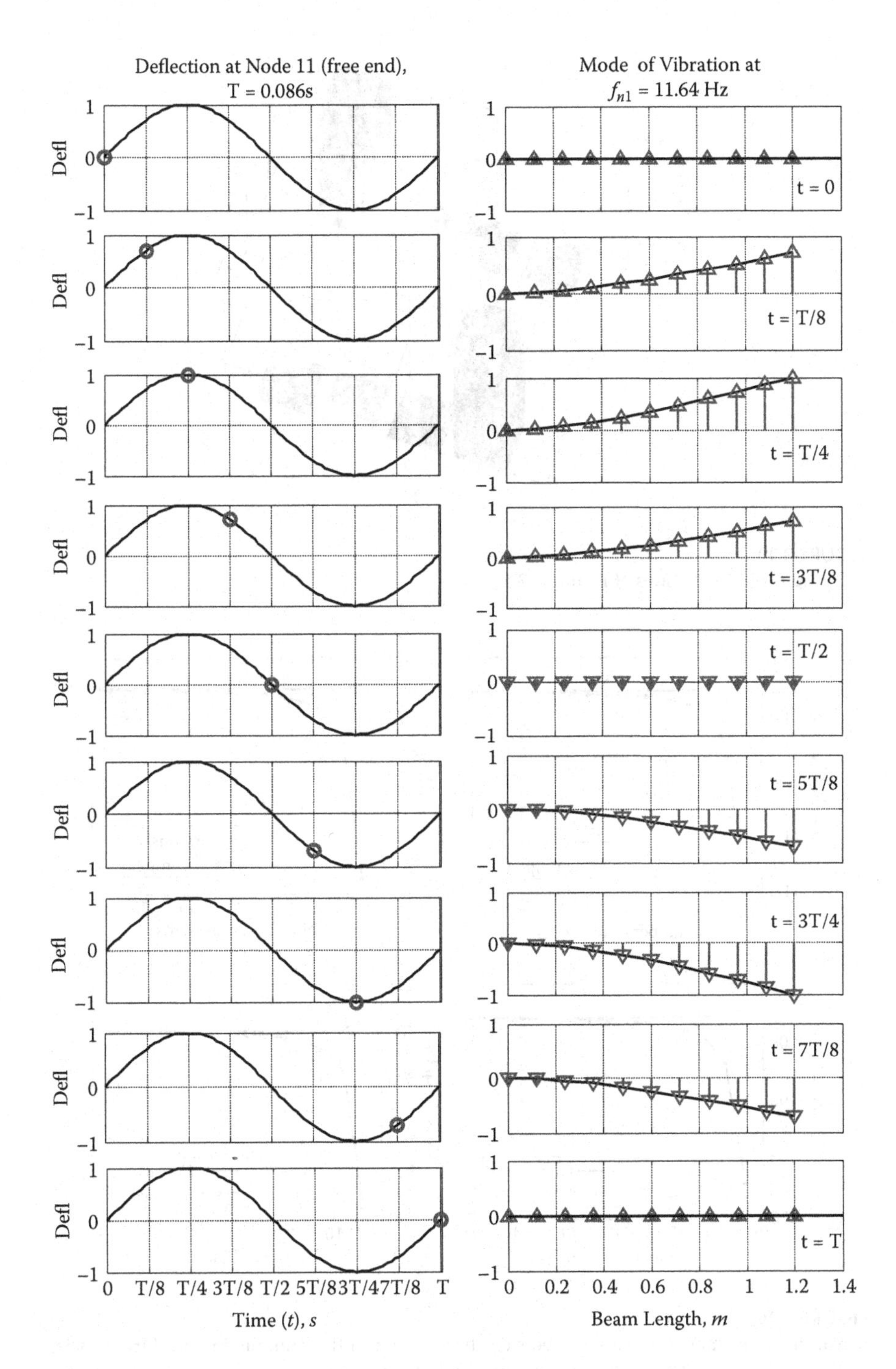

FIGURE 3.18
Mode of vibration of the cantilever beam at mode 1.

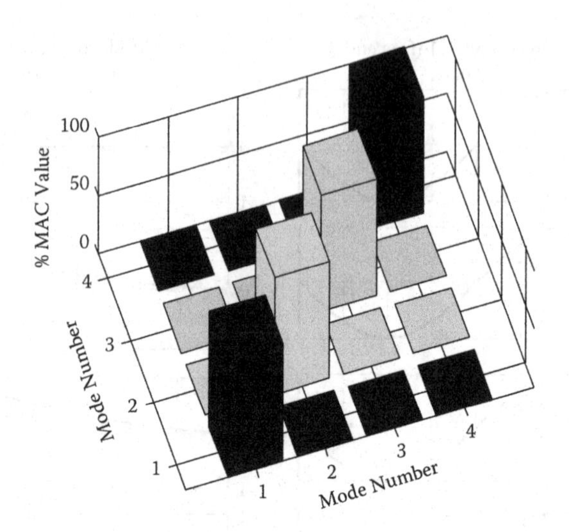

FIGURE 3.19
Bar graphs of MAC values of Example 3.4.

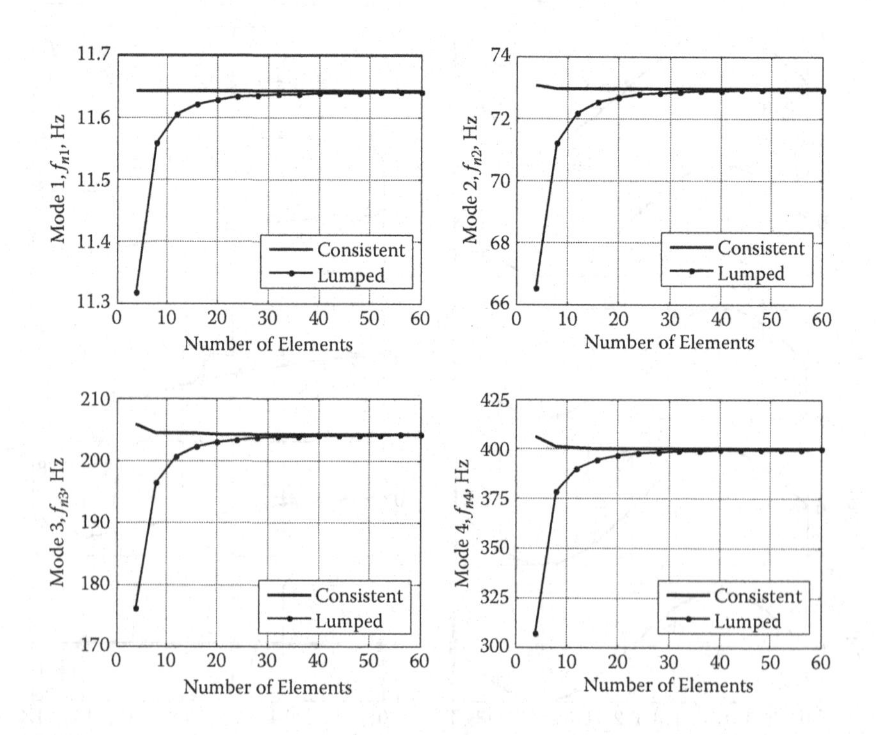

FIGURE 3.20
Sensitivity of element size and numbers on the accuracy of the computed natural frequencies (solid line for consistent mass, line with dots for lumped mass).

frequencies show large errors when a smaller number of elements are used for the lumped mass matrix. Using a large number of elements for the lumped mass matrix yields much better results. However, the use of 10 or more elements provides much more accurate results when the consistent mass matrix is used. This study simply indicates that the consistent mass matrix is a better choice. The use of a large number of small-size elements for any system is always better in the FE model, but this often increases the computational time and cost for large and complex systems. Hence, an optimized selection of element size and number of elements is important. This can be done initially by comparing the computed natural frequencies with different numbers of elements to find out the optimum element size and number of elements for further analysis.

3.10 Damping Modeling

Damping generally causes the dissipation of energy in the vibrating system, which depends on a number of different factors, including joints, frictions, surrounding medium, material intermolecular force, strength of the applied loads, etc. The damping matrix cannot be constructed using the material properties and geometrical dimensions like the mass and stiffness matrices. In fact, the construction of damping matrix is a really difficult part in the mathematical modeling. However, the use of damping in vibration and dynamics analysis is an integral part of any system; hence, the damping matrix is essential. Different approaches are discussed in the literature based on the experimental observations for the damping matrix, **C**, but the mass and stiffness proportional damping is a simple method and generally accepted approach for many applications. It can be expressed mathematically as

$$\mathbf{C} = a\mathbf{M} + b\mathbf{K} \tag{3.48}$$

where a and b are the proportional coefficients for the mass and stiffness matrices, respectively. Equation (3.48) for a SDOF system can be written as

$$c = am + bk \tag{3.49}$$

$$\Rightarrow \frac{c}{m} = a + b\frac{k}{m}$$

$$\Rightarrow 2\zeta_n\omega_n = a + b\omega_n^2 \tag{3.50}$$

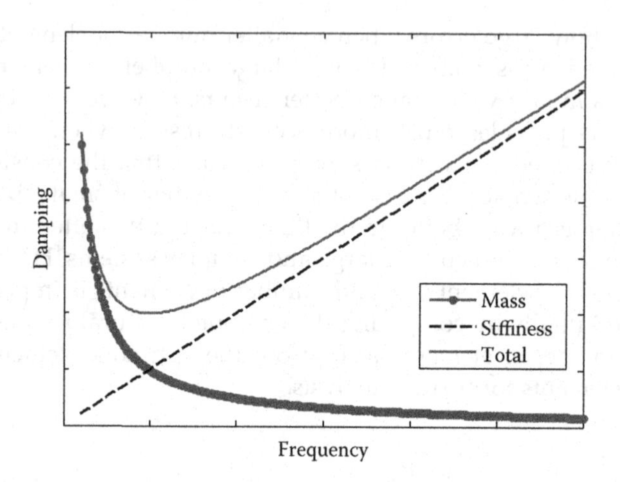

FIGURE 3.21
Damping variation with frequency in the mass and stiffness proportional damping.

It is obvious from Equation (3.50) that the mass-dependent damping, $\zeta_{n,m} = \frac{a}{2\omega_n}$, decreases with the increase in natural frequency, and stiffness-dependent damping, $\zeta_{n,k} = \frac{b}{2}\omega_n$, increases linearly with frequency. Overall, resultant damping, $\zeta_n = \zeta_{n,m} + \zeta_{n,k}$, decreases first and then increases, with the increase in the frequency that is shown in Figure 3.21. This proportional damping often represents the actual dynamic behavior of many systems.

If the mode shape vectors are $\varphi = [\varphi_1\ \varphi_2 \ldots \varphi_r \ldots \varphi_s]$ for s number of modes, then multiplying Equation (3.48) by φ^T and φ gives

$$\varphi^T\mathbf{C}\varphi = a\varphi^T\mathbf{M}\varphi + b\varphi^T\mathbf{K}\varphi$$

$$\varphi^T\mathbf{C}\varphi = a\mathbf{I} + b\varphi^T\mathbf{K}\varphi \tag{3.51}$$

where **I** is the identity matrix of size $s \times s$. Matrices $\varphi^T\mathbf{C}\varphi$ and $\varphi^T\mathbf{K}\varphi$ are given by

$$\varphi^T\mathbf{C}\varphi = \begin{bmatrix} 2\zeta_{n_1}\omega_{n_1} & 0 & \cdots & 0 & \cdots & 0 \\ 0 & 2\zeta_{n_2}\omega_{n_2} & \cdots & 0 & \cdots & 0 \\ \vdots & \vdots & \ddots & \vdots & \cdots & \vdots \\ 0 & 0 & 0 & 2\zeta_{n_r}\omega_{n_r} & 0 & 0 \\ \vdots & \vdots & \vdots & \vdots & \ddots & \vdots \\ 0 & 0 & 0 & 0 & \cdots & 2\zeta_{n_s}\omega_{n_s} \end{bmatrix} \tag{3.52}$$

$$\varphi^T \mathbf{K} \varphi = \begin{bmatrix} \omega_{n1}^2 & 0 & \cdots & 0 & \cdots & 0 \\ 0 & \omega_{n2}^2 & \cdots & 0 & \cdots & 0 \\ \vdots & \vdots & \ddots & \vdots & \cdots & \vdots \\ 0 & 0 & 0 & \omega_{nr}^2 & 0 & 0 \\ \vdots & \vdots & \vdots & \vdots & \ddots & \vdots \\ 0 & 0 & 0 & 0 & \cdots & \omega_{ns}^2 \end{bmatrix} \tag{3.53}$$

3.11 Summary

A simplified concept of DOFs and FE modeling is introduced. Formal methods for the derivation of mass, stiffness, and damping matrices are discussed. The modal analysis to compute the natural frequencies and mode shapes is explained through eigenvalue and eigenvector analysis. A few examples are used to aid these concepts for better understanding.

References

Cook, R.D., Malkus, D.S., Plesha, M.E. 1989. *Concept and Applications of Finite Element Analysis*. Wiley, New York.

Fenner, R.T. 1986. *Engineering Elasticity Application of Numerical and Analytical Techniques*. Ellis Horwood Ltd., West Sussex, England.

Thomson, W.T., Dahleh, M.D. 1998. *Theory of Vibration with Applications*, 5th ed. Prentice Hall, Englewood Cliffs, NJ.

Zienkiewicz, O.C., Taylor, R.L. 1989. *The Finite Element Method: Basic Formulations and Linear Problems*, vol. 1, 4th ed. McGraw-Hill Book Company, London, U.K.

4

Force Response Analysis

4.1 Introduction

The response (acceleration, velocity, and displacement) estimation for the vibration load is important to understand the dynamic behavior of structures, machines, etc. The dynamic equation of a system is given by

$$\mathbf{M}\ddot{\mathbf{r}}(t) + \mathbf{C}\dot{\mathbf{r}}(t) + \mathbf{K}\mathbf{r}(t) = \mathbf{F}(t) \tag{4.1}$$

where **M**, **C**, and **K** are the system mass, damping, and stiffness matrices, respectively. If the number of DOFs at a node is six and the total number of nodes is equal to n, then the system matrices size will be $6n \times 6n$. The response vectors, $\mathbf{r}(t)$, and the force vectors, $\mathbf{F}(t)$, at time t can be written as

$$\mathbf{r}(t) = \mathbf{r}_t = \begin{Bmatrix} x_{1\_t} \\ y_{1\_t} \\ z_{1\_t} \\ \theta_{x1\_t} \\ \theta_{y1\_t} \\ \theta_{z1\_t} \\ x_{2\_t} \\ y_{2\_t} \\ z_{2\_t} \\ \theta_{x2\_t} \\ \theta_{y2\_t} \\ \theta_{z2\_t} \\ \vdots \\ x_{n\_t} \\ y_{n\_t} \\ z_{n\_t} \\ \theta_{xn\_t} \\ \theta_{yn\_t} \\ \theta_{zn\_t} \end{Bmatrix} \text{ and } \mathbf{F}(t) = \mathbf{F}_t = \begin{Bmatrix} F_{x1\_t} \\ F_{y1\_t} \\ F_{z1\_t} \\ M_{x1\_t} \\ M_{y1\_t} \\ M_{z1\_t} \\ F_{x2\_t} \\ F_{y2\_t} \\ F_{z2\_t} \\ M_{x2\_t} \\ M_{y2\_t} \\ M_{z2\_t} \\ \vdots \\ F_{xn\_t} \\ F_{yn\_t} \\ F_{zn\_t} \\ M_{xn\_t} \\ M_{yn\_t} \\ M_{zn\_t} \end{Bmatrix} \tag{4.2}$$

If the time increment is Δt and the total number of time steps is q, then the time data are $t = [0 \quad \Delta t \quad 2\Delta t \quad \cdots \quad (q-1)\Delta t]^T$ and the corresponding force and response vectors are $\mathbf{F}(t) = \mathbf{F}_t = [\mathbf{F}_0 \quad \mathbf{F}_{\Delta t} \quad \mathbf{F}_{2\Delta t} \quad \cdots \quad \mathbf{F}_{(q-1)\Delta t}]^T$ and $\mathbf{r}(t) = \mathbf{r}_t = [\mathbf{r}_0 \quad \mathbf{r}_{\Delta t} \quad \mathbf{r}_{2\Delta t} \quad \cdots \quad \mathbf{r}_{(q-1)\Delta t}]^T$, respectively. The responses can be calculated either by the *direct integration method* or the *mode superposition method*. Both methods are discussed here.

4.2 Direct Integration (DI) Method

There are a number of integration schemes (methods) available for solving Equation (4.1) to calculate the response vectors; a few are listed here (Bathe and Wilson, 1978):

1. Central difference method
2. Runge-Kutta method
3. Wilson-θ method
4. Houbolt method
5. Newmark-β method

The Newmark-β method is perhaps the more popular and acceptable method for response estimation in dynamic analysis because it provides an unconditional stable result. Hence, this method is only discussed here in the following steps.

Step 1: Choose the time step, Δt. The selection of the time step depends on the highest frequency (f_{max}) to be included in the response. It is reasonable to select a time step equal to 1/10 of the minimum time period $\left(T_{min} = \frac{1}{f_{max}}\right)$ of the maximum frequency or smaller, i.e., $\Delta t \leq \frac{T_{min}}{10}$ for the better solution. Let's say the highest frequency required in the responses for a system is up to 100 Hz, i.e., f_{max}= 100 Hz; then choose $\Delta t \leq 1$ ms. If the chosen $\Delta t = 1$ ms, then the sampling rate (f_s) becomes equal to 1000 samples/s for the computed responses.

Step 2: Assume the initial condition at time $t = 0$ for the response vectors $(\mathbf{r}_0, \dot{\mathbf{r}}_0, \ddot{\mathbf{r}}_0)$ for all DOFs ($6n$). Generally, $(\mathbf{r}_0 = 0, \dot{\mathbf{r}}_0 = 0, \text{ and } \ddot{\mathbf{r}}_0 = 0)$ if nothing is known.

Step 3: Estimate the following at any time $t_k = (k-1)\Delta t$:

$$\hat{\mathbf{r}}_{t_k} = \mathbf{r}_{t_{k-1}} + \Delta t \dot{\mathbf{r}}_{t_{k-1}} + \frac{\Delta t}{2}(1-2\beta)\Delta t \ddot{\mathbf{r}}_{t_{k-1}} \tag{4.3}$$

$$\hat{\dot{\mathbf{r}}}_{t_k} = \dot{\mathbf{r}}_{t_{k-1}} + (1-\alpha)\Delta t \ddot{\mathbf{r}}_{t_{k-1}} \tag{4.4}$$

and

$$\hat{\mathbf{K}} = \mathbf{M} + \alpha \Delta t \mathbf{C} + \beta \Delta t^2 \mathbf{K} \tag{4.5}$$

Step 4: Estimate the acceleration $(\ddot{\mathbf{r}}_{t_k})$, velocity $(\dot{\mathbf{r}}_{t_k})$, and displacement $(\ddot{\mathbf{r}}_{t_k})$ at time t_k using the following equations:

$$\ddot{\mathbf{r}}_{t_k} = \hat{\mathbf{K}}^{-1}\left(\mathbf{F}_{t_k} - \mathbf{C}\hat{\dot{\mathbf{r}}}_{t_k} - \mathbf{K}\hat{\mathbf{r}}_{t_k}\right) \tag{4.6}$$

$$\dot{\mathbf{r}}_{t_k} = \hat{\dot{\mathbf{r}}}_{t_k} + \alpha \Delta t \ddot{\mathbf{r}}_{t_k} \tag{4.7}$$

$$\mathbf{r}_{t_k} = \hat{\mathbf{r}}_{t_k} + \beta \Delta t^2 \ddot{\mathbf{r}}_{t_k} \tag{4.8}$$

where $\alpha \geq \frac{1}{2}$ and $\beta \leq \frac{1}{4}$. Generally, the use of $\alpha = \frac{1}{2}$ and $\beta = \frac{1}{4}$ provides solution stability.

Step 5: Repeat steps 3 to 4 to complete the response estimation at the time steps $t_2 = \Delta t$, $t_3 = 2\Delta t$, $t_4 = 3\Delta t$, …, $t_q = (q-1)\Delta t$.

Step 6: Once the response data computation is completed for all time steps, then computed data should be low-pass filtered at the frequency, f_{max}, to remove any unexpected computational error beyond this frequency.

4.2.1 Example 4.1: A SDOF System

A SDOF system, shown in Figure 4.1, is considered here for simple illustration. Let us assume that the mass, $\mathbf{M} = \mathrm{m} = 1$ kg, the stiffness, $\mathbf{K} = \mathrm{k} = 986.9604$ Nm/s, and the damping, $\mathbf{C} = \mathrm{c} = 5$ Ns/m for the SDOF system, which yields the natural frequency $f_n = 5$ Hz, the normalized mode shape, $\varphi = \phi = 1$, and the damping ratio, $\zeta = 7.96\%$. The applied force is assumed to be the sinusoidal force at the natural frequency, i.e., $\mathrm{F}(t) = \sin(2\pi f n t)$ for 256 s. It is also assumed

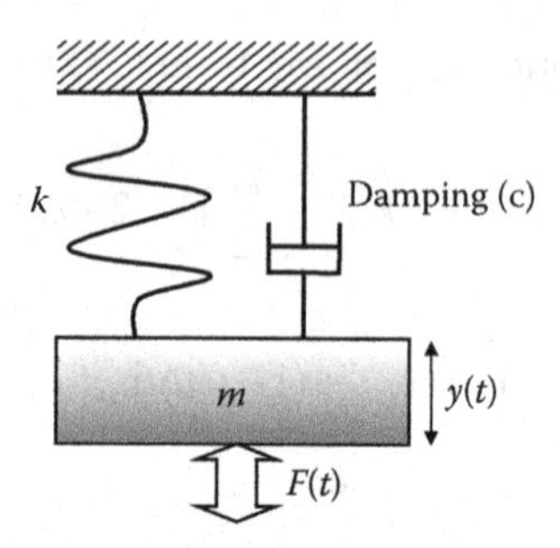

FIGURE 4.1
SDOF system of Example 4.1.

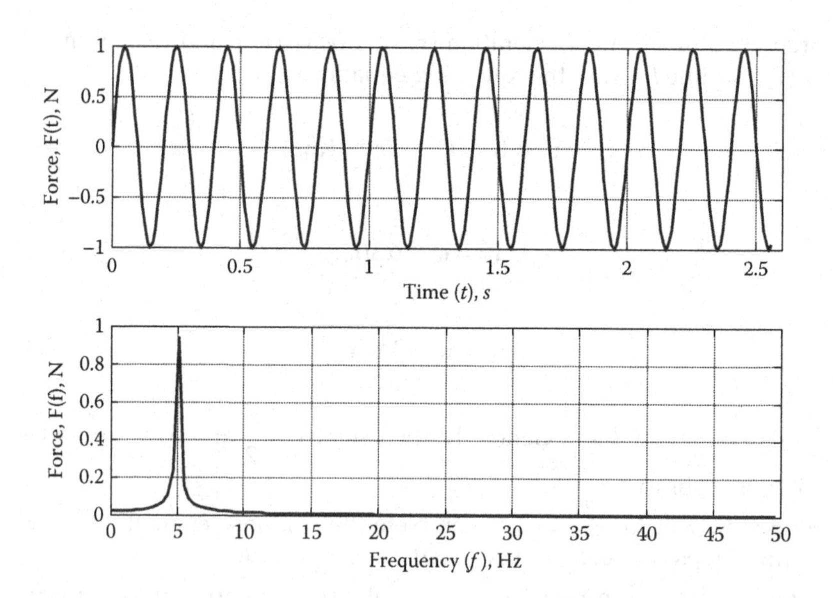

FIGURE 4.2
Applied force in time and frequency domains.

that $f_{max} = 10$ Hz is required for the present case. Hence, the time step 0.01 s has been selected so that the time $t = [000.1 \quad 0.02 \quad 0.03 \cdots 256]^T$ s and the corresponding force vectors $\mathbf{F}(t) = [0 \quad 0.3090 \quad 0.5878 \quad 0.8090 \cdots -0.9511]^T$ N are shown in Figure 4.2.

The initial responses at time $t_1 = 0$ s were $\mathbf{r}_0 = y_0 = y_{t_1} = 0$, $\dot{\mathbf{r}}_0 = \dot{y}_0 = \dot{y}_{t_1} = 0$, and $\ddot{\mathbf{r}}_0 = \ddot{y}_0 = \ddot{y}_{t_1} = 0$. The constants are $\alpha = \frac{1}{2}$ and $\beta = \frac{1}{4}$.

At time $t_2 = 0.01$s:

$$\text{Equation (4.3)},\ \hat{y}_{t_2} = y_{t_1} + (0.01)\dot{y}_{t_1} + \frac{(0.01)}{2}(1-2\beta)(0.01)\ddot{y}_{t_1}$$

$$\Rightarrow \hat{y}_{t2} = 0 + (0.01)(0) + \frac{(0.01)}{2}(1-2\beta)(0.01)(0) = 0$$

$$\text{Equation (4.4), } \dot{\hat{y}}_{t2} = \dot{y}_{t1} + (1-\alpha)(0.01)\ddot{y}_{t1}$$

$$\Rightarrow \dot{\hat{y}}_{t2} = 0 + (1-\alpha)(0.01)(0) = 0$$

$$\text{Equation (4.5), } \hat{\mathbf{K}} = \mathbf{M} + \alpha\Delta t\mathbf{C} + \beta\Delta t^2\mathbf{K}$$

$$\Rightarrow \hat{\mathbf{K}} = 1 + \alpha(0.01)\mathbf{C} + \beta(0.01)^2\mathbf{K} = 1.0497$$

The response estimations at time $t_2 = 0.01$ s:

$$\text{Acceleration, Equation (4.6), } a_{t_2} = \ddot{y}_{t_2} = \hat{\mathbf{K}}^{-1}\left(\mathbf{F}_{t_2} - \mathbf{C}\dot{\hat{y}}_{t_2} - \mathbf{K}\hat{y}_{t_2}\right)$$

$$\Rightarrow a_{t_2} = \ddot{y}_{t_2} = 1.0497^{-1}\left(0.3090 - 5\times 0 - 986.9604\times 0\right) = 0.2944 \text{ m/s}^2$$

$$\text{Velocity, Equation (4.7), } v_{t_2} = \dot{y}_{t_2} = \dot{\hat{y}}_{t_2} + \alpha\Delta t\ddot{y}_{t_2}$$

$$\Rightarrow v_{t_2} = \dot{y}_{t_2} = 0 + \frac{1}{2}\times 0.01\times 0.2944 = 0.001472 \text{ m/s}$$

$$\text{Displacement, Equation (4.8), } y_{t2} = \hat{y}_{t2} + \beta\Delta t^2\ddot{y}_{t2}$$

$$\Rightarrow y_{t_2} = 0 + \frac{1}{4}\times 0.01^2\times 0.2944 = 7.4\text{e-}6 \text{ m}$$

Similarly, calculate the responses at $t = t_3 \quad t_4 \quad \ldots \quad t_q$, where $t_q = (q-1)\Delta t$ = 256 s. The force and the calculated responses are shown in Figure 4.3 and also listed in Table 4.1.

4.2.2 Example 4.2: A Two-DOF System

A two-DOF system is shown in Figure 4.4. The masses $m_1 = m_2 = 1$ kg, the stiffnesses $k_1 = k_2 = 1000$ N/m, and the damping values $c_1 = c_2 = 0.5$ Ns/m have been assumed for the system. Hence, the system mass, damping, and stiffness matrices can be written as

$$\mathbf{M} = \begin{bmatrix} m_1 & 0 \\ 0 & m_2 \end{bmatrix}, \mathbf{C} = \begin{bmatrix} c_1 & -c_1 \\ -c_1 & c_1 + c_2 \end{bmatrix}, \text{ and } \mathbf{K} = \begin{bmatrix} k_1 & -k_1 \\ -k_1 & k_1 + k_2 \end{bmatrix} \quad (4.9)$$

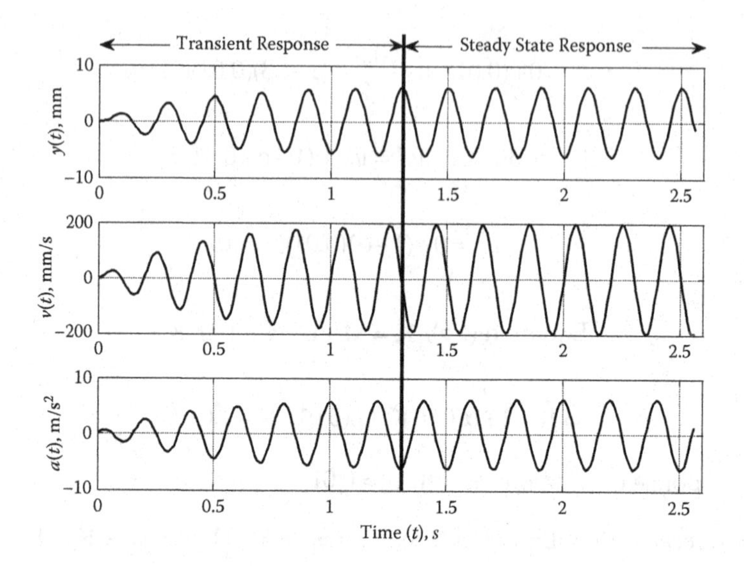

FIGURE 4.3
The calculated responses for Example 4.1.

$$\Rightarrow \mathbf{M} = \begin{bmatrix} 1 & 0 \\ 0 & 1 \end{bmatrix} \text{kg}, \ \mathbf{C} = \begin{bmatrix} 0.5 & -0.5 \\ -0.5 & 1 \end{bmatrix} \text{Ns/m}$$

$$\text{and } \mathbf{K} = \begin{bmatrix} 1000 & -1000 \\ -1000 & 2000 \end{bmatrix} \tag{4.10}$$

which yields the natural frequencies and the normalized mode shapes:

$$\{f_n\} = \begin{Bmatrix} f_{n1} \\ f_{n2} \end{Bmatrix} = \begin{Bmatrix} 3.1105 \\ 8.1434 \end{Bmatrix} \text{Hz}$$

and

$$\varphi = \begin{bmatrix} \varphi_1 & \varphi_2 \end{bmatrix} = \begin{bmatrix} \phi_{y1,1} & \phi_{y1,2} \\ \phi_{y2,1} & \phi_{y2,2} \end{bmatrix} = \begin{bmatrix} -0.8507 & -0.5257 \\ -0.5257 & 0.8507 \end{bmatrix} \tag{4.11}$$

The force is assumed to be a sweep-sine excitation at node 2, and there is no force at node 1. The mathematical expression for the applied force vectors is given by Equation (4.12):

$$\mathbf{F}(t) = \begin{Bmatrix} F_1(t) \\ F_2(t) \end{Bmatrix} = \begin{Bmatrix} F_{1\_t} \\ F_{2\_t} \end{Bmatrix} = \begin{Bmatrix} 0 \\ \sin(2\pi f t) \end{Bmatrix} \tag{4.12}$$

TABLE 4.1
Applied Force and the Calculated Responses

Time, t, s	Force, F(t), N	Displacement, y(t), mm	Velocity, v(t), mm/s	Acceleration, a(t), m/s^2
0	0	0	0	0
0.0100	0.3090	0.0074	1.4720	0.2944
0.0200	0.5878	0.0424	5.5353	0.5183
0.0300	0.8090	0.1264	11.2662	0.6279
0.0400	0.9511	0.2697	17.3953	0.5979
0.0500	1.0000	0.4692	22.5066	0.4244
0.0600	0.9511	0.7080	25.2582	0.1260
0.0700	0.8090	0.9573	24.5941	–0.2588
0.0800	0.5878	1.1799	19.9187	–0.6763
0.0900	0.3090	1.3355	11.2115	–1.0651
0.1000	0.0000	1.3869	–0.9349	–1.3641
0.1100	–0.3090	1.3054	–15.3588	–1.5206
0.1200	–0.5878	1.0764	–30.4514	–1.4979
0.1300	–0.8090	0.7024	–44.3435	–1.2805
0.1400	–0.9511	0.2050	–55.1349	–0.8777
0.1500	–1.0000	–0.3763	–61.1379	–0.3229
0.1600	–0.9511	–0.9876	–61.1064	0.3292
0.1700	–0.8090	–1.5652	–54.4211	1.0079
0.1800	–0.5878	–2.0433	–41.2070	1.6350
0.1900	–0.3090	–2.3612	–22.3660	2.1332
0.2000	–0.0000	–2.4706	0.4803	2.4360
0.2100	0.3090	–2.3426	25.1371	2.4953
0.2200	0.5878	–1.9716	49.0557	2.2884
0.2300	0.8090	–1.3783	69.6042	1.8213
0.2400	0.9511	–0.6085	84.3598	1.1298
0.2500	1.0000	0.2703	91.3902	0.2763
0.2600	0.9511	1.1747	89.4927	–0.6558
0.2700	0.8090	2.0140	78.3613	–1.5705
0.2800	0.5878	2.6991	58.6618	–2.3694
0.2900	0.3090	3.1524	32.0033	–2.9623
0.3000	0.0000	3.3165	0.8056	–3.2772
⋮	⋮	⋮	⋮	⋮
⋮	⋮	⋮	⋮	⋮
2.4900	0.3090	5.7369	80.2653	–5.7544
2.5000	–0.0000	6.2392	20.1994	–6.2588
2.5100	–0.3090	6.1309	–41.8485	–6.1508
2.5200	–0.5878	5.4227	–99.8059	–5.4407
2.5300	–0.8090	4.1836	–147.9999	–4.1981
2.5400	–0.9511	2.5351	–181.7129	–2.5445
2.5500	–1.0000	0.6383	–197.6441	–0.6417
2.5600	–0.9511	–1.3211	–194.2329	1.3240

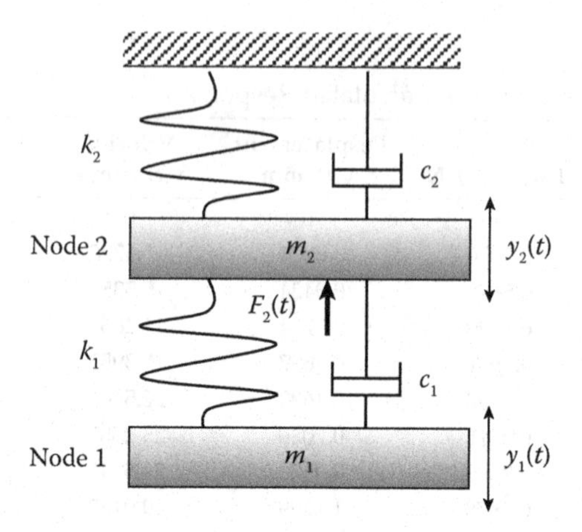

FIGURE 4.4
Two-DOF system of Example 4.2.

where F_{1_t} and F_{2_t} are the applied force vectors at nodes 1 and 2, respectively, at the time t, and f is the exciting frequency varying linearly from 1.5 Hz at time $t_1 = 0$ s to 9.5 Hz at time $t_q = (q-1)\Delta t = 10.24$ s, where the time step $\Delta t = 0.01$ s. This means that the excitation frequency of the applied force will pass through natural frequencies, $f_{n1} = 3.1105$ Hz and $f_{n2} = 8.1434$ Hz, at the times 2.0614 and 8.5036 s, respectively. Hence, the system will resonate for a short period when the exciting frequency passes through the natural frequencies. The applied forces are shown in Figure 4.5.

Now the following responses have been computed, which are explained in the following steps:

$$\text{Displacement, } \mathbf{r}_t = \begin{Bmatrix} y_1(t) \\ y_2(t) \end{Bmatrix} = \begin{Bmatrix} y_{1\_t} \\ y_{2\_t} \end{Bmatrix} \tag{4.13}$$

$$\text{Velocity, } \dot{\mathbf{r}}_t = \begin{Bmatrix} \dot{y}_1(t) = v_1(t) \\ \dot{y}_2(t) = v_1(t) \end{Bmatrix} = \begin{Bmatrix} \dot{y}_{1\_t} = v_{1\_t} \\ \dot{y}_{2\_t} = v_{2\_t} \end{Bmatrix} \tag{4.14}$$

$$\text{Acceleration, } \ddot{\mathbf{r}}_t = \begin{Bmatrix} \ddot{y}_1(t) = a_1(t) \\ \ddot{y}_2(t) = a_1(t) \end{Bmatrix} = \begin{Bmatrix} \ddot{y}_{1\_t} = a_{1\_t} \\ \ddot{y}_{2\_t} = a_{2\_t} \end{Bmatrix} \tag{4.15}$$

The initial responses at time $t_1 = 0$ s:

$$\text{Initial displacement, } \mathbf{r}_0 = \begin{Bmatrix} y_1(0) \\ y_2(0) \end{Bmatrix} = \begin{Bmatrix} y_1(t_1) \\ y_2(t_1) \end{Bmatrix} = \begin{Bmatrix} y_{1\_t_1} \\ y_{2\_t_1} \end{Bmatrix} = \begin{Bmatrix} 0 \\ 0 \end{Bmatrix}$$

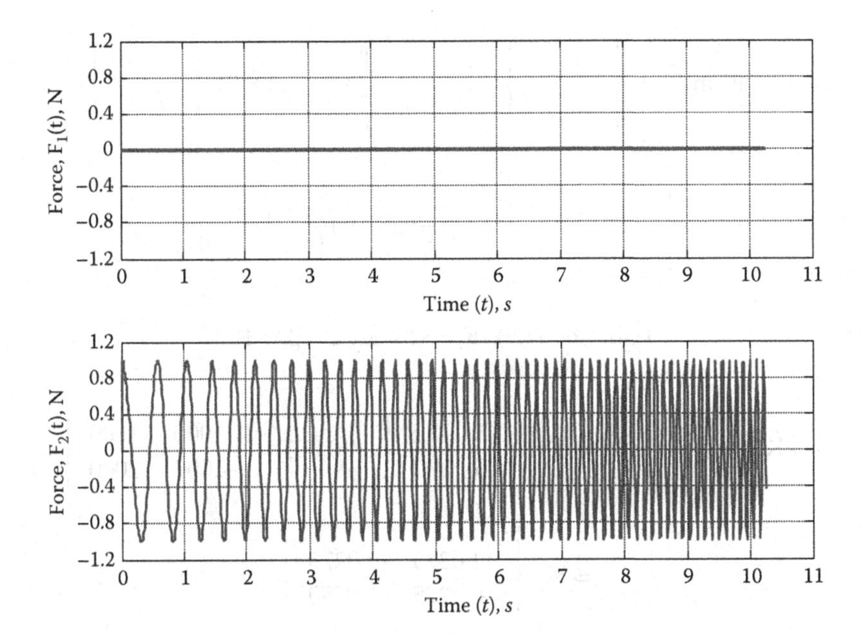

FIGURE 4.5
The applied force at nodes 1 and 2 for Example 4.2.

$$\text{Initial velocity, } \dot{\mathbf{r}}_0 = \begin{Bmatrix} v_1(0) \\ v_2(0) \end{Bmatrix} = \begin{Bmatrix} v_1(t_1) \\ v_2(t_1) \end{Bmatrix} = \begin{Bmatrix} v_{1\_t_1} \\ v_{2\_t_1} \end{Bmatrix} = \begin{Bmatrix} 0 \\ 0 \end{Bmatrix}$$

$$\text{Initial acceleration, } \ddot{\mathbf{r}}_0 = \begin{Bmatrix} a_1(0) \\ a_2(0) \end{Bmatrix} = \begin{Bmatrix} a_1(t_1) \\ a_2(t_1) \end{Bmatrix} = \begin{Bmatrix} a_{1\_t_1} \\ a_{2\_t_1} \end{Bmatrix} = \begin{Bmatrix} 0 \\ 0 \end{Bmatrix}$$

At time $t_2 = 0.01$ s:

$$\text{Equation (4.3), } \begin{Bmatrix} \hat{y}_{1\_t_2} \\ \hat{y}_{2\_t_2} \end{Bmatrix} = \begin{Bmatrix} y_{1\_t_1} \\ y_{2\_t_1} \end{Bmatrix} + (0.01)\begin{Bmatrix} v_{1\_t_1} \\ v_{2\_t_1} \end{Bmatrix} + \frac{(0.01)}{2}(1-2\beta)(0.01)\begin{Bmatrix} a_{1\_t_1} \\ a_{2\_t_1} \end{Bmatrix}$$

$$\Rightarrow \begin{Bmatrix} \hat{y}_{1\_t_2} \\ \hat{y}_{2\_t_2} \end{Bmatrix} = \begin{Bmatrix} 0 \\ 0 \end{Bmatrix} + (0.01)\begin{Bmatrix} 0 \\ 0 \end{Bmatrix} + \frac{(0.01)}{2}(1-2\beta)(0.01)\begin{Bmatrix} 0 \\ 0 \end{Bmatrix} = \begin{Bmatrix} 0 \\ 0 \end{Bmatrix}$$

Equation (4.4), $\begin{Bmatrix} \dot{\hat{y}}_{1\_t_2} \\ \dot{\hat{y}}_{2\_t_2} \end{Bmatrix} = \begin{Bmatrix} v_{1\_t_1} \\ v_{2\_t_1} \end{Bmatrix} + (1-\alpha)(0.01) \begin{Bmatrix} a_{1\_t_1} \\ a_{2\_t_1} \end{Bmatrix}$

$$\Rightarrow \begin{Bmatrix} \dot{\hat{y}}_{1\_t_2} \\ \dot{\hat{y}}_{2\_t_2} \end{Bmatrix} = \begin{Bmatrix} 0 \\ 0 \end{Bmatrix} + (1-\alpha)(0.01) \begin{Bmatrix} 0 \\ 0 \end{Bmatrix} = \begin{Bmatrix} 0 \\ 0 \end{Bmatrix}$$

Equation (4.5), $\hat{\mathbf{K}} = \mathbf{M} + \alpha\Delta t\mathbf{C} + \beta\Delta t^2\mathbf{K}$

$$\Rightarrow \hat{\mathbf{K}} = \begin{bmatrix} 1 & 0 \\ 0 & 1 \end{bmatrix} + \alpha(0.01) \begin{bmatrix} 0.5 & -0.5 \\ -0.5 & 1 \end{bmatrix} + \beta(0.001)^2 \begin{bmatrix} 1000 & -1000 \\ -1000 & 2000 \end{bmatrix}$$

$$\Rightarrow \hat{\mathbf{K}} = \begin{bmatrix} 1.0275 & -0.0275 \\ -0.0275 & 1.0550 \end{bmatrix}$$

where $\alpha = \dfrac{1}{2}$ and $\beta = \dfrac{1}{4}$.

The response estimations at time $t_2 = 0.01$ s:

Acceleration, Equation (4.6), $\begin{Bmatrix} a_{1\_t_2} \\ a_{2\_t_2} \end{Bmatrix} = \hat{\mathbf{K}}^{-1} \left(\mathbf{F}_{t_2} - \mathbf{C} \begin{Bmatrix} \dot{\hat{y}}_{1\_t_2} \\ \dot{\hat{y}}_{2\_t_2} \end{Bmatrix} - \mathbf{K} \begin{Bmatrix} \hat{y}_{1\_t_2} \\ \hat{y}_{2\_t_2} \end{Bmatrix} \right)$

$$\Rightarrow \begin{Bmatrix} a_{1\_t_2} \\ a_{2\_t_2} \end{Bmatrix} = \begin{bmatrix} 1.0275 & -0.0275 \\ -0.0275 & 1.0550 \end{bmatrix}^{-1} \left(\begin{Bmatrix} F_{1\_t2} \\ F_{2\_t2} \end{Bmatrix} - \mathbf{C} \begin{Bmatrix} 0 \\ 0 \end{Bmatrix} - \mathbf{K} \begin{Bmatrix} 0 \\ 0 \end{Bmatrix} \right)$$

$$= \begin{Bmatrix} 2.5247e-2 \\ 9.4331e-1 \end{Bmatrix} \text{m/s}^2$$

Velocity, Equation (4.7), $\begin{Bmatrix} v_{1\_t_2} \\ v_{2\_t_2} \end{Bmatrix} = \begin{Bmatrix} \dot{\hat{y}}_{1\_t_2} \\ \dot{\hat{y}}_{2\_t_2} \end{Bmatrix} + \alpha\Delta t \begin{Bmatrix} a_{1\_t_2} \\ a_{2\_t_2} \end{Bmatrix}$

$$\Rightarrow \begin{Bmatrix} v_{1\_t_2} \\ v_{2\_t_2} \end{Bmatrix} = \begin{Bmatrix} 0 \\ 0 \end{Bmatrix} + \alpha\Delta t \begin{Bmatrix} 2.5247e-2 \\ 9.4331e-1 \end{Bmatrix} = \begin{Bmatrix} 1.2637e-4 \\ 4.7215e-3 \end{Bmatrix} \text{m/s}$$

Displacement, Equation (4.8), $\begin{Bmatrix} y_{1\_t_2} \\ y_{2\_t_2} \end{Bmatrix} = \begin{Bmatrix} \hat{y}_{1\_t_2} \\ \hat{y}_{2\_t_2} \end{Bmatrix} + \beta \Delta t^2 \begin{Bmatrix} a_{1\_t_2} \\ a_{2\_t_2} \end{Bmatrix}$

$$\Rightarrow \begin{Bmatrix} y_{1\_t_2} \\ y_{2\_t_2} \end{Bmatrix} = \begin{Bmatrix} 0 \\ 0 \end{Bmatrix} + \beta \Delta t^2 \begin{Bmatrix} 2.5247e-2 \\ 9.4331e-1 \end{Bmatrix} = \begin{Bmatrix} 6.3183e-7 \\ 2.3607e-5 \end{Bmatrix} \text{m}$$

The estimation at the next time, $t_3 = 2\Delta t = 0.02$ s:

Equation (4.3), $\begin{Bmatrix} \hat{y}_{1\_t_3} \\ \hat{y}_{2\_t_3} \end{Bmatrix} = \begin{Bmatrix} y_{1\_t_2} \\ y_{2\_t_2} \end{Bmatrix} + (0.01) \begin{Bmatrix} v_{1\_t_2} \\ v_{2\_t_2} \end{Bmatrix}$

$$+ \frac{(0.01)}{2}(1-2\beta)(0.01) \begin{Bmatrix} a_{1\_t_2} \\ a_{2\_t_2} \end{Bmatrix}$$

$$\Rightarrow \begin{Bmatrix} \hat{y}_{1\_t_3} \\ \hat{y}_{2\_t_3} \end{Bmatrix} = \begin{Bmatrix} 6.3183e-7 \\ 2.3607e-5 \end{Bmatrix} + (0.01) \begin{Bmatrix} 1.2637e-4 \\ 4.7215e-3 \end{Bmatrix}$$

$$+ \frac{(0.01)}{2}(1-2\beta)(0.01) \begin{Bmatrix} 2.5247e-2 \\ 9.4331e-1 \end{Bmatrix}$$

$$\Rightarrow \begin{Bmatrix} \hat{y}_{1\_t_3} \\ \hat{y}_{2\_t_3} \end{Bmatrix} = \begin{Bmatrix} 2.5273e-6 \\ 9.4430e-5 \end{Bmatrix}$$

Equation (4.4), $\begin{Bmatrix} \hat{\dot{y}}_{1\_t_3} \\ \hat{\dot{y}}_{2\_t_3} \end{Bmatrix} = \begin{Bmatrix} v_{1\_t_2} \\ v_{2\_t_2} \end{Bmatrix} + (1-\alpha)(0.01) \begin{Bmatrix} a_{1\_t_2} \\ a_{1\_t_2} \end{Bmatrix}$

$$\Rightarrow \begin{Bmatrix} \hat{\dot{y}}_{1\_t_3} \\ \hat{\dot{y}}_{2\_t_3} \end{Bmatrix} = \begin{Bmatrix} 1.2637e-4 \\ 4.7215e-3 \end{Bmatrix} + (1-\alpha)(0.01) \begin{Bmatrix} 2.5247e-2 \\ 9.4331e-1 \end{Bmatrix} = \begin{Bmatrix} 2.5273e-4 \\ 9.4430e-3 \end{Bmatrix}$$

Acceleration, Equation (4.6), $\begin{Bmatrix} a_{1\_t_3} \\ a_{2\_t_3} \end{Bmatrix} = \hat{\mathbf{K}}^{-1} \left(\mathbf{F}_{t_3} - \mathbf{C} \begin{Bmatrix} \hat{\dot{y}}_{1\_t_3} \\ \hat{\dot{y}}_{2\_t_3} \end{Bmatrix} - \mathbf{K} \begin{Bmatrix} \hat{y}_{1\_t_3} \\ \hat{y}_{2\_t_3} \end{Bmatrix} \right)$

$$\Rightarrow \begin{Bmatrix} a_{1\_t3} \\ a_{2\_t3} \end{Bmatrix} = \begin{bmatrix} 1.0275 & -0.0275 \\ -0.0275 & 1.0550 \end{bmatrix}^{-1}$$

$$\left(\begin{Bmatrix} F_{1\_t3} \\ F_{2\_t3} \end{Bmatrix} - \mathbf{C} \begin{Bmatrix} 2.5273e-4 \\ 9.4430e-3 \end{Bmatrix} - \mathbf{K} \begin{Bmatrix} 2.5273e-6 \\ 9.4430e-5 \end{Bmatrix} \right)$$

$$\Rightarrow \begin{Bmatrix} a_{1\_t3} \\ a_{2\_t3} \end{Bmatrix} = \begin{Bmatrix} 1.1353e-1 \\ 7.4729e-1 \end{Bmatrix} \text{m/s}^2$$

Velocity, Equation (4.7), $\begin{Bmatrix} v_{1\_t3} \\ v_{2\_t3} \end{Bmatrix} = \begin{Bmatrix} \dot{\hat{y}}_{1\_t3} \\ \dot{\hat{y}}_{2\_t3} \end{Bmatrix} + \alpha\Delta t \begin{Bmatrix} a_{1\_t3} \\ a_{2\_t3} \end{Bmatrix}$

$$\Rightarrow \begin{Bmatrix} v_{1\_t3} \\ v_{2\_t3} \end{Bmatrix} = \begin{Bmatrix} 2.5273e-4 \\ 9.4430e-3 \end{Bmatrix}$$

$$+\alpha\Delta t \begin{Bmatrix} 1.1353e-1 \\ 7.4729e-1 \end{Bmatrix} = \begin{Bmatrix} 8.2246e-4 \\ 1.3185e-2 \end{Bmatrix} \text{m/s}$$

Displacement, Equation (4.8), $\begin{Bmatrix} y_{1\_t3} \\ y_{2\_t3} \end{Bmatrix} = \begin{Bmatrix} \hat{y}_{1\_t3} \\ \hat{y}_{2\_t3} \end{Bmatrix} + \beta\Delta t^2 \begin{Bmatrix} a_{1\_t3} \\ a_{2\_t3} \end{Bmatrix}$

$$\Rightarrow \begin{Bmatrix} y_{1\_t3} \\ y_{2\_t3} \end{Bmatrix} = \begin{Bmatrix} 2.5273e-6 \\ 9.4430e-5 \end{Bmatrix}$$

$$+\beta\Delta t^2 \begin{Bmatrix} 1.1353e-1 \\ 7.4729e-1 \end{Bmatrix} = \begin{Bmatrix} 5.3760e-6 \\ 1.1314e-4 \end{Bmatrix} \text{m}$$

Similarly, calculate the responses at $t = t_4 \quad t_5 \quad \ldots \quad t_q$, where $t_q = (q-1)\Delta t$ = 10.24 s. The calculated acceleration responses are shown in Figure 4.6 and also listed in Table 4.2. The amplitude spectra of the force at node 2 and the acceleration responses at nodes 1 and 2 are shown in Figure 4.7. Note that the Fourier transformation and spectrum calculation are discussed in Chapter 6. The presence of two peaks in the acceleration spectra is related to the natural frequencies, which indicate the occurrence of resonance at both

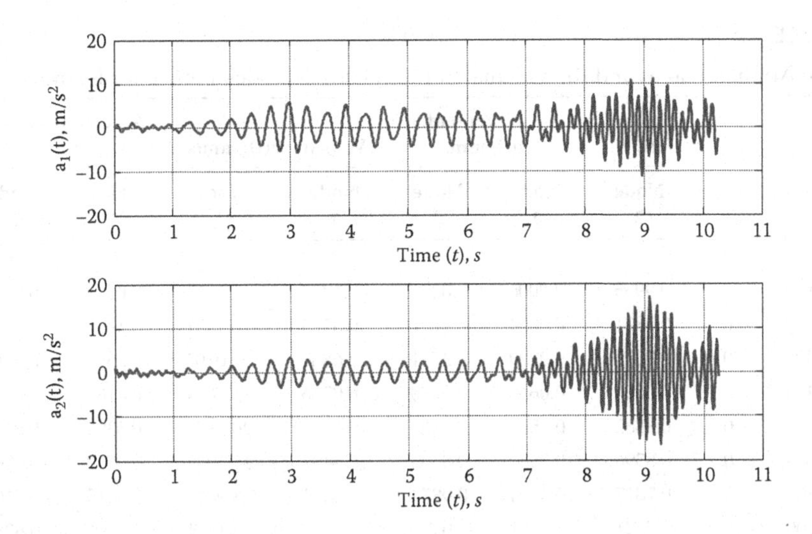

FIGURE 4.6
The calculated acceleration responses at nodes 1 and 2 for Example 4.2.

natural frequencies due to the sweep-sine excitation in the frequency band of 1.5 to 9.5 Hz. In fact, these resonances are also seen in the time domain responses shown in Figure 4.6.

4.2.3 Example 4.3: A Cantilever Beam

It is an example of a cantilever beam made of steel shown in Figure 4.8. The physical properties are length (L = 1.2 m), width (w = 20 mm), and depth (d = 20 mm), and the material properties elasticity (E = 200 GPa) and density (ρ = 7800 kg/m^3). The FE model using two-node Euler-Bernoulli beam elements for the cantilever beam is shown in Figure 4.9. The beam has been divided into 10 equal elements of length 120 mm. Only two DOFs—one for bending deflection in the y-direction and the other for associated bending rotation about the z-direction—are used for the simple simulation. However, all six DOFs can be used at each node if required. The mass and stiffness matrices are constructed as discussed in Chapter 3. Both displacement and rotational DOFs at node 1 are removed from the mass and stiffness matrices to simulate the zero displacement and rotation, respectively, at the clamped location required for the cantilever beam. The eigenvalue and eigenvector analysis is carried out to compute the natural frequencies and mode shapes of the beam. The first three computed bending modes are 23.28, 145.92, and 408.66 Hz. The damping matrix is constructed using Equation (3.46), $\mathbf{C} = a\mathbf{M} + b\mathbf{K}$, where the proportional coefficients for the mass and stiffness matrices, a and b, are assumed to be 2.664 and 3.0812e-5, respectively. This yields the damping for the first three modes as 1, 1.54, and 4%, respectively.

TABLE 4.2

The Applied Forces and the Calculated Responses at Nodes 1 and 2 for Example 4.2

	Force, F(*t*), N		Displacement, y(*t*), mm		Velocity, v(*t*), mm/s		Acceleration, a(*t*), m/s^2	
Time, *t*, s	Node 1	Node 2	Node 1	Node 2	Node 1	Node 2	Node 1	Node 2
0	0	0	0	0	0	0	0	0
0.0100	0	0.9955	0.0006	0.0236	0.1264	4.7215	0.0253	0.9443
0.0200	0	0.9821	0.0054	0.1131	0.8225	13.1851	0.1139	0.7484
0.0300	0	0.9597	0.0229	0.2741	2.6889	19.0107	0.2593	0.4167
0.0400	0	1.0000	0.0668	0.4770	6.0756	21.5654	0.4180	0.0942
0.0500	0	0.8882	0.1515	0.6863	10.8635	20.2973	0.5396	−0.3478
0.0600	0	0.8396	0.2880	0.8653	16.4457	15.5062	0.5769	−0.6104
0.0700	0	0.7827	0.4794	0.9874	21.8374	8.9013	0.5015	−0.7106
0.0800	0	0.7181	0.7181	1.0427	25.9081	2.1571	0.3126	−0.6383
0.0900	0	0.6463	0.9859	1.0376	27.6527	−3.1644	0.0363	−0.4260
0.1000	0	0.5678	1.2564	0.9918	26.4303	−5.9962	−0.2808	−0.1403
0.1100	0	0.4833	1.4991	0.9318	22.1194	−6.0181	−0.5814	0.1359
0.1200	0	0.3935	1.6855	0.8831	15.1531	−3.7176	−0.8118	0.3242
0.1300	0	0.2994	1.7934	0.8632	6.4267	−0.2479	−0.9335	0.3698
0.1400	0	0.2018	1.8110	0.8764	−2.8992	2.8795	−0.9317	0.2557
0.1500	0	0.1017	1.7383	0.9118	−11.6507	4.1901	−0.8186	0.0064
0.1600	0	0.0000	1.5856	0.9459	−18.8884	2.6310	−0.6289	−0.3182
0.1700	0	−0.1022	1.3707	0.9483	−24.0903	−2.1497	−0.4114	−0.6379
0.1800	0	−0.2037	1.1141	0.8890	−27.2289	−9.6975	−0.2163	−0.8716
0.1900	0	−0.3036	0.8343	0.7463	−28.7255	−18.8433	−0.0830	−0.9575
⋮	⋮	⋮	⋮	⋮	⋮	⋮	⋮	⋮
⋮	⋮	⋮	⋮	⋮	⋮	⋮	⋮	⋮
10.1400	0	−0.7022	−3.6015	1.2069	58.0410	164.2854	4.8616	−6.8529
10.1500	0	−0.9801	−2.7691	2.4609	108.4442	86.5046	5.2190	−8.7032
10.1600	0	−0.9239	−1.4473	2.9057	155.9205	2.4534	4.2762	−8.1070
10.1700	0	−0.5521	0.2737	2.5948	188.2750	−64.6280	2.1947	−5.3093
10.1800	0	0.0088	2.1974	1.7865	196.4603	−97.0320	−0.5576	−1.1715
10.1900	0	0.5671	4.0647	0.8639	177.0070	−87.4895	−3.3331	3.0800
10.2000	0	0.9312	5.6144	0.2213	132.9410	−41.0362	−5.4802	6.2106
10.2100	0	0.9754	6.6439	0.1494	72.9521	26.6684	−6.5176	7.3303
10.2200	0	0.6838	7.0540	0.7529	9.0710	94.0322	−6.2586	6.1425
10.2300	0	0.1562	6.8670	1.9222	−46.4801	139.8231	−4.8516	3.0157
10.2400	0	−0.4258	6.2127	3.3674	−84.3808	149.2055	−2.7285	−1.1392

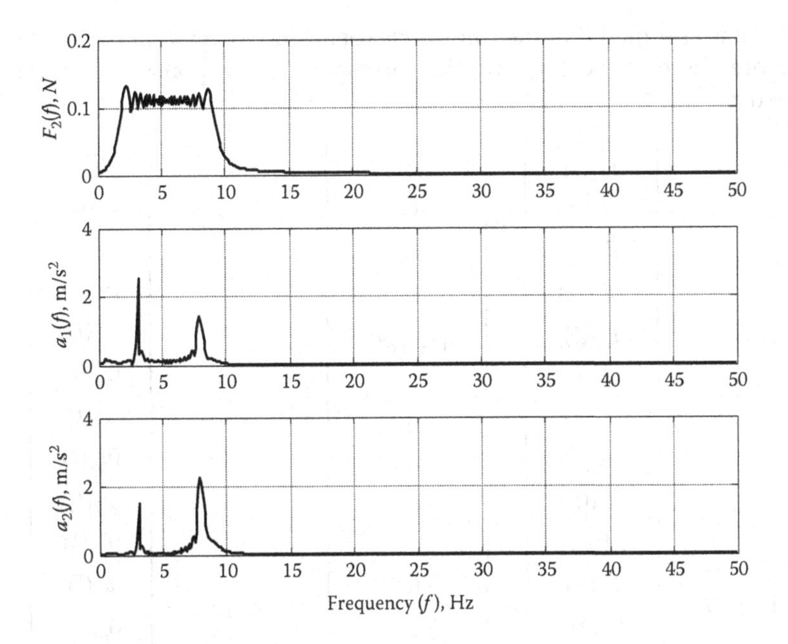

FIGURE 4.7
The amplitude spectra of the applied force at node 2 and the calculated acceleration responses at nodes 1 and 2 for Example 4.2.

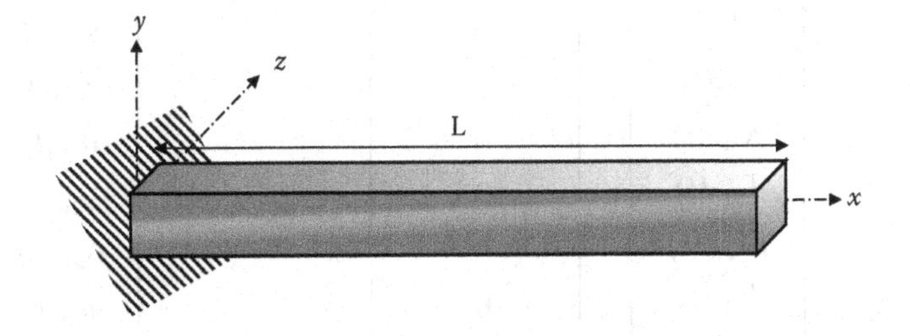

FIGURE 4.8
The cantilever beam of Example 4.3.

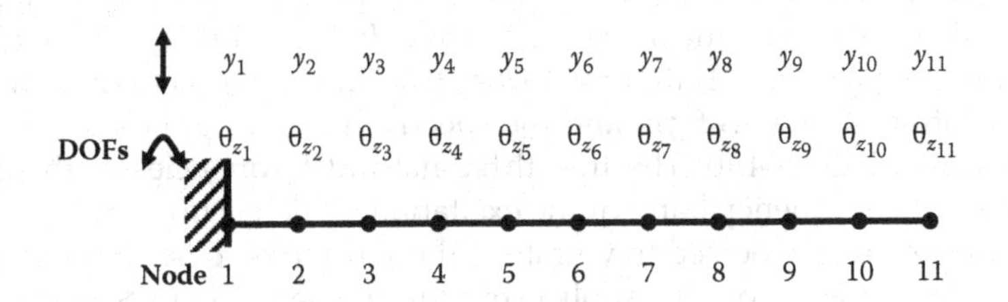

FIGURE 4.9
The FE model of the cantilever beam of Example 4.3 ($y1$ to $y11$: the bending deflection in y-direction; θ_{z1} to θ_{z11}: the bending rotation about z-axis for the nodes 1 to 11, respectively).

Now it is assumed that the transverse (bending) force is applied at node 3; therefore, the force vectors and the corresponding response vectors can be written as

$$\mathbf{F}(t)=\mathbf{F}_t=\begin{Bmatrix} F_{y1}(t)\\ M_{z1}(t)\\ F_{y2}(t)\\ M_{z2}(t)\\ F_{y3}(t)\\ M_{z3}(t)\\ F_{y4}(t)\\ M_{z4}(t)\\ F_{y5}(t)\\ M_{z5}(t)\\ F_{y6}(t)\\ M_{z6}(t)\\ F_{y7}(t)\\ M_{z7}(t)\\ F_{y8}(t)\\ M_{z8}(t)\\ F_{y9}(t)\\ M_{z9}(t)\\ F_{y10}(t)\\ M_{z10}(t)\\ F_{y11}(t)\\ M_{z11}(t) \end{Bmatrix}=\begin{Bmatrix} 0\\0\\0\\0\\ \sin(2\pi ft)\\0\\0\\0\\0\\0\\0\\0\\0\\0\\0\\0\\0\\0\\0\\0\\0\\0 \end{Bmatrix} \text{ and } \mathbf{r}(t)=\mathbf{r}_t=\begin{Bmatrix} y_1(t)\\ \theta_{y1}(t)\\ y_2(t)\\ \theta_{y2}(t)\\ y_3(t)\\ \theta_{y3}(t)\\ y_4(t)\\ \theta_{y4}(t)\\ y_5(t)\\ \theta_{y5}(t)\\ y_6(t)\\ \theta_{y6}(t)\\ y_7(t)\\ \theta_{y7}(t)\\ y_8(t)\\ \theta_{y8}(t)\\ y_9(t)\\ \theta_{y9}(t)\\ y_{10}(t)\\ \theta_{y10}(t)\\ y_{11}(t)\\ \theta_{y11}(t) \end{Bmatrix}$$

where the applied force $F_{y3}(t) = \sin(2\pi ft)$; f is the frequency of the applied force, which is varying linearly from 0 to 512 Hz from time $t_1 = 0$ s to $t_q = 8$ s at the time steps of $\Delta t = 1/10240 = 0.97656$ μs. This is also known as sweep-sine excitation. The waveform and the spectrum of the applied force, $F_{y3}(t)$, are shown in Figure 4.10. The first three natural frequencies of the beam are within the frequency band of the excitation, which is up to 512 Hz. As a result, the beam is expected to vibrate at these three modes. Now the steps 1–6 of direct integration are applied exactly the same as in Sections 4.1.1 and 4.1.2 to compute the displacement, velocity, and acceleration responses. The computed bending displacement at the beam's free end (node 11, DOFy_{11}) and the corresponding velocity and acceleration responses are also

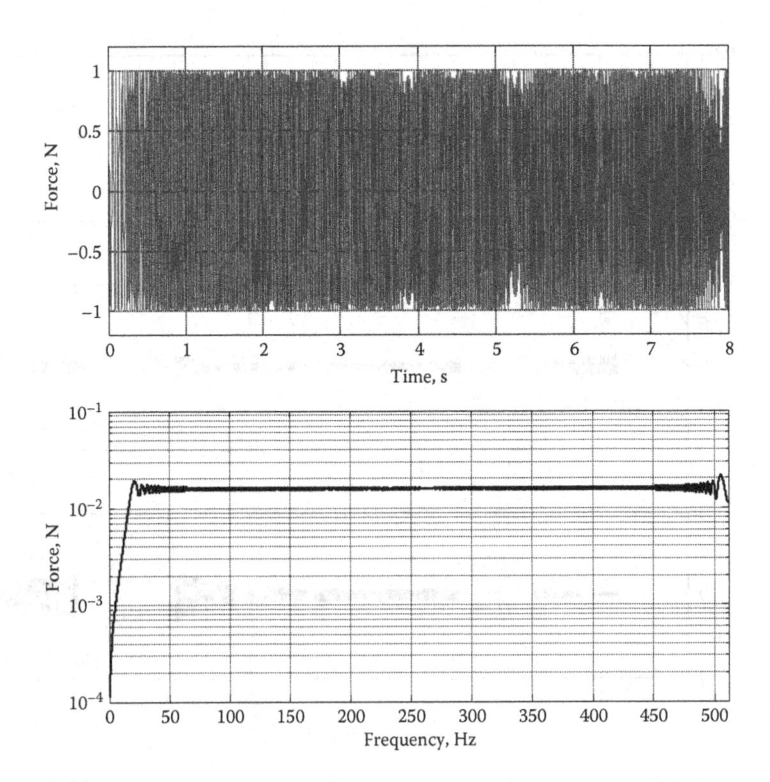

FIGURE 4.10
Time waveform and its spectrum of the applied force for Example 4.3.

shown in Figure 4.11. The three zones of amplification in the time waveform of the responses indicate the responses corresponding to the excitation frequencies that are close to these three natural frequencies. The amplitude spectrum for the acceleration response in Figure 4.11(c) is also shown in Figure 4.12, where three frequency peaks corresponding to the first three modes are clearly observed.

4.3 Mode Superposition (MS) Method

The DI method often requires exorbitant computation time and huge memory if the size of the mass, stiffness, and damping matrices is large, as the method directly involves the system matrices in the response estimation. The mode superposition (MS) method generally overcomes these limitations to a great extent. It is because the later method involves only a limited number of modes in the response estimation, unlike the earlier method, which uses all modes due to direct use of the system matrices. Perhaps the use of a limited but

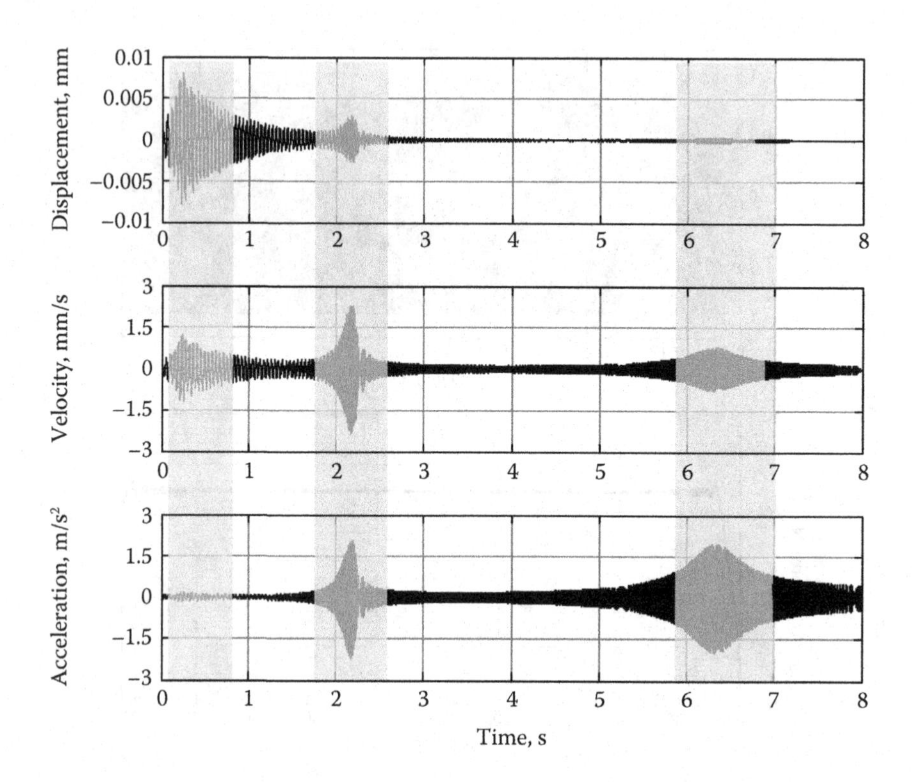

FIGURE 4.11
Computed responses (displacement, velocity, and acceleration) at the free end of the cantilever beam of Example 4.3.

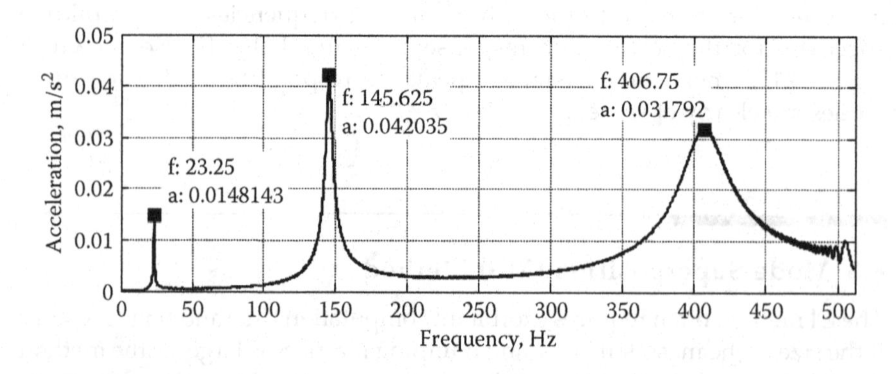

FIGURE 4.12
Acceleration spectrum of the free end acceleration response of the cantilever beam of Example 4.3.

required number of modes in the response analysis justifies the application of the MS method for any system to meet practical requirements. For example, the excitation given in Example 4.3 is limited to the first two modes of the cantilever beam, which means the use of just these two modes in the response estimation would be sufficient for the response calculation. Therefore, the MS is likely to be a very efficient approach compared to the DI for Example 4.3.

The steps involved in the MS for the response estimation are discussed here.

Step 1: Carry out the modal analysis to compute the natural frequencies and the normalized mode shapes. Let's assume that s is the number of modes required to be included in the response estimation. The normalized mode shape matrix, φ, for the s number of modes corresponding to the DOFs of the response vectors, $\mathbf{r}(t)$, in the Equation (4.2) can be written as

$$\varphi = \begin{bmatrix} \varphi_1 & \varphi_2 & \cdots & \varphi_r & \cdots & \varphi_s \end{bmatrix} \tag{4.16}$$

$$= \begin{bmatrix} \phi_{x1,1} & \phi_{x1,2} & \cdots & \phi_{x1,r} & \cdots & \phi_{x1,s} \\ \phi_{y1,1} & \phi_{y1,2} & \cdots & \phi_{y1,r} & \cdots & \phi_{y1,s} \\ \phi_{z1,1} & \phi_{z1,2} & \cdots & \phi_{z1,r} & \cdots & \phi_{z1,s} \\ \phi_{\theta_{x1},1} & \phi_{\theta_{x1},2} & \cdots & \phi_{\theta_{x1},r} & \cdots & \phi_{\theta_{x1},s} \\ \phi_{\theta_{y1},1} & \phi_{\theta_{y1},2} & \cdots & \phi_{\theta_{y1},r} & \cdots & \phi_{\theta_{y1},s} \\ \phi_{\theta_{z1},1} & \phi_{\theta_{z1},2} & \cdots & \phi_{\theta_{z1},r} & \cdots & \phi_{\theta_{z1},s} \\ \phi_{x2,1} & \phi_{x2,2} & \cdots & \phi_{x2,r} & \cdots & \phi_{x2,s} \\ \phi_{y2,1} & \phi_{y2,2} & \cdots & \phi_{y2,r} & \cdots & \phi_{y2,s} \\ \phi_{z2,1} & \phi_{z2,2} & \cdots & \phi_{z2,r} & \cdots & \phi_{z2,s} \\ \phi_{\theta_{x2},1} & \phi_{\theta_{x2},2} & \cdots & \phi_{\theta_{x2},r} & \cdots & \phi_{\theta_{x2},s} \\ \phi_{\theta_{y2},1} & \phi_{\theta_{y2},2} & \cdots & \phi_{\theta_{y2},r} & \cdots & \phi_{\theta_{y2},s} \\ \phi_{\theta_{z2},1} & \phi_{\theta_{z2},2} & \cdots & \phi_{\theta_{z2},r} & \cdots & \phi_{\theta_{z2},s} \\ \vdots & \vdots & \ddots & \vdots & \ddots & \vdots \\ \phi_{xn,1} & \phi_{xn,2} & \cdots & \phi_{xn,r} & \cdots & \phi_{xn,s} \\ \phi_{yn,1} & \phi_{yn,2} & \cdots & \phi_{yn,r} & \cdots & \phi_{yn,s} \\ \phi_{zn,1} & \phi_{zn,2} & \cdots & \phi_{zn,r} & \cdots & \phi_{zn,s} \\ \phi_{\theta_{xn},1} & \phi_{\theta_{xn},2} & \cdots & \phi_{\theta_{xn},r} & \cdots & \phi_{\theta_{xn},s} \\ \phi_{\theta_{yn},1} & \phi_{\theta_{yn},2} & \cdots & \phi_{\theta_{yn},r} & \cdots & \phi_{\theta_{yn},s} \\ \phi_{\theta_{zn},1} & \phi_{\theta_{zn},2} & \cdots & \phi_{\theta_{zn},r} & \cdots & \phi_{\theta_{zn},s} \end{bmatrix}$$

Step 2: The dynamic response of any system is basically the response contributed by the modes so the displacement response vectors, $\mathbf{r}(t)$, can be expressed

$$\mathbf{r}(t) = \varphi \mathbf{p}(t) \tag{4.17}$$

where $\mathbf{p}(t)$ is the displacement response vector in modal domain, which is written as $\mathbf{p}(t) = [p_1(t) \;\; p_2(t) \;\; \cdots \;\; p_s(t)]^T$, where $p_1(t) \;\; p_2(t) \;\; \cdots \;\; p_s(t)$ are the modal response vectors at time t for the first, second, $\cdots$, sth modes, respectively. Hence, the nodal displacement response vectors at time t can be estimated as

$$\mathbf{r}(t) = \mathbf{r}_t = \begin{Bmatrix} x_{1\_t} \\ y_{1\_t} \\ z_{1\_t} \\ \theta_{x1\_t} \\ \theta_{y1\_t} \\ \theta_{z1\_t} \\ x_{2\_t} \\ y_{2\_t} \\ z_{2\_t} \\ \theta_{x2\_t} \\ \theta_{y2\_t} \\ \theta_{z2\_t} \\ \vdots \\ x_{n\_t} \\ y_{n\_t} \\ z_{n\_t} \\ \theta_{xn\_t} \\ \theta_{yn\_t} \\ \theta_{zn\_t} \end{Bmatrix} = \begin{bmatrix} \phi_{x1,1} & \phi_{x1,2} & \cdots & \phi_{x1,r} & \cdots & \phi_{x1,s} \\ \phi_{y1,1} & \phi_{y1,2} & \cdots & \phi_{y1,r} & \cdots & \phi_{y1,s} \\ \phi_{z1,1} & \phi_{z1,2} & \cdots & \phi_{z1,r} & \cdots & \phi_{z1,s} \\ \phi_{\theta_{x1},1} & \phi_{\theta_{x1},2} & \cdots & \phi_{\theta_{x1},r} & \cdots & \phi_{\theta_{x1},s} \\ \phi_{\theta_{y1},1} & \phi_{\theta_{y1},2} & \cdots & \phi_{\theta_{y1},r} & \cdots & \phi_{\theta_{y1},s} \\ \phi_{\theta_{z1},1} & \phi_{\theta_{z1},2} & \cdots & \phi_{\theta_{z1},r} & \cdots & \phi_{\theta_{z1},s} \\ \phi_{x2,1} & \phi_{x2,2} & \cdots & \phi_{x2,r} & \cdots & \phi_{x2,s} \\ \phi_{y2,1} & \phi_{y2,2} & \cdots & \phi_{y2,r} & \cdots & \phi_{y2,s} \\ \phi_{z2,1} & \phi_{z2,2} & \cdots & \phi_{z2,r} & \cdots & \phi_{z2,s} \\ \phi_{\theta_{x2},1} & \phi_{\theta_{x2},2} & \cdots & \phi_{\theta_{x2},r} & \cdots & \phi_{\theta_{x2},s} \\ \phi_{\theta_{y2},1} & \phi_{\theta_{y2},2} & \cdots & \phi_{\theta_{y2},r} & \cdots & \phi_{\theta_{y2},s} \\ \phi_{\theta_{z2},1} & \phi_{\theta_{z2},2} & \cdots & \phi_{\theta_{z2},r} & \cdots & \phi_{\theta_{z2},s} \\ \vdots & \vdots & \ddots & \vdots & \ddots & \vdots \\ \phi_{xn,1} & \phi_{xn,2} & \cdots & \phi_{xn,r} & \cdots & \phi_{xn,s} \\ \phi_{yn,1} & \phi_{yn,2} & \cdots & \phi_{yn,r} & \cdots & \phi_{yn,s} \\ \phi_{zn,1} & \phi_{zn,2} & \cdots & \phi_{zn,r} & \cdots & \phi_{zn,s} \\ \phi_{\theta_{xn},1} & \phi_{\theta_{xn},2} & \cdots & \phi_{\theta_{xn},r} & \cdots & \phi_{\theta_{xn},s} \\ \phi_{\theta_{yn},1} & \phi_{\theta_{yn},2} & \cdots & \phi_{\theta_{yn},r} & \cdots & \phi_{\theta_{yn},s} \\ \phi_{\theta_{zn},1} & \phi_{\theta_{zn},2} & \cdots & \phi_{\theta_{zn},r} & \cdots & \phi_{\theta_{zn},s} \end{bmatrix} \begin{Bmatrix} p_1(t) \\ p_2(t) \\ \vdots \\ p_r(t) \\ \vdots \\ p_s(t) \end{Bmatrix} \tag{4.18}$$

Similarly, the nodal velocity and acceleration vectors at time t can be calculated as

$$\mathbf{v}(t) = \dot{\mathbf{r}}(t) = \varphi \dot{\mathbf{p}}(t) \tag{4.19}$$

$$\mathbf{a}(t) = \ddot{\mathbf{r}}(t) = \varphi \ddot{\mathbf{p}}(t) \tag{4.20}$$

Step 3: The estimation of the modal vectors—$\mathbf{p}(t)$, $\dot{\mathbf{p}}(t)$, and $\ddot{\mathbf{p}}(t)$—is essentially required for the estimation of the displacement, velocity, and acceleration vectors of the complete system. This requires the conversion of the equilibrium Equation (4.1) into the modal domain. Substituting Equations (4.17), (4.19), and (4.20) into Equation (4.1) and premultiplying Equation (4.1) by φ^T yields the following equation:

$$\varphi^T \mathbf{M}\varphi\ddot{\mathbf{p}}(t) + \varphi^T \mathbf{C}\varphi\dot{\mathbf{p}}(t) + \varphi^T \mathbf{K}\varphi\mathbf{p}(t) = \varphi^T \mathbf{F}(t) \tag{4.21}$$

$$\Rightarrow \mathbf{I}\ddot{\mathbf{p}}(t) + \begin{bmatrix} 2\zeta_1\omega_{n_1} & 0 & \cdots & 0 & \cdots & 0 \\ 0 & 2\zeta_2\omega_{n_2} & \cdots & 0 & \cdots & 0 \\ \vdots & \vdots & \ddots & \vdots & \ddots & \vdots \\ 0 & 0 & \cdots & 2\zeta_r\omega_{n_r} & \cdots & 0 \\ \vdots & \vdots & \ddots & \vdots & \ddots & \vdots \\ 0 & 0 & \cdots & 0 & \cdots & 2\zeta_n\omega_{n_s} \end{bmatrix}$$
$$\dot{\mathbf{p}}(t) + \begin{bmatrix} \omega_{n_1}^2 & 0 & \cdots & 0 & \cdots & 0 \\ 0 & \omega_{n_2}^2 & \cdots & 0 & \cdots & 0 \\ \vdots & \vdots & \ddots & \vdots & \ddots & \vdots \\ 0 & 0 & \cdots & \omega_{n_r}^2 & \cdots & 0 \\ \vdots & \vdots & \ddots & \vdots & \ddots & \vdots \\ 0 & 0 & \cdots & 0 & \cdots & \omega_{n_s}^2 \end{bmatrix} \mathbf{p}(t) = \varphi^T \mathbf{F}(t) \tag{4.22}$$

$$\Rightarrow \begin{bmatrix} 1 & 0 & \cdots & 0 & \cdots & 0 \\ 0 & 1 & \cdots & 0 & \cdots & 0 \\ \vdots & \vdots & \ddots & \vdots & \ddots & \vdots \\ 0 & 0 & \cdots & 1 & \cdots & 0 \\ \vdots & \vdots & \ddots & \vdots & \ddots & \vdots \\ 0 & 0 & \cdots & 0 & \cdots & 1 \end{bmatrix} \begin{Bmatrix} \ddot{p}_1(t) \\ \ddot{p}_2(t) \\ \vdots \\ \ddot{p}_r(t) \\ \vdots \\ \ddot{p}_s(t) \end{Bmatrix}$$

$$+ \begin{bmatrix} 2\zeta_1\omega_{n_1} & 0 & \cdots & 0 & \cdots & 0 \\ 0 & 2\zeta_2\omega_{n_2} & \cdots & 0 & \cdots & 0 \\ \vdots & \vdots & \ddots & \vdots & \ddots & \vdots \\ 0 & 0 & \cdots & 2\zeta_r\omega_{n_r} & \cdots & 0 \\ \vdots & \vdots & \ddots & \vdots & \ddots & \vdots \\ 0 & 0 & \cdots & 0 & \cdots & 2\zeta_n\omega_{n_s} \end{bmatrix} \begin{Bmatrix} \dot{p}_1(t) \\ \dot{p}_2(t) \\ \vdots \\ \dot{p}_r(t) \\ \vdots \\ \dot{p}_s(t) \end{Bmatrix}$$

$$+\begin{bmatrix} \omega_{n1}^2 & 0 & \cdots & 0 & \cdots & 0 \\ 0 & \omega_{n2}^2 & \cdots & 0 & \cdots & 0 \\ \vdots & \vdots & \ddots & \vdots & \ddots & \vdots \\ 0 & 0 & \cdots & \omega_{nr}^2 & \cdots & 0 \\ \vdots & \vdots & \ddots & \vdots & \ddots & \vdots \\ 0 & 0 & \cdots & 0 & \cdots & \omega_{ns}^2 \end{bmatrix} \begin{Bmatrix} p_1(t) \\ p_2(t) \\ \vdots \\ p_r(t) \\ \vdots \\ p_s(t) \end{Bmatrix} = \begin{Bmatrix} \varphi_1^T \\ \varphi_2^T \\ \vdots \\ \varphi_r^T \\ \vdots \\ \varphi_s^T \end{Bmatrix} \mathbf{F}(t) \quad (4.23)$$

Hence, the decoupled equation for each mode can be extracted from Equation (4.23). For example, the equation for the rth mode can be written as

$$\ddot{p}_r(t)+2\zeta_r\omega_{n_r}\dot{p}_r(t)+\omega_{n_r}^2 p_r(t)=\varphi_r^T\mathbf{F}(t) \quad (4.24)$$

where ζ_r and $\omega_{n_r}\left(=2\pi f_{n_r}\right)$ are the damping ratio and the natural frequency (rad/s) of the rth mode, respectively.

Step 4: Steps 1 to 4 of the Newmark-β integration method discussed in Section 4.1 can be used for solving the modal Equation (4.23) to estimate the modal response vectors—$p_r(t)$, $\dot{p}_r(t)$, $\ddot{p}_r(t)$—just by replacing the system matrices (**M**, **C**, **K**) by ($\varphi^T \mathbf{M}\varphi$, $\varphi^T \mathbf{C}\varphi$, $\varphi^T \mathbf{K}\varphi$), respectively, and the force vectors, **F**(t), by $\varphi^T \mathbf{F}(t)$ so that Equation (4.1) becomes comparable to Equation (4.23).

Step 5: Repeat the estimation of the modal response vectors for all modes selected for the response calculation. The example in Section 4.1.2 has two natural frequencies, so either only one mode or both modes can be used for the response estimation. Since the applied exciting frequency consists of both modes, it is recommended to use both modes for the response estimation. However, the other example in Section 4.1.3 is a cantilever beam, which has a number of modes, but the applied force consists of frequencies up to the first three modes. In this case, it is recommended to use only the first three modes for the response estimation. The use of more modes in the latter case unnecessarily increases the computational time and cost without any noticeable advantage.

Step 6: Finally, the nodal displacement, velocity, and acceleration can be computed using Equations (4.18), (4.19), and (4.20), respectively.

4.3.1 Example 4.4: A Two-DOF System

It is actually the same as Example 4.2 of Section 4.1.2, but the responses are now estimated using the MS method. Since the applied force at node 2 has the frequency band of excitation that covers both modes of the system, both modes are included for the response calculation. The response computations are explained in the following steps.

Modal analysis using the mass and stiffness matrices (Equation (4.10)) yields the natural frequencies and the normalized mode shapes as

$$\{f_n\} = \begin{Bmatrix} f_{n1} \\ f_{n2} \end{Bmatrix} = \begin{Bmatrix} 3.1105 \\ 8.1434 \end{Bmatrix} \text{Hz}$$

$$\text{and } \varphi = \begin{bmatrix} \varphi_1 & \varphi_2 \end{bmatrix} = \begin{bmatrix} \phi_{y1,1} & \phi_{y1,2} \\ \phi_{y2,1} & \phi_{y2,2} \end{bmatrix} = \begin{bmatrix} -0.8507 & -0.5257 \\ -0.5257 & 0.8507 \end{bmatrix} \tag{4.25}$$

The damping ratios ζ_1 and ζ_2 have been calculated as

$$\varphi^T \mathbf{C} \varphi = \begin{bmatrix} \phi_{y1,1} & \phi_{y2,1} \\ \phi_{y1,2} & \phi_{y2,2} \end{bmatrix} \begin{bmatrix} 0.5 & -0.5 \\ -0.5 & 1 \end{bmatrix} \begin{bmatrix} \phi_{y1,1} & \phi_{y1,2} \\ \phi_{y2,1} & \phi_{y2,2} \end{bmatrix}$$

$$\Rightarrow \begin{bmatrix} \varphi_1^T \\ \varphi_2^T \end{bmatrix} \mathbf{C} \begin{bmatrix} \varphi_1 & \varphi_2 \end{bmatrix}$$

$$= \begin{bmatrix} -0.8507 & -0.5257 \\ -0.5257 & 0.8507 \end{bmatrix} \begin{bmatrix} 0.5 & -0.5 \\ -0.5 & 1 \end{bmatrix} \begin{bmatrix} -0.8507 & -0.5257 \\ -0.5257 & 0.8507 \end{bmatrix}$$

$$\Rightarrow \begin{bmatrix} 2\zeta_1\omega_{n1} & 0 \\ 0 & 2\zeta_2\omega_{n2} \end{bmatrix} = \begin{bmatrix} 0.1910 & 0 \\ 0 & 1.3090 \end{bmatrix}$$

$$\Rightarrow \begin{bmatrix} \zeta_1 & 0 \\ 0 & \zeta_2 \end{bmatrix} = \begin{bmatrix} 0.0049 & 0 \\ 0 & 0.0128 \end{bmatrix}$$

Once again, the time step $\Delta t = 0.01$ s has been assumed. Now the system equilibrium equation in the modal domain can be written as (Equation (4.23))

$$\begin{bmatrix} 1 & 0 \\ 0 & 1 \end{bmatrix} \begin{Bmatrix} \ddot{p}_1(t) \\ \ddot{p}_2(t) \end{Bmatrix} + \begin{bmatrix} 2\zeta_1\omega_{n1} & 0 \\ 0 & 2\zeta_2\omega_{n2} \end{bmatrix} \begin{Bmatrix} \dot{p}_1(t) \\ \dot{p}_2(t) \end{Bmatrix}$$

$$+ \begin{bmatrix} \omega_{n1}^2 & 0 \\ 0 & \omega_{n2}^2 \end{bmatrix} \begin{Bmatrix} p_1(t) \\ p_2(t) \end{Bmatrix} = \begin{bmatrix} \phi_{y1,1} & \phi_{y2,1} \\ \phi_{y1,2} & \phi_{y2,2} \end{bmatrix} \begin{Bmatrix} F_1(t) \\ F_2(t) \end{Bmatrix} \tag{4.26}$$

At time $t_1 = 0$ s the initial modal responses are assumed as:

$$\text{Initial modal displacement, } \begin{Bmatrix} p_1(t_1) \\ p_2(t_1) \end{Bmatrix} = \begin{Bmatrix} p_1(0) \\ p_2(0) \end{Bmatrix} = \begin{Bmatrix} 0 \\ 0 \end{Bmatrix}$$

$$\text{Initial modal velocity, } \begin{Bmatrix} \dot{p}_1(t_1) \\ \dot{p}_2(t_1) \end{Bmatrix} = \begin{Bmatrix} \dot{p}_1(0) \\ \dot{p}_2(0) \end{Bmatrix} = \begin{Bmatrix} 0 \\ 0 \end{Bmatrix}$$

$$\text{Initial modal acceleration, } \begin{Bmatrix} \ddot{p}_1(t_1) \\ \ddot{p}_2(t_1) \end{Bmatrix} = \begin{Bmatrix} \ddot{p}_1(0) \\ \ddot{p}_2(0) \end{Bmatrix} = \begin{Bmatrix} 0 \\ 0 \end{Bmatrix}$$

Hence, the nodal responses at time $t_1 = 0$ s:

$$\text{Displacement, } \begin{Bmatrix} y_1(t_1) \\ y_2(t_1) \end{Bmatrix} = \begin{bmatrix} \phi_{y1,1} & \phi_{y1,2} \\ \phi_{y2,1} & \phi_{y2,2} \end{bmatrix} \begin{Bmatrix} p_1(t_1) \\ p_2(t_1) \end{Bmatrix}$$

$$= \begin{bmatrix} -0.8507 & -0.5257 \\ -0.5257 & 0.8507 \end{bmatrix} \begin{Bmatrix} 0 \\ 0 \end{Bmatrix} = \begin{Bmatrix} 0 \\ 0 \end{Bmatrix}$$

$$\text{Velocity, } \begin{Bmatrix} v_1(t_1) \\ v_2(t_1) \end{Bmatrix} = \begin{bmatrix} \phi_{y1,1} & \phi_{y1,2} \\ \phi_{y2,1} & \phi_{y2,2} \end{bmatrix} \begin{Bmatrix} \dot{p}_1(t_1) \\ \dot{p}_2(t_1) \end{Bmatrix}$$

$$= \begin{bmatrix} -0.8507 & -0.5257 \\ -0.5257 & 0.8507 \end{bmatrix} \begin{Bmatrix} 0 \\ 0 \end{Bmatrix} = \begin{Bmatrix} 0 \\ 0 \end{Bmatrix}$$

$$\text{Acceleration, } \begin{Bmatrix} a_1(t_1) \\ a_2(t_1) \end{Bmatrix} = \begin{bmatrix} \phi_{y1,1} & \phi_{y1,2} \\ \phi_{y2,1} & \phi_{y2,2} \end{bmatrix} \begin{Bmatrix} \ddot{p}_1(t_1) \\ \ddot{p}_2(t_1) \end{Bmatrix}$$

$$= \begin{bmatrix} -0.8507 & -0.5257 \\ -0.5257 & 0.8507 \end{bmatrix} \begin{Bmatrix} 0 \\ 0 \end{Bmatrix} = \begin{Bmatrix} 0 \\ 0 \end{Bmatrix}$$

At time $t_2 = 0.01$ s:

$$\text{Equation (4.3), } \begin{Bmatrix} \hat{p}_1(t_2) \\ \hat{p}_2(t_2) \end{Bmatrix} = \begin{Bmatrix} p_1(t_1) \\ p_2(t_1) \end{Bmatrix} + (0.01) \begin{Bmatrix} \dot{p}_1(t_1) \\ \dot{p}_2(t_1) \end{Bmatrix}$$

$$+ \frac{(0.01)}{2}(1-2\beta)(0.01) \begin{Bmatrix} \ddot{p}_1(t_1) \\ \ddot{p}_2(t_1) \end{Bmatrix}$$

$$\Rightarrow\begin{Bmatrix} \hat{p}_1(t_2) \\ \hat{p}_2(t_2) \end{Bmatrix} = \begin{Bmatrix} 0 \\ 0 \end{Bmatrix} + (0.01)\begin{Bmatrix} 0 \\ 0 \end{Bmatrix} + \frac{(0.01)}{2}(1-2\beta)(0.01)\begin{Bmatrix} 0 \\ 0 \end{Bmatrix} = \begin{Bmatrix} 0 \\ 0 \end{Bmatrix}$$

Equation (4.4), $\begin{Bmatrix} \dot{\hat{p}}_1(t_2) \\ \dot{\hat{p}}_2(t_2) \end{Bmatrix} = \begin{Bmatrix} \dot{p}_1(t_1) \\ \dot{p}_2(t_1) \end{Bmatrix} + (1-\alpha)(0.01)\begin{Bmatrix} \ddot{p}_1(t_1) \\ \ddot{p}_2(t_1) \end{Bmatrix}$

$$\Rightarrow\begin{Bmatrix} \dot{\hat{p}}_1(t_2) \\ \dot{\hat{p}}_2(t_2) \end{Bmatrix} = \begin{Bmatrix} 0 \\ 0 \end{Bmatrix} + (1-\alpha)(0.01)\begin{Bmatrix} 0 \\ 0 \end{Bmatrix} = \begin{Bmatrix} 0 \\ 0 \end{Bmatrix}$$

Equation (4.5), $\hat{\mathbf{K}} = \varphi^T\mathbf{M}\varphi + \alpha\Delta t(\varphi^T\mathbf{C}\varphi) + \beta\Delta t^2(\varphi^T\mathbf{K}\varphi)$

$$\Rightarrow \hat{\mathbf{K}} = \begin{bmatrix} 1 & 0 \\ 0 & 1 \end{bmatrix} + \alpha(0.01)\begin{bmatrix} 0.1910 & 0 \\ 0 & 1.3090 \end{bmatrix} + \beta(0.01)^2\begin{bmatrix} 381.966 & 0 \\ 0 & 2618.034 \end{bmatrix}$$

$$\Rightarrow \hat{\mathbf{K}} = \begin{bmatrix} 1.0105 & 3.969e-18 \\ 5.684e-18 & 1.0720 \end{bmatrix}$$

where $\alpha = \dfrac{1}{2}$ and $\beta = \dfrac{1}{4}$.

Modal responses at time $t_2 = 0.01$ s:

Modal acceleration, Equation (4.6),

$$\begin{Bmatrix} \ddot{p}_1(t_2) \\ \ddot{p}_2(t_2) \end{Bmatrix} = \hat{\mathbf{K}}^{-1}\left(\varphi^T\mathbf{F}_{t_2} - \varphi^T\mathbf{C}\varphi\begin{Bmatrix} \dot{\hat{p}}_1(t_2) \\ \dot{\hat{p}}_2(t_2) \end{Bmatrix} - \varphi^T\mathbf{K}\varphi\begin{Bmatrix} \hat{p}_1(t_2) \\ \hat{p}_2(t_2) \end{Bmatrix}\right)$$

$$\Rightarrow \begin{Bmatrix} \ddot{p}_1(t_2) \\ \ddot{p}_2(t_2) \end{Bmatrix} = \begin{bmatrix} 1.0105 & 3.969e-18 \\ 5.684e-18 & 1.0720 \end{bmatrix}^{-1}$$

$$\left(\varphi^T \begin{Bmatrix} F_{1\_t_2} \\ F_{2\_t_2} \end{Bmatrix} - \varphi^T \mathbf{C} \varphi \begin{Bmatrix} 0 \\ 0 \end{Bmatrix} - \varphi^T \mathbf{K} \varphi \begin{Bmatrix} 0 \\ 0 \end{Bmatrix} \right)$$

$$\Rightarrow \begin{Bmatrix} \ddot{p}_1(t_2) \\ \ddot{p}_2(t_2) \end{Bmatrix} = \begin{Bmatrix} -0.5179 \\ 0.7900 \end{Bmatrix} \text{m/s}^2$$

Modal velocity, Equation (4.7), $\begin{Bmatrix} \dot{p}_1(t_2) \\ \dot{p}_2(t_2) \end{Bmatrix} = \begin{Bmatrix} \dot{\hat{p}}_1(t_2) \\ \dot{\hat{p}}_2(t_2) \end{Bmatrix}$

$$+ \alpha \Delta t \begin{Bmatrix} \ddot{p}_1(t_2) \\ \ddot{p}_2(t_2) \end{Bmatrix}$$

$$\Rightarrow \begin{Bmatrix} \dot{p}_1(t_2) \\ \dot{p}_2(t_2) \end{Bmatrix} = \begin{Bmatrix} 0 \\ 0 \end{Bmatrix} + \alpha \Delta t \begin{Bmatrix} -0.5179 \\ 0.7900 \end{Bmatrix}$$

$$= \begin{Bmatrix} -2.5897e-3 \\ 3.9499e-3 \end{Bmatrix} \text{m/s}$$

Modal displacement, Equation (4.8), $\begin{Bmatrix} p_1(t_2) \\ p_2(t_2) \end{Bmatrix} = \begin{Bmatrix} \hat{p}_1(t_2) \\ \hat{p}_2(t_2) \end{Bmatrix}$

$$+ \beta \Delta t^2 \begin{Bmatrix} \ddot{p}_1(t_2) \\ \ddot{p}_2(t_2) \end{Bmatrix}$$

$$\Rightarrow \begin{Bmatrix} p_1(t_2) \\ p_2(t_2) \end{Bmatrix} = \begin{Bmatrix} 0 \\ 0 \end{Bmatrix} + \beta \Delta t^2 \begin{Bmatrix} -0.5179 \\ 0.7900 \end{Bmatrix}$$

$$= \begin{Bmatrix} -12.9e-6 \\ 19.7e-6 \end{Bmatrix} \text{m}$$

Hence, the nodal responses at time $t_2 = 0.01$ s:

$$\text{Displacement,} \begin{Bmatrix} y_1(t_2) \\ y_2(t_2) \end{Bmatrix} = \begin{bmatrix} \phi_{y1,1} & \phi_{y1,2} \\ \phi_{y2,1} & \phi_{y2,2} \end{bmatrix} \begin{Bmatrix} p_1(t_2) \\ p_2(t_2) \end{Bmatrix}$$

$$= \begin{bmatrix} -0.8507 & -0.5257 \\ -0.5257 & 0.8507 \end{bmatrix} \begin{Bmatrix} -12.9e-6 \\ 19.7e-6 \end{Bmatrix}$$

$$= \begin{Bmatrix} 6.3183e-7 \\ 2.3607e-5 \end{Bmatrix} \text{m}$$

$$\text{Velocity,} \begin{Bmatrix} v_1(t_2) \\ v_2(t_2) \end{Bmatrix} = \begin{bmatrix} \phi_{y1,1} & \phi_{y1,2} \\ \phi_{y2,1} & \phi_{y2,2} \end{bmatrix} \begin{Bmatrix} \dot{p}_1(t_2) \\ \dot{p}_2(t_2) \end{Bmatrix}$$

$$= \begin{bmatrix} -0.8507 & -0.5257 \\ -0.5257 & 0.8507 \end{bmatrix} \begin{Bmatrix} -2.5897e-3 \\ 3.9499e-3 \end{Bmatrix}$$

$$= \begin{Bmatrix} 1.2637e-4 \\ 4.7215e-3 \end{Bmatrix} \text{m/s}$$

$$\text{Acceleration,} \begin{Bmatrix} a_1(t_2) \\ a_2(t_2) \end{Bmatrix} = \begin{bmatrix} \phi_{y1,1} & \phi_{y1,2} \\ \phi_{y2,1} & \phi_{y2,2} \end{bmatrix} \begin{Bmatrix} \ddot{p}_1(t_2) \\ \ddot{p}_2(t_2) \end{Bmatrix}$$

$$= \begin{bmatrix} -0.8507 & -0.5257 \\ -0.5257 & 0.8507 \end{bmatrix} \begin{Bmatrix} -0.5179 \\ 0.7900 \end{Bmatrix}$$

$$= \begin{Bmatrix} 2.5247e-2 \\ 9.4331e-1 \end{Bmatrix} \text{m/s}^2$$

In similar fashion, the modal responses at $t = t_3 \quad t_4 \quad \ldots \quad t_q$, and then the nodal responses, are computed. The calculated modal and nodal responses are listed in Table 4.3. The calculated nodal responses by the MS method are exactly the same as the responses computed by the DI method (see Table 4.2). The estimated modal responses in terms of displacement (p_1 and p_2), velocity $\left(\dot{p}_1 = {dp_1}/{dt} \text{ and } \dot{p}_2 = {dp_2}/{dt}\right)$, and acceleration $\ddot{p}_1 = {d^2 p_1}/{dt}$ and $\ddot{p}_2 = {d^2 p_2}/{dt}$ are also shown in Figures 4.13 and 4.14, and their acceleration response spectra in Figure 4.15 indicate the presence of resonance in their modal responses related to the first and second modes, respectively.

TABLE 4.3

Modal and Nodal Responses for Example 4.4 Computed by the MS Method

			Modal Responses						Nodal Responses					
	Force, F(*t*), N		Displacement, mm		Velocity, mm/s		Acceleration, m/s²		Displacement, y(*t*), mm		Velocity, v(*t*), mm/s		Acceleration, a(*t*), m/s²	
Time, *t*, s	Node 1	Node 2	Mode 1, $p_1(t)$	Mode 2, $p_2(t)$	Mode 1, $\dot{p}_1(t)$	Mode 2, $\dot{p}_2(t)$	Mode 1, $\ddot{p}_1(t)$	Mode 2, $\ddot{p}_2(t)$	Node 1	Node 2	Node 1	Node 2	Node 1	Node 2
0	0	0	0	0	0	0	0	0	0	0	0	0	0	0
0.0100	0	0.9955	−0.0129	0.0197	−2.5897	3.9499	−0.5179	0.7900	0.0006	0.0236	0.1264	4.7215	0.0253	0.9443
0.0200	0	0.9821	−0.0641	0.0934	−7.6314	10.7835	−0.4904	0.5767	0.0054	0.1131	0.8225	13.1851	0.1139	0.7484
0.0300	0	0.9597	−0.1636	0.2211	−12.2819	14.7578	−0.4397	0.2181	0.0229	0.2741	2.6889	19.0107	0.2593	0.4167
0.0400	0	1.0000	−0.3076	0.3707	−16.5058	15.1505	−0.4051	−0.1396	0.0668	0.4770	6.0756	21.5654	0.4180	0.0942
0.0500	0	0.8882	−0.4896	0.5042	−19.9119	11.5547	−0.2761	−0.5796	0.1515	0.6863	10.8635	20.2973	0.5396	−0.3478
0.0600	0	0.8396	−0.6999	0.5847	−22.1416	4.5444	−0.1698	−0.8225	0.2880	0.8653	16.4457	15.5062	0.5769	−0.6104
0.0700	0	0.7827	−0.9269	0.5879	−23.2558	−3.9087	−0.0530	−0.8681	0.4794	0.9874	21.8374	8.9013	0.5015	−0.7106
0.0800	0	0.7181	−1.1590	0.5094	−23.1728	−11.7857	0.0696	−0.7073	0.7181	1.0427	25.9081	2.1571	0.3126	−0.6383
0.0900	0	0.6463	−1.3842	0.3643	−21.8592	−17.2297	0.1931	−0.3815	0.9859	1.0376	27.6527	−3.1644	0.0363	−0.4260
0.1000	0	0.5678	−1.5902	0.1832	−19.3306	−18.9959	0.3126	0.0282	1.2564	0.9918	26.4303	−5.9962	−0.2808	−0.1403
0.1100	0	0.4833	−1.7651	0.0045	−15.6520	−16.7482	0.4231	0.4213	1.4991	0.9318	22.1194	−6.0181	−0.5814	0.1359
0.1200	0	0.3935	−1.8980	−0.1349	−10.9356	−11.1288	0.5202	0.7026	1.6855	0.8831	15.1531	−3.7176	−0.8118	0.3242
0.1300	0	0.2994	−1.9794	−0.2085	−5.3365	−3.5896	0.5996	0.8053	1.7934	0.8632	6.4267	−0.2479	−0.9335	0.3698
0.1400	0	0.2018	−2.0013	−0.2066	0.9524	3.9736	0.6581	0.7073	1.8110	0.8764	−2.8992	2.8795	−0.9317	0.2557
0.1500	0	0.1017	−1.9580	−0.1383	7.7079	9.6894	0.6930	0.4358	1.7383	0.9118	−11.6507	4.1901	−0.8186	0.0064

0.1600	0	0.0000	−1.8460	−0.0290	14.6842	12.1683	0.7023	0.0600	1.5856	0.9459	−18.8884	2.6310	−0.6289	−0.3182
0.1700	0	−0.1022	−1.6645	0.0860	21.6226	10.8364	0.6854	−0.3263	1.3707	0.9483	−24.0903	−2.1497	−0.4114	−0.6379
0.1800	0	−0.2037	−1.4151	0.1705	28.2606	6.0659	0.6422	−0.6278	1.1141	0.8890	−27.2289	−9.6975	−0.2163	−0.8716
0.1900	0	−0.3036	−1.1021	0.1962	34.3419	−0.9272	0.5740	−0.7708	0.8343	0.7463	−28.7255	−18.8433	−0.0830	−0.9575
⋮	⋮	⋮	⋮	⋮	⋮	⋮	⋮	⋮	⋮	⋮	⋮	⋮	⋮	⋮
⋮	⋮	⋮	⋮	⋮	⋮	⋮	⋮	⋮	⋮	⋮	⋮	⋮	⋮	⋮
10.1700	0	−0.5521	−1.5970	2.0634	−126.179	−153.958	0.9243	−5.6701	0.2737	2.5948	188.2750	−64.6280	2.1947	−5.3093
10.1800	0	0.0088	−2.8084	0.3645	−116.106	−185.826	1.0902	−0.7034	2.1974	1.7865	196.4603	−97.0320	−0.5576	−1.1715
10.1900	0	0.5671	−3.9118	−1.4021	−104.575	−167.481	1.2160	4.3723	4.0647	0.8639	177.0070	−87.4895	−3.3331	3.0800
10.2000	0	0.9312	−4.8923	−2.7635	−91.5124	−104.799	1.3966	8.1642	5.6144	0.2213	132.9410	−41.0362	−5.4802	6.2106
10.2100	0	0.9754	−5.7302	−3.3658	−76.0771	−15.6677	1.6905	9.6620	6.6439	0.1494	72.9521	26.6684	−6.5176	7.3303
10.2200	0	0.6838	−6.3963	−3.0680	−57.1519	75.2197	2.0946	8.5155	7.0540	0.7529	9.0710	94.0322	−6.2586	6.1425
10.2300	0	0.1562	−6.8520	−1.9751	−33.9710	143.3767	2.5416	5.1159	6.8670	1.9222	−46.4801	139.8231	−4.8516	3.0157
10.2400	0	−0.4258	−7.0551	−0.4018	−6.6633	171.2834	2.9199	0.4654	6.2127	3.3674	−84.3808	149.2055	−2.7285	−1.1392

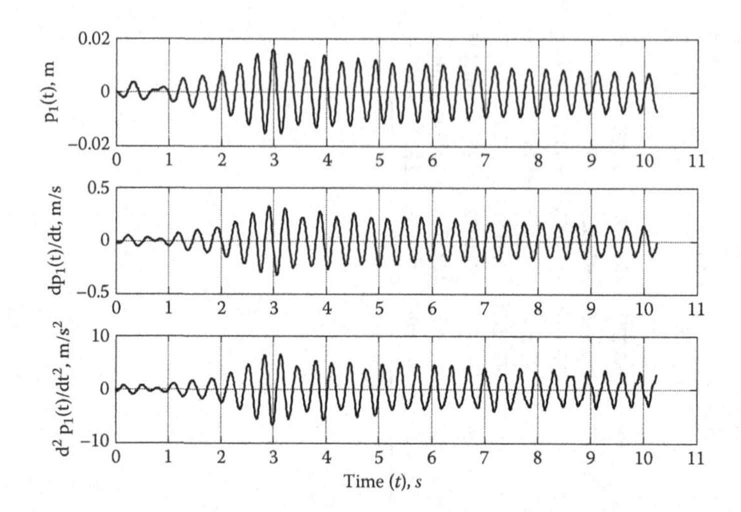

FIGURE 4.13
Modal responses (displacement, velocity, and acceleration) for mode 1 for Example 4.4.

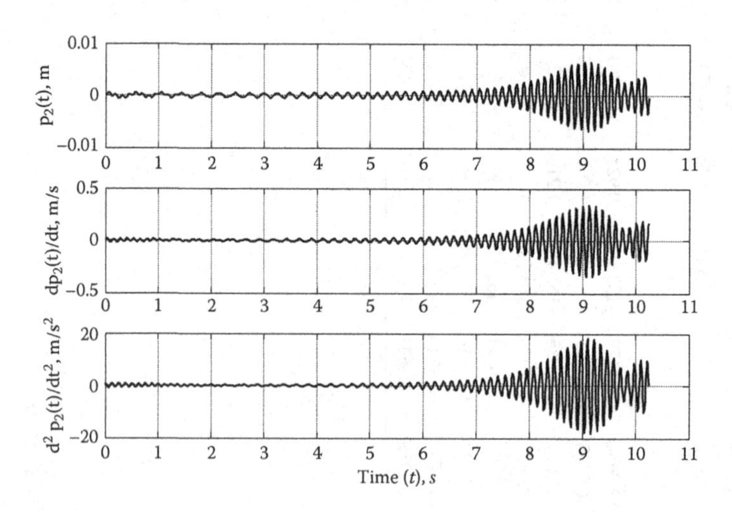

FIGURE 4.14
Modal responses (displacement, velocity, and acceleration) for mode 2 for Example 4.4.

4.4 Excitation at the Base

It is a typical phenomenon in case of earthquakes where the bases of structures, buildings, etc., are subjected to seismic loads, often referred to as ground motion. Hence, the response and stress estimation for such loadings are important to understand the dynamics and, most important, safety of the

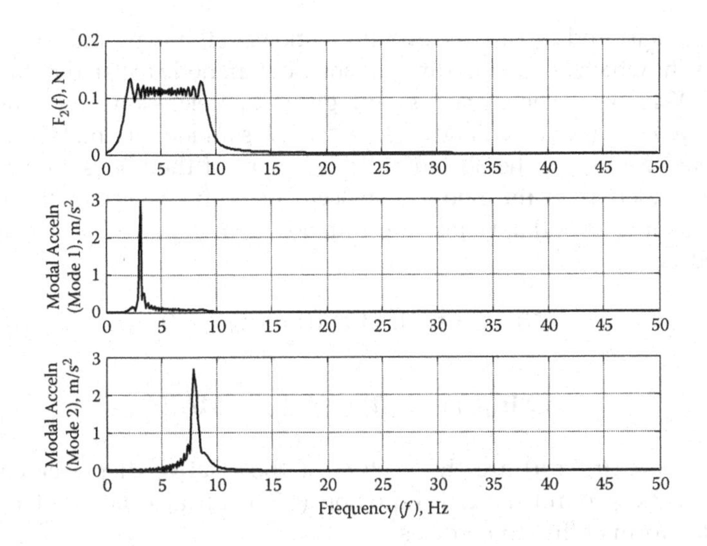

FIGURE 4.15
Applied force spectrum and spectra of the modal acceleration responses for Example 4.4.

structures and systems. Equation (4.1) for the motion for the base loading is written as

$$\mathbf{M}\ddot{\mathbf{r}}(t)+\mathbf{C}(\dot{\mathbf{r}}(t)-\dot{\mathbf{r}}_g(t))+\mathbf{K}(\mathbf{r}(t)-\mathbf{r}_g(t))=0 \tag{4.27}$$

and the vectors of the ground motion are given as

$$\mathbf{r}_g(t)=\mathbf{I}\begin{Bmatrix} x_g(t) \\ y_g(t) \\ z_g(t) \\ 0 \\ 0 \\ 0 \\ x_g(t) \\ y_g(t) \\ z_g(t) \\ 0 \\ 0 \\ 0 \\ \vdots \\ x_g(t) \\ y_g(t) \\ z_g(t) \\ 0 \\ 0 \\ 0 \end{Bmatrix} \tag{4.28}$$

where $x_g(t)$, $y_g(t)$, and $z_g(t)$ are the ground motions in the x-, y-, and z-directions, respectively, which are acting simultaneously at all nodes related to the translational DOFs only. The corresponding ground accelerations are denoted as $a_{xg}(t) = \ddot{x}_g(t)$, $a_{yg}(t) = \ddot{y}_g(t)$, and $a_{zg}(t) = \ddot{z}_g(t)$. $\boldsymbol{I}$ is an identity matrix. The displacement vectors are the absolute displacement of the nodes; however, for the stress calculation the nodal displacement with respect to the ground displacement is essential. Hence, the equation of motion in Equation (4.27) is modified as

$$\mathbf{M}(\ddot{\mathbf{r}}_{rel}(t) + \ddot{\mathbf{r}}_g(t)) + \mathbf{C}\dot{\mathbf{r}}_{rel}(t) + \mathbf{K}\mathbf{r}_{rel}(t) = 0 \tag{4.29}$$

$$\Rightarrow \mathbf{M}\ddot{\mathbf{r}}_{rel}(t) + \mathbf{C}\dot{\mathbf{r}}_{rel}(t) + \mathbf{K}\mathbf{r}_{rel}(t) = -\mathbf{M}\ddot{\mathbf{r}}_g(t) \tag{4.30}$$

where $\mathbf{r}_{rel}(t) = \mathbf{r}(t) - \mathbf{r}_g(t)$ are the vectors of the relative displacement with respect to the ground motion. Equation (4.30) can also be written in the decoupled form at the rth mode as

$$\ddot{p}_r(t) + +2\zeta_r \dot{p}_r(t) + \omega_{n_r}^2 p_r(t) = -\varphi_r^T \mathbf{M}\ddot{\mathbf{r}}_g(t) \tag{4.31}$$

where the relative displacement vectors $\mathbf{r}_{rel}(t) = \varphi\mathbf{p}(t)$. Figure 4.16 illustrates the concept of the ground motion subjected to a simple structure in the x-direction only. Equations (4.30) and (4.31) can be solved using the DI and MS methods, respectively.

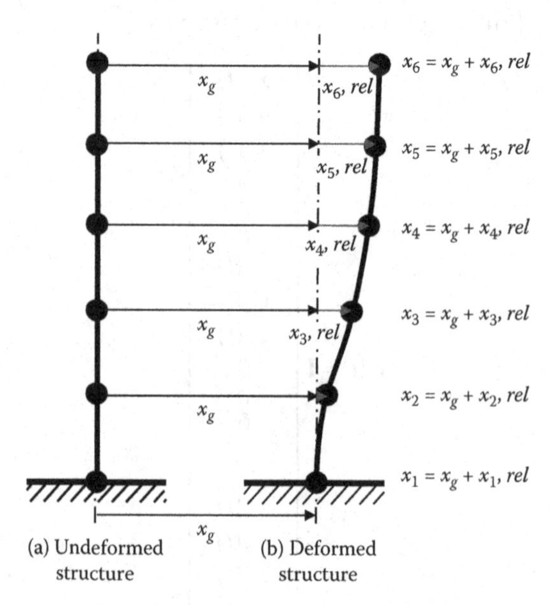

FIGURE 4.16
Structure subjected to ground motion in x-direction.

4.5 Summary

This chapter explained the numerical integration method for solving the differential equation of motions in an illustrative step-by-step approach through a few simple examples. The Newmark-β method for the integration is discussed in the chapter. Both the DI and MS methods generally used in the response computation are explained through a few simple examples. The formulation for the ground motion is also discussed, which is useful for seismic analysis.

Reference

Bathe, K.J., Wilson, E.L. 1978. *Numerical Methods in Finite Element Analysis.* Prentice Hall Ltd., New Delhi, India.

5

Introduction to Vibration Instruments

5.1 Vibration Measurement

In order to aid understanding, a single degree of freedom (SDOF) system consists of a spring of stiffness (k), a mass (m), and a damper, again considered here and shown in Figure 5.1. The equilibrium dynamic equation at time t is written as

$$F_i(t) + F_c(t) + F_s(t) = F(t) \tag{5.1}$$

$$\Rightarrow \text{Inertia force} + \text{Damping force} + \text{Stiffness force} = \text{External force}$$

$$\Rightarrow m\, a(t) + c\, v(t) + k\, y(t) = F(t) \tag{5.2}$$

where $a(t)$, $v(t)$, and $y(t)$ are, respectively, the acceleration, velocity, and displacement of the SDOF system at time t. In general, any vibrating object consists of three types of vibration responses: acceleration, velocity, and displacement responses. All these parameters are interrelated, which implies that the measurement of one parameter may be used to generate any of the other parameters through integration or differentiation.

In practice, vibration measurement is generally grouped into two main types:

1. **Measurement of only vibration responses due to natural excitation:** For example, vibration of civil buildings, mechanical structures, etc., due to wind flow and earthquakes; bridge vibration due to movement of vehicles; pipe vibration due to fluid flow; machinery vibration due to shaft rotation; etc.
2. **Measurement of both excitation forces and vibration responses:** In many experimental studies, it is sometimes essential to give known excitation (disturbance) to structures by some external means (shaker, hammer, etc.). In such cases, measurements of both excitation forces and the structural responses are required.

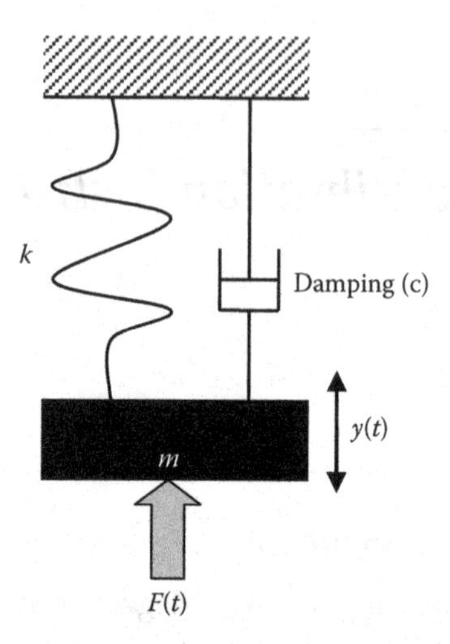

FIGURE 5.1
A spring-mass system.

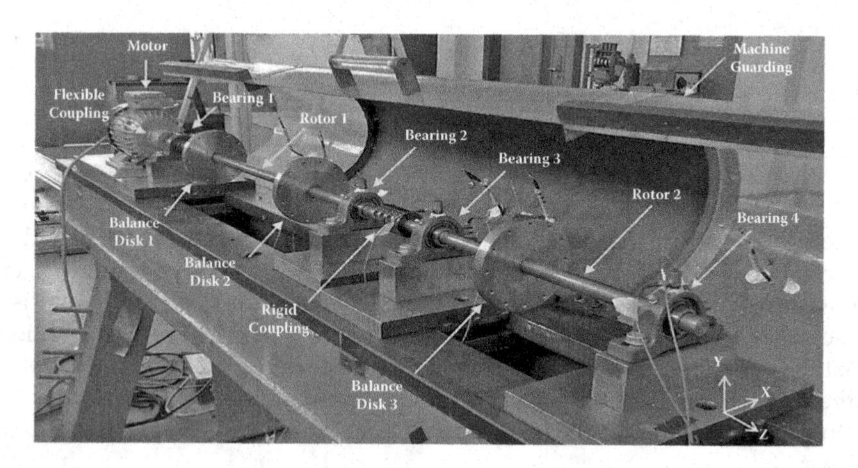

FIGURE 5.2
Photograph of a rotating rig with the vibration sensors on bearing housing.

5.1.1 Typical Measurement Setups

A few typical examples of experimental setups for vibration measurements are discussed here, so as to enhance better understanding. The former setup illustrates the vibration response measurement on a rotating rig, while the latter illustrates the measurements of both applied force and vibration responses at a number of locations on a beam.

A typical experimental setup for the vibration and temperature measurements on the bearing housings of a rotating rig during normal machine operation is shown in Figure 5.2. Figure 5.3, however, presents the schematic

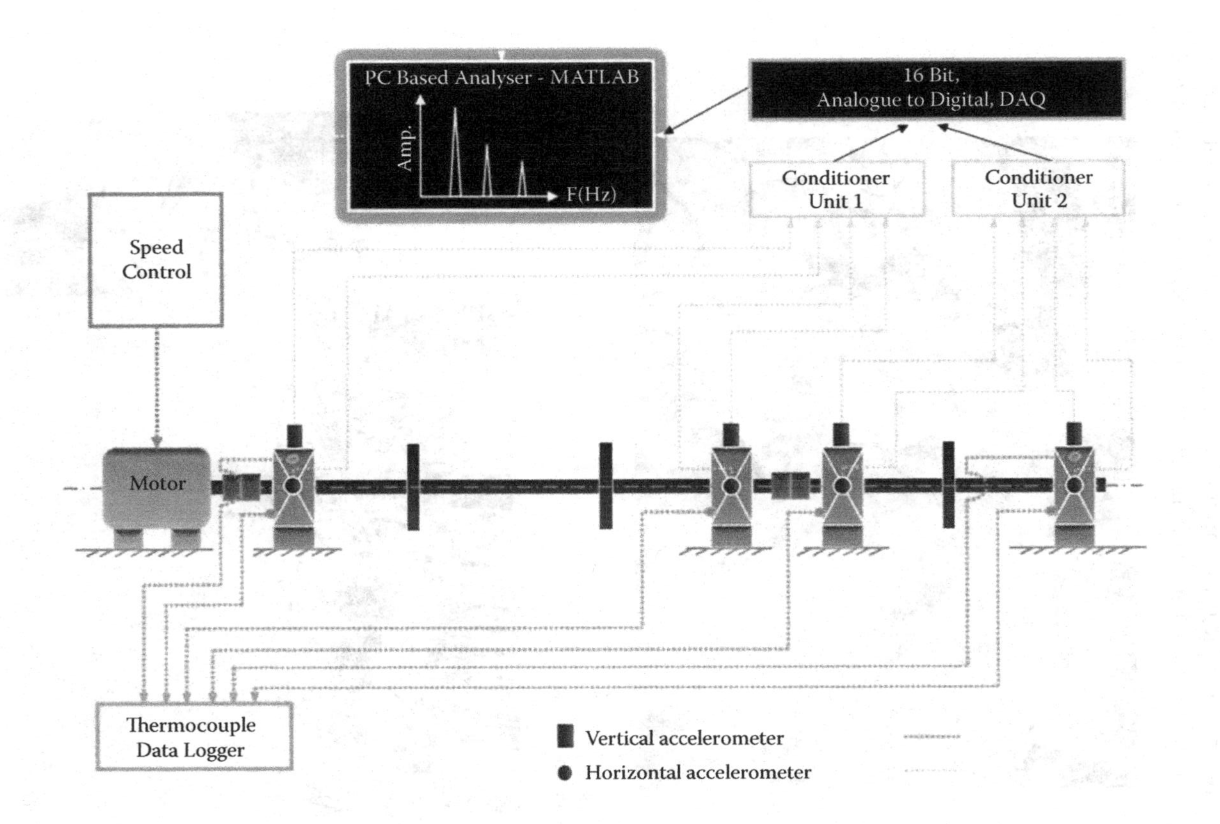

FIGURE 5.3
Schematic of vibration response measurement setup on a rotating rig.

diagram of the same rig, showing the vibration and temperature sensors, sensor power supply/conditioner units, data logger, and data acquisition (DAQ) system for collecting the measured data into the computer. This only represents a typical example of the response measurement that was highlighted in number 1 above. Similarly, Figures 5.4 and 5.5 show typical

FIGURE 5.4
Beam excited by a shaker.

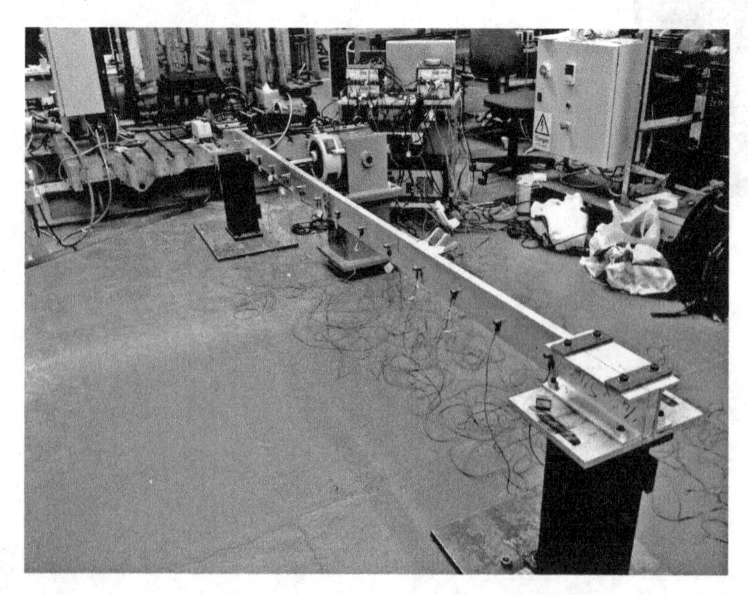

FIGURE 5.5
Schematic of a typical force and response measurement along the beam length.

examples of the force and vibration response measurements discussed in number 2 above. Figure 5.4 shows a beam that is excited at a known frequency through the aid of a shaker with an attached force sensor, while the vibration measurements all along the beam are measured using the accelerometers (vibration sensors) in Figure 5.5. Each of the elements shown in Figures 5.2 to 5.5 is systematically explained in this chapter for the purpose of clarity.

5.1.2 Steps Involved in the Collection of Data

It is vital that you understand or have knowledge of the following prior to setting up a measurement/monitoring system (from a vibration engineer's point of view):

1. Specifications (specs) of object (machine, structure, etc.) on which vibration measurements need to be carried out.
2. Knowledge of sensors' working principle (not the exact electrical/electronic circuits).
3. Knowledge of sensors' specifications, so as to aid appropriate selection. Understand the specs of sensors so that the appropriate selection is possible.
4. Knowledge of how to collect the data into a PC.
5. Signal processing for the correlation of theory and data display.

Figure 5.6 typically shows the experimental setup and data collection procedure in an abstract form, which involves the following:

1. A number of different sensors depending on the requirements. P indicates the measuring parameter from the sensor (e.g., P will be displacement in mm for the displacement sensor). S_0 indicates the sensor output, possibly in voltage, depending upon the sensor sensitivity, S.
2. Conditioner/power supply unit for each sensor, which supplies power to the sensor and provides an analogue voltage output with appropriate gain or attenuation (G).
3. Analogue voltage output from each conditioner unit is then connected to the data acquisition (DAQ) device, which converts the analogue signals from different sensors into a digital form.
4. Select appropriate DAQ device in terms of the number of bits (b) for analogue-to-digital conversion (ADC), DAQ sensitivity, S_d, sampling frequency, f_s, and the antialiasing filter.
5. V_o indicates the recorded voltage digital data from each sensor into the PC.

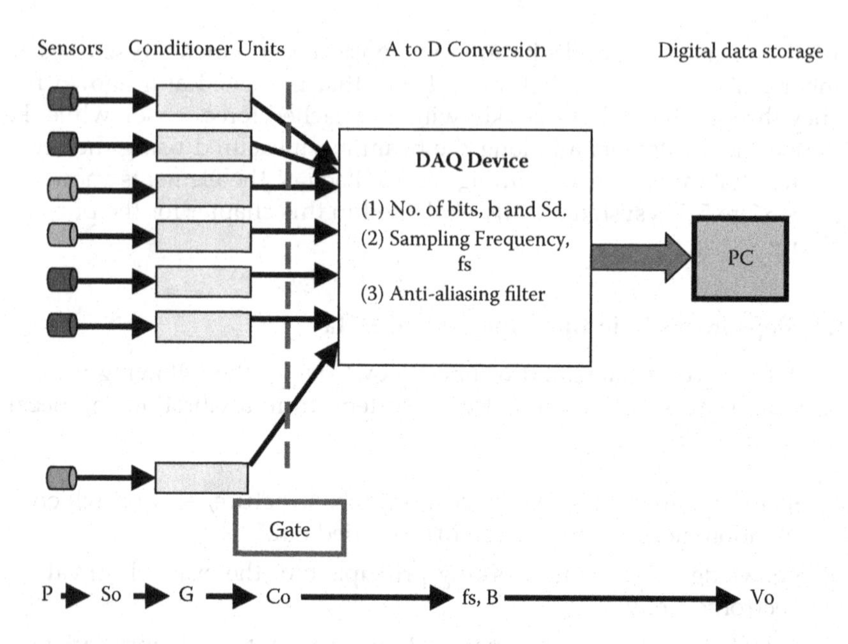

FIGURE 5.6
Typical abstract form of any measurement procedure.

Once the data are recorded, signal processing can then be performed to extract the appropriate information that will be well correlated with the theory.

5.1.3 Instrument Calibration and Specifications

To further stress the importance of appropriate sensor selection, an example of a cantilever beam is considered here. The objective is to experimentally measure the deflection when a point load at the free end of a cantilever beam is slowly and gradually increased, but within the elastic limit. Hence, the following sensors and actuators are required:

1. Displacement sensor
2. A load cell for applying the load

Since it is absolutely impossible to select any of the above instruments at random, prior ideas about the following could prove helpful:

1. Maximum possible deflection at the free end (at $x = L$) within the elastic limit
2. How much load will be applied to get maximum deflection
3. The step size for the gradual increase in load from 0 to maximum possible load

You will be required to perform initial calculations using the beam theory to determine the maximum load (P_{max}) and the maximum deflection (∂_{max}) due to the load, P_{max}, so as to get a fair idea of the values for these parameters. Based on these estimated values, it is often advised to select displacement sensor, strain gauge, and a load cell with upper limits equal to at least 20% higher than the estimated values. Once the upper range of the sensors and instruments is decided, it is also important to consider their measurement resolution. For example, if it is decided to increase the load from 0 to P_{max} in a very small step of 10 mN, then the increase in the displacement also needs to be estimated. Hence, the chosen instruments must have better resolution than the required step increase.

Hence, the following considerations are vital for the selection of the appropriate sensor(s):

- **What kind of output do you require?** This will depend on the parameter (e.g., temperature, pressure, etc.) to measure.
- **What level of accuracy is required?** The measurement error should be within 5 or 1% or less. This needs to be judged based on applications and requirements.
- **Resolution:** A step resolution that the sensor can accurately measure.
- **Is the sensor sensitive to environmental conditions?** It is also important to know whether the sensor is sensitive to environment conditions such as variations in temperature.
- **Range of measurements:** The upper and lower limits for the measurement.
- **What happens to the sensor if the measured parameter exceeds the limit?** It is required to know what happens to the sensor if the measurement exceeds the limit. Will this cause damage to the sensor or does the sensor have an in-built protection capability?
- **Mounting arrangement?** The size, weight, and mounting arrangements are important considerations, so that this should not affect the normal setup or required function of the system on which experiments are to be conducted.
- **Portable type or permanent mounting?** Sometimes, this also becomes an essential consideration.

5.1.3.1 Specifications of Instruments/Sensors

The consideration of the above aspects is only possible if the following technical specifications are carefully looked into.

Technical parts:

- **Calibration chart:** It is a standard practice that newly manufactured sensors have to pass through a series of tests to check their performance and quality. One of the important tests is calibration.

During this test, the performance of the sensor is tested against a known input or by comparing its output to the output from a well-accepted reference sensor. Such a calibration chart is often considered a valuable document for any sensor.

- **Sensitivity (electrical parameter/measuring parameter):** Nowadays, all kinds of sensors are associated with an electrical/electronic circuit that provides electrical output, generally in voltage. Hence, the ratio of the electrical output to the parameter measured by the sensor is generally known as the sensitivity of the sensor. For example, a sensitivity of 10 mV/mm for a displacement sensor means that an output of 10 mV represents a displacement of 1 mm.
- **Range of measurements (lower and upper):** The lower and upper bounds of the parameter to be accurately measured by a sensor are called the range of measurements. The difference between the measurable upper and lower limits is called full-scale (FS) reading of the instrument.
- **Linearity or accuracy:** The accuracy or linearity of an instrument is the measure of how close the output reading of the instrument is to the actual value. In practice, it is common to quote the inaccuracy figure rather than the accuracy figure. Inaccuracy is the extent to which the reading might be wrong. It is often quoted as a percentage of the FS. For example, a pressure gauge of range 0–10 bar has a quoted inaccuracy of ±1% FS, which means that the maximum error is expected to be 0.1 bar.
- **Repeatability/reproducibility:** Repeatability describes the closeness of the output reading when the same input is applied repetitively over a short period of time, while keeping all measurement conditions (instrument, observer, location, usage, etc.) the same. Reproducibility, on the other hand, describes the closeness of the output readings for the same input when there are changes in the method of the measurement, observer, measuring instrument, measurement locations, conditions of use, and time of measurement. The degree of repeatability or reproducibility in measurements from an instrument is an alternative way of expressing its precision.
- **Stability:** The output of a sensor should remain constant when a constant input is applied over long periods of time compared with the time of taking a reading, under fixed conditions of use.
- **Resolution:** The smallest change of input to an instrument that can be detected with certainty.
- **How sensitive is the sensor under different environmental conditions?** Calibration is usually conducted under a fixed condition, such as constant temperature, humidity, etc. Hence, it is

good to know the effect of different environment conditions on the accuracy of the measurements.

- **Overrange protection:** It is likely that selection of a sensor/instrument may not be accurate, and in case the measuring parameter is more than the limit of the sensor, the sensor may be damaged and cannot be used further. It is good to have a range protection system within the sensor. Sometimes, the measured parameter could exceed the limits of the sensor, due to inaccurate sensor/instrument selection. This can cause permanent damage to the sensor/instrument. It is therefore desired to have a protection system within the sensor.
- **Shock withstanding capacity:** Sensors should have in-built shock protection capabilities, so as to prevent damage/faults when accidentally dropped from their mounting locations.

Other parts:

- **Signal conditioner: output range, power supply, amplification, etc.:** Each sensor requires some kind of power supply to activate its circuit and receive the voltage output from the sensor. This unit is often called the signal conditioner unit. This is further discussed in Section 5.2.
- **Size, weight, mounting guide, etc.:** The size and weight of the sensors are also important considerations when performing experiments. It is good to refer to the mounting guidelines from the product catalogue as well as the manufacturer's notes.

5.2 Response Measuring Transducers

There are several types of transducers (sensors) commercially available in the market, and the selection of the appropriate transducer will depend on the measurement requirements. Hence, this section explains the concepts and basic working principles of the most commonly used transducers, the displacement, velocity, and acceleration measuring sensors.

5.2.1 Sensor/Transducer Unit

A complete sensor unit consists of a sensor and a signal conditioner, as shown in Figure 5.7. The functions of the conditioner unit are:

1. Supply the required power to the sensor.
2. Probably equipped with built-in filter to remove selectable frequency bands from the measured signal by the sensor.

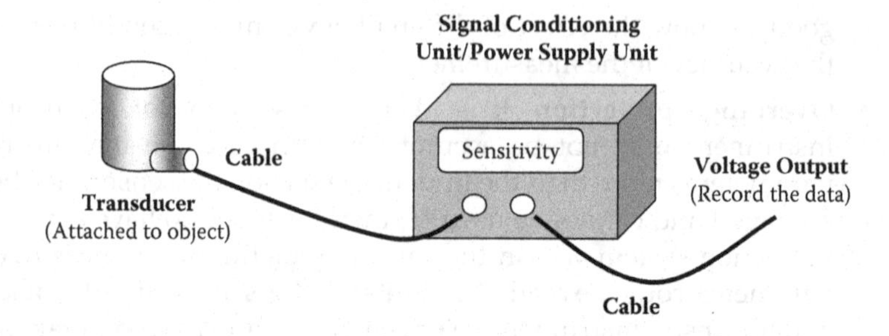

FIGURE 5.7
Vibration transducer (sensor) with a signal conditioning unit/power supply unit.

3. Amplification or attenuation (*G*) of the signal from the sensor can also be possible.
4. Analogue output voltage.

It is important to note that the amplification or attenuation is generally defined in terms of gain (*G*), and usually either a linear or dB scale is used for this purpose. The dB unit is calculated in logarithmic scale as $20\log_{10}\left(\frac{reference2}{reference1}\right)$ dB.

Here reference 1 is output voltage from the sensor, while reference 2 is the output from the conditioner based on the gain (*G*) set on the conditioner unit. For example, $G = 0$ dB means that $0 = 20\log_{10}\left(\frac{reference2}{reference1}\right)$ $\Rightarrow \frac{reference2}{reference1} = 10^{(0/20)} = 1$, i.e., $G = 1$ (on linear scale).

Another example: $G = -20$ dB, i.e.,

$$\Rightarrow 20\log_{10}\left(\frac{reference2}{reference1}\right) = 20 \Rightarrow \frac{reference2}{reference1} = 10^{(-20/20)} = 0.1$$

$G = 0.1$ (on linear scale); i.e., the signal conditioner has attenuated the output from the sensor by 1/10.

Broadly, transducers are classified in the following three categories (Sections 5.3 to 5.5).

5.3 Displacement Transducers

They are used for measuring the displacement of the vibrating object and are of two types: *proximity probe* and *optical sensor*.

5.3.1 Proximity or Eddy Current Probe

It is used for measuring the gap or displacement of a moving metal object relative to the probe. A proximity probe assembly consists of a probe containing a coil as shown in Figures 5.8(a, b). The coil is excited by an input current and voltage at a high frequency, typically 1 MHz. Because of the high frequency of excitation, eddy currents are introduced only on the surface of the target object. The eddy current alters the induction of the probe coil, and this change can be translated into the voltage output, which is proportional to the distance between the probe and the vibrating object.

This transducer is popularly used for measuring shaft relative displacement with respect to the bearing pedestal in rotating machinery. Another typical application is in the measurement of the reference signal (tacho signal) of a rotating shaft with a key phasor, with respect to the bearing pedestal (or nonrotating part) in rotating machinery. Refer to Chapter 9 for further details.

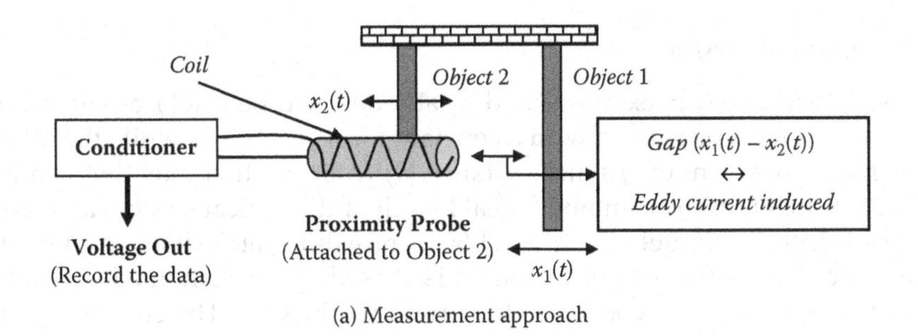

(a) Measurement approach

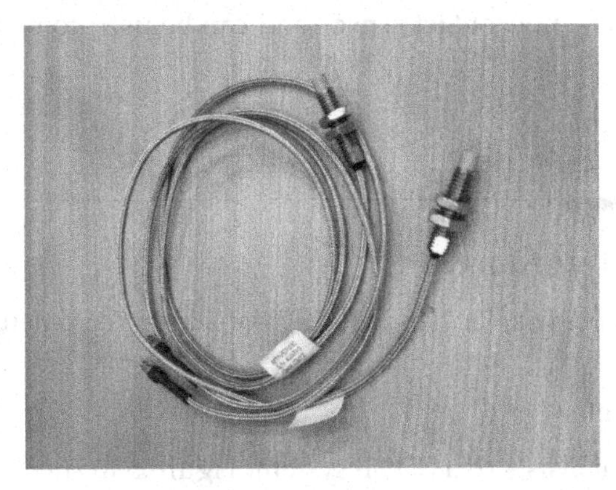

(b) Photograph of 2 proximity probes

FIGURE 5.8
Proximity probe and its working principle. (a) Measurement approach. (b) Photograph of two proximity probes.

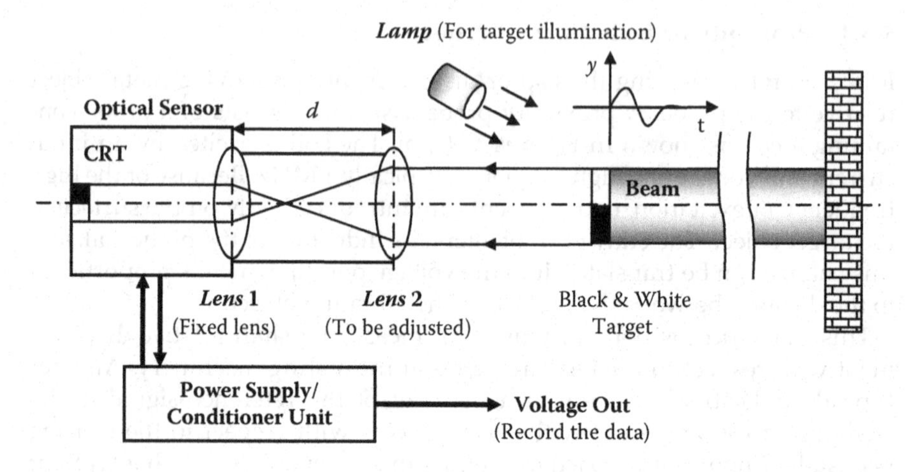

FIGURE 5.9
Optical sensor and its working principle.

5.3.2 Optical Sensor

It consists of two convex lenses and a cathode ray tube (CRT), as shown in Figure 5.9. One lens is fixed with a constant focal length (f_1), while the other lens has a focal length (f_2) and distance (d) from the first lens that can be adjusted such that the combined focal length of the optical sensor can focus on the target. The target is usually a black-and-white patch stuck on a vibrating object. The movement of the target is sensed by the combined lens, and the change in image is converted to current by the CRT. The current signal is then converted into voltage through the power supply/signal conditioner unit. This is also a noncontact sensor and can be used to measure absolute displacement.

5.4 Velocity Transducers

There are two types of velocity measuring transducers commercially available.

5.4.1 Seismometer

A seismometer consists of a SDOF system. Figure 5.10 shows a simple configuration of a seismometer mounted on a vibrating object. The equation of motion can be written as

$$m\ddot{y}_e(t) + c(\dot{y}_e(t) - \dot{y}(t)) + k(y_e(t) - y(t)) = 0 \tag{5.3}$$

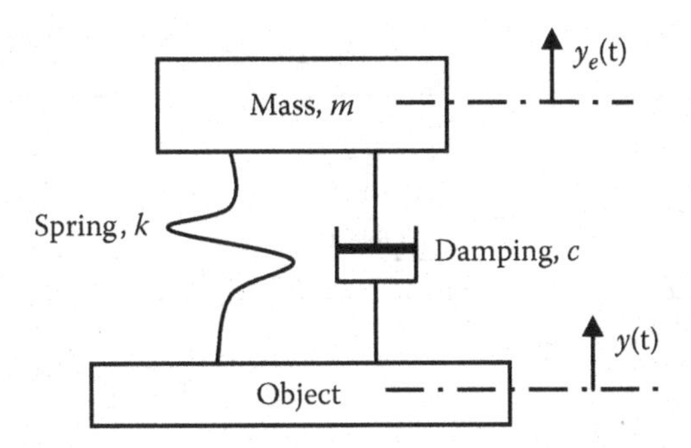

FIGURE 5.10
A SDOF system attached to a vibrating object.

where $y(t)$ and $y_e(t)$ are the displacements of the vibrating object and the seismometer mass, m. Equation (5.3) can further be written as

$$\Rightarrow m\ddot{y}_{rel}(t) + c\dot{y}_{rel}(t) + ky_{rel}(t) = -m\ddot{y}(t) \tag{5.4}$$

where $y_{rel}(t) = y_e(t) - y(t)$ is the relative displacement of the sensor seismic mass with respect to the displacement of the vibrating object. The solution of Equation (5.4) is then written as

$$y_{rel}(t) = \frac{-m\ddot{y}(t)/k}{\sqrt{(1-r^2)^2 + (2\zeta r)^2}} = \frac{r^2 y(t)}{\sqrt{(1-r^2)^2 + (2\zeta r)^2}} \tag{5.5}$$

where $r = \frac{\omega}{\omega_n} = \frac{f}{f_n}$; f is the frequency of the vibrating object and f_n is the natural frequency of the seismometer sensor. The equation for the relative displacement, i.e., Equation (5.5), is then given by

$$\frac{y_{rel}(t)}{y(t)} = \frac{r^2}{\sqrt{(1-r^2)^2 + (2\zeta r)^2}} \tag{5.6}$$

Now let's assume that $r = \frac{\omega}{\omega_n} >>>> 1$, meaning that $\frac{1}{r} \approx 0$. Hence, $y_{rel}(t) = y(t)$; i.e., the movement of the sensor mass becomes equal to 0, $y_e(t) = 0$. This means that the spring stiffness and mass weight are adjusted such that the displacement of the transducer mass, $y_e(t)$, is always equal to zero irrespective of the displacement amplitude, $y(t)$, of the vibrating object. Hence, the spring absorbs the displacement of the vibrating object as pictorially illustrated in Figure 5.11. The relative displacement, $y_{rel}(t)$, is nothing, but the output (OP) of the transducer, and the object vibration displacement, $y(t)$, is the input (IP) to the transducer (or sensor), which needs to be measured. Figure 5.12 shows a typical

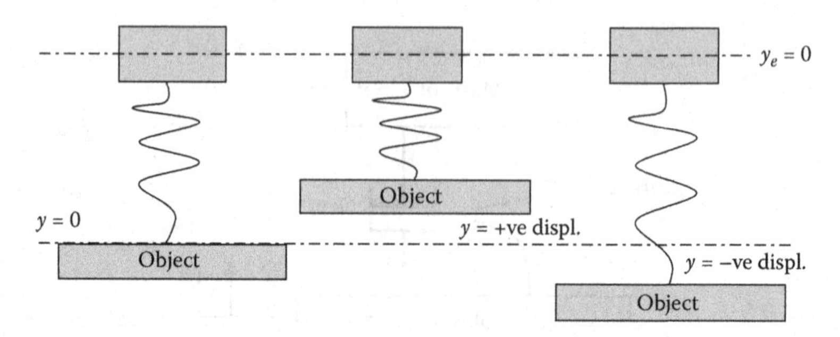

FIGURE 5.11
Zero movement of seismometer mass.

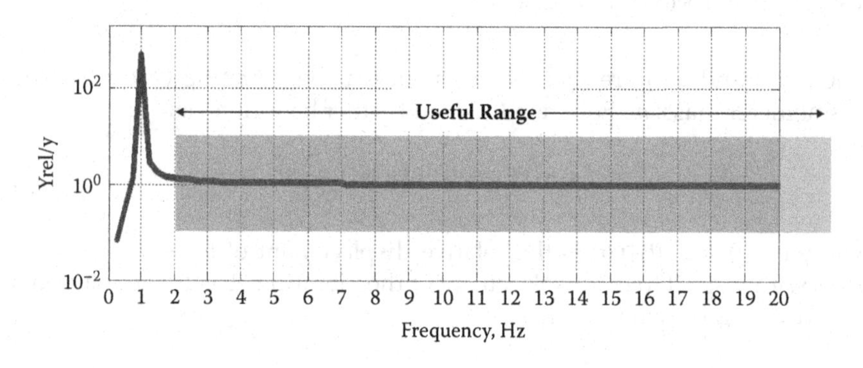

FIGURE 5.12
Typical frequency response curve for a seismometer (velocity sensor).

plot of the transfer function $(OP/IP) = \frac{y_{rel}}{y}$ versus the frequency ratio, f. It can be seen that this ratio (OP/IP) is nearly equal to 1 for the frequency, $f = 2$Hz and above. Hence, the useful range of frequency is above the sensor natural frequency. Based on this condition, the natural frequency of the sensor should be as small as possible (may be less than 1 Hz), and it should measure the displacement of the vibrating object and not the velocity. This is only possible when the seismometer consists of a soft spring with a relatively heavy mass so as to have a low natural frequency. A typical arrangement for measuring earthquake motion during a seismic event in earlier years based on this vibration principle is shown in Figure 5.13. This was done much before the electronic era, when the measurement system was totally mechanical. The system consisted of a soft spring with a heavy mass to ensure that the natural frequency was nearly zero with negligible damping and a drum rotating at constant speed. The time axis (i.e., x-axis) on the graph paper attached to the drum is calibrated with the drum speed, and the pen attached with seismic mass is used to plot the vibration of the earth during the event of an earthquake. This is why this transducer is known as a seismometer.

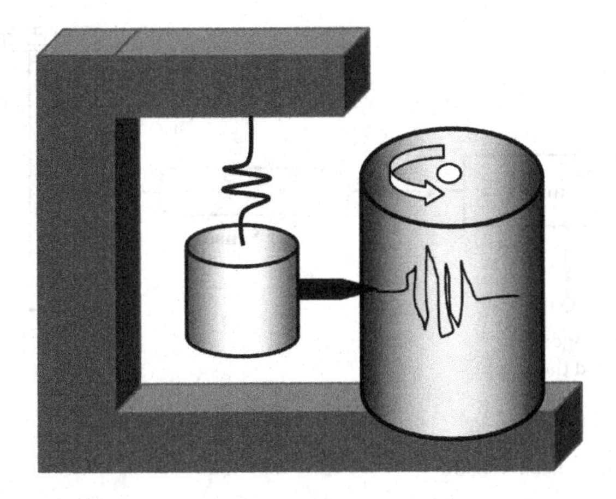

FIGURE 5.13
A typical ancient earthquake measurement approach.

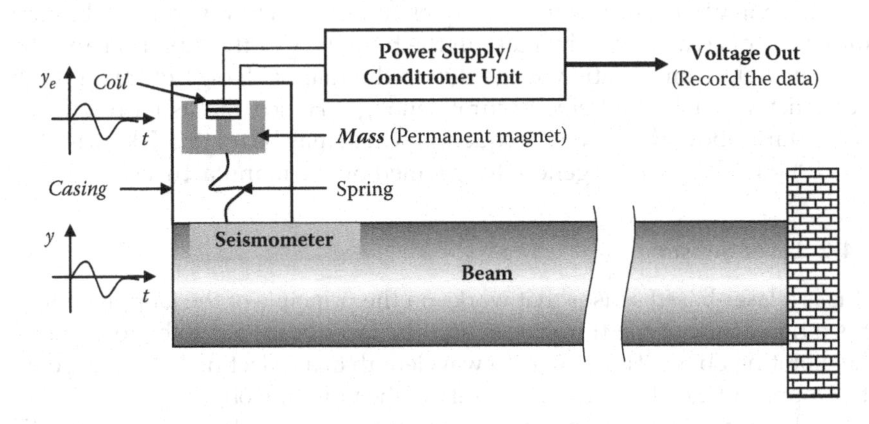

FIGURE 5.14
Seismometer and its working principle.

However, the OP measurement approach of the transducer has significantly changed with time, owing to the development of electrical theories. This is schematically shown in Figure 5.14. It is directly attached to the vibrating object whose vibration velocity has to be measured. To get the equivalent electrical signal from the transducer, the mass of the transducer is made of a permanent magnet with a wire coil attached to the transducer casing. Therefore, during the object's vibration, the coil will generate an electric flux and the rate of change in the electrical flux generates the electromagnetic voltage. This rate of change of flux, $\dot{\varphi}(t)$, is proportional to the rate of change in displacement, $\dot{y}(t)$, i.e., velocity. Hence, the output voltage is directly proportional to the velocity of the vibrating object and not its displacement.

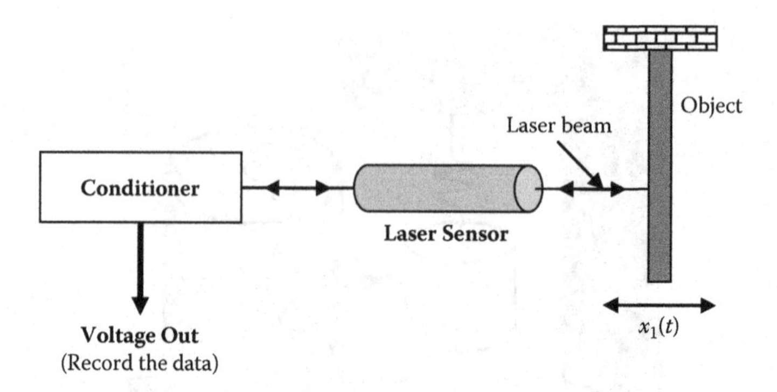

FIGURE 5.15
Measurement using a laser sensor.

According to vibration theories (see Figure 5.12), seismometers generally work above their natural frequencies due to their bulkiness and should measure vibration up to infinite frequency ranges. However, the vibration amplitude decreases with increase in the frequency of the vibration and the generated electromagnetic voltage due to the rate of change of flux at high frequency possibly not being accurate enough. Hence, the usual frequency range starts above the natural frequency, which may be up to 2–3 kHz for the $(OP/IP) = \frac{y_{rel}}{y} = 1$. This is generally specified by the manufacturer.

5.4.2 Laser Sensor

This is a laser-based sensor that works on the principle of the Doppler effect. It is a non-contact-type transducer, and the laser beam has to be focused on the vibrating object. The change in wavelength of a reflecting beam, based on the Doppler principle, gives the velocity of the vibrating object. A simple schematic diagram of measurements using a laser sensor is shown in Figure 5.15. Some typical advantages of the laser vibrometer (sensor) are (1) measurement is possible from distance (no accessibility constraints), (2) its noncontact nature ensures that the weight of the sensor is not added to that of the vibrating object, (3) a single sensor can replace several contact-type vibration sensors by simply scanning the vibration of an entire area of the object, and (4) vibration measurement on small vibrating objects (up to micro-level) is also possible.

5.5 Acceleration Transducers

Several types of acceleration transducers (accelerometers) are available, and they are all contact types (i.e., they are mounted directly on the vibrating object for acceleration measurement). The piezoelectric accelerometers are the most popular and most commonly used in practice. The working principle is schematically illustrated in Figure 5.16. It consists of a piezoelectric crystal (acting as a mechanical spring) attached to a mass, and the complete system is housed in a casing. It is like the spring-mass system shown in Figure 5.10. The displacement, $y(t)$, of the vibrating object causes displacement, $y_e(t)$, of the accelerometer (seismic) mass so that the relative displacement, $y_{rel}(t) = y_e(t) - y(t)$, leads to a deformation in the piezoelectric crystal. This deformation will generate an equivalent amount of charge due to the piezoelectric property of the crystal, which is then converted to voltage signal $v(t)$. Theoretically, it has been known that such a relative displacement is directly proportional to the acceleration of the vibrating object. Hence, the generated charge (or voltage) is directly proportional to the acceleration and this is explained here.

It is similar to the SDOF shown in Figure 5.10, but the theory is slightly different from the seismometer explained in Section 5.4.1. Equation (5.5) is reproduced here to explain the principle for an accelerometer.

$$y_{rel}(t) = \frac{-m\ddot{y}(t)/k}{\sqrt{\left(1-r^2\right)^2+\left(2\zeta r\right)^2}} = \frac{r^2 y(t)}{\sqrt{\left(1-r^2\right)^2+\left(2\zeta r\right)^2}} \tag{5.7}$$

If the ratio of the frequency $r = \dfrac{\omega}{\omega_n} = \dfrac{f}{f_n} <<< 1 \approx 0$, then

$$y_{rel}(t) = r^2 y(t) = \frac{\omega^2 y(t)}{\omega_n^2} \propto \textit{Acceleration of object} \tag{5.8}$$

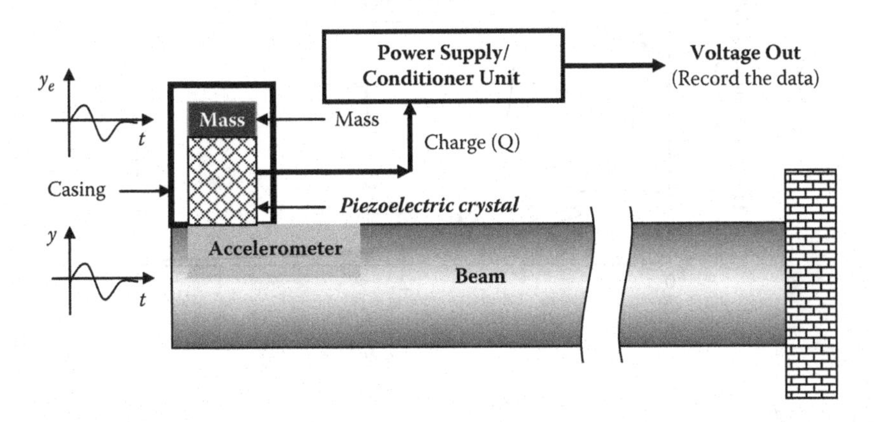

FIGURE 5.16
Accelerometer and its use.

In Equation (5.8), $\omega^2 y(t)$ is the acceleration of the vibrating object to be measured, which is nothing but input (IP) to the accelerometer, while ω_n^2 is the constant term for the accelerometer. The output (OP) of an accelerometer, i.e., the relative displacement, $y_{rel}(t)$, is given by

$$y_{rel}(t) = \frac{1/\omega_n^2}{\sqrt{\left(1-r^2\right)^2 + \left(2\zeta r\right)^2}}\,\omega^2 y(t) \tag{5.9}$$

In the nondimensional form, OP/IP,

$$\frac{\omega_n^2 y_{rel}(t)}{\omega^2 y(t)} = \frac{1}{\sqrt{\left(1-r^2\right)^2 + \left(2\zeta r\right)^2}} \tag{5.10}$$

Figure 5.17 shows the characteristic curve of an accelerometer in terms of the nondimensional form of the transfer function (OP/IP) for different values of damping ratios. It is obvious the (OP/IP) plot in Figure 5.17 is nearly equal to 1 for a very small value of the frequency ratio, r. This means that the working frequency range of an accelerometer may be a small fraction of its natural frequency, f_n. It is generally 20% of f_n when the damping ratio is equal to 0.7.

Figure 5.18 also shows the typical construction of an accelerometer, where the stiffness is given by a piezoelectric crystal, hence the relative

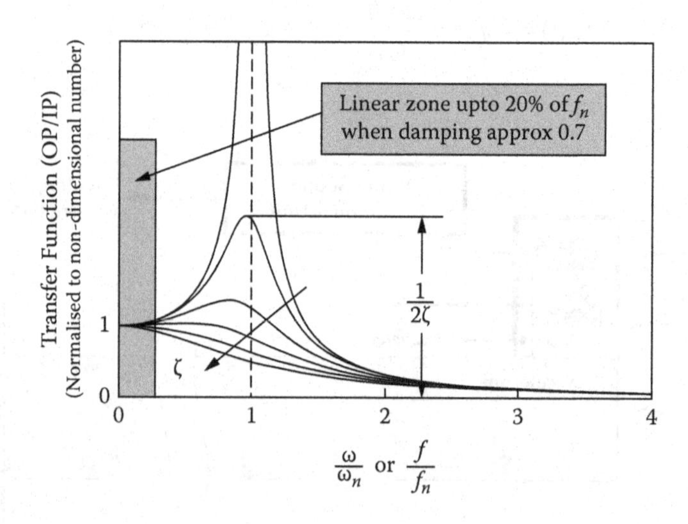

FIGURE 5.17
Characteristic curve of an accelerometer in nondimensional form of the transfer function (OP/IP).

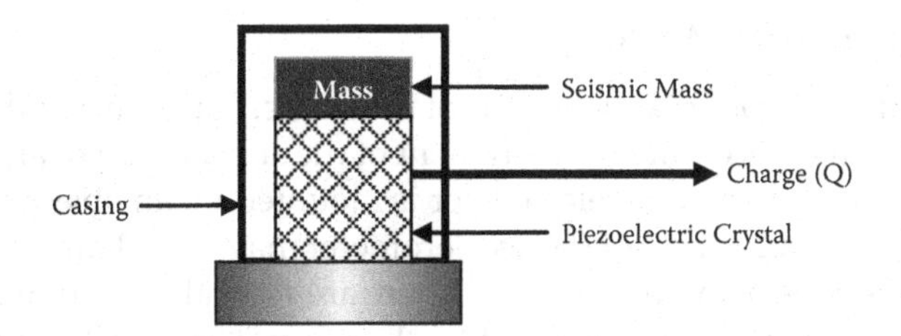

FIGURE 5.18
Simplified construction of an accelerometer.

displacement, $y_{rel}(t)$, in the spring (i.e., in the piezoelectric crystal) that generates an equivalent amount of the charge, $Q(t)$.

$$\textit{Relative Displacement, } y_{rel}(t) \propto Q(t) \propto \textit{Acceleration of object} \quad (5.11)$$

Hence, the charge, Q, is the measure of the acceleration, which can then be converted into the output voltage.

5.5.1 Technical Specifications of Accelerometer

Technical specifications of accelerometer:

- Sensitivity: pC/g, pC/(m/s^2), mV/g, mV/(m/s^2)
- Frequency range of measurement
- Resonance frequency
- Maximum acceleration measurement limit
- Amplitude linearity
- Phase distortion

These parameters must be understand carefully while selecting the accelerometer to meet the requirements. Typical specs of an accelerometer are listed here as a sample.

Sensitivity = 100 mV/g
Measurement range = ±50 g
Frequency range (±5%) = 0.5 Hz to 10 kHz
Resonance frequency ≥50 kHz
Resolution = 0.00015 g
Overload limit (shock) = ±5000 g (peak)
Operating temperature range = −65 to +200°F
Excitation voltage = 18 to 30 V (DC)
Constant excitation current = 2 to 20 mA
Size = approximately 11 mm (diameter) × 15 mm (height)
Weight = 6 g (approximately)

5.5.2 Accelerometer Mounting

It is known from theory that up to 20% of the accelerometer natural frequency, f_n, is only useful for accurate measurement of the acceleration of any vibrating object. However, when an accelerometer is mounted on an object, the mounting resonance frequency, f_m, of the accelerometer may vary. This will generally depend on the type of mounting used. There are several mounting techniques (wax, adhesive, magnet, or stud), with each type possessing its own stiffness. The mounting stiffness may get added to the accelerometer stiffness in series. Hence, the effective stiffness, k_{eff}, for the accelerometer can be given by

$$\frac{1}{k_{eff}} = \frac{1}{k} + \frac{1}{k_m} \tag{5.12}$$

where k_m is the stiffness of the mounting type, which may be equal to, greater than, or less than the stiffness, k, of the accelerometer. Hence, the effective stiffness, k_{eff}, may be less than or equal to k, and therefore the mounting natural frequency, $f_m = \frac{1}{2\pi}\sqrt{\frac{k_{eff}}{m}} \le f_n$. This means that the working frequency range of the accelerometer may be up to $0.2f_m$, which may be less than $0.2f_n$. This is illustrated in Figure 5.19. Figure 5.19 compares the change in the natural frequency of an accelerometer due to two different mounting arrangements. This simply indicates that even if the sensor is good, it may not provide accurate measurements and reduce the linear frequency range if the appropriate mounting is not used, depending upon the applications.

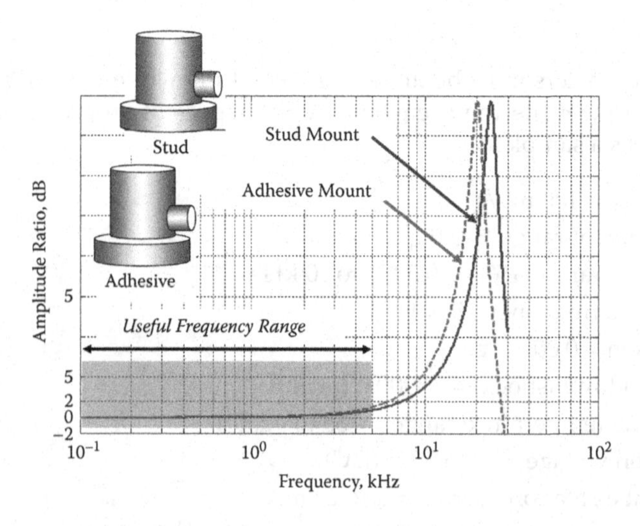

FIGURE 5.19
Mounting natural frequency of an accelerometer.

5.6 External Excitation Instruments

The source of excitation to cause the vibration in any object/system could be either natural or externally applied excitation. There are three types of external excitation instruments used in practice, which are briefly explained here.

5.6.1 Instrumented Hammer

As the name suggests, this hammer can generate an impulsive excitation to structures, and then the structures follow free decay of vibration. In general, it appears as a normal hammer that consists of a force transducer for measuring how much force is applied to a structure, and an impacting head. A schematic of the instrumented hammer is shown in Figure 5.20. A photograph of a small instrumented hammer is shown in Figure 5.21. The force transducer contains a piezoelectric crystal that generates charge due to deformation in the crystal as a result of the impact applied. Depending on the requirements, the impact head may be changed from soft to hard. Instrumented hammers of different sizes and impact capacities are commercially available from different manufacturers. The frequency band of excitation will be small if the soft impact head is used, while a relatively high frequency band of excitation will be obtained with harder impact heads, as shown in Figure 5.22.

5.6.2 Portable Shaker

As discussed earlier, the frequency band of excitation can only be controlled through the impact head hardness. This control is, however, qualitative in nature, and may not be adequate if a highly precise frequency range of excitation is required. Hence, for such cases, portable shakers of different ratings (loads, frequency bands, etc.) are often used. This type of shaker is generally attached to the object to be excited. A simple schematic of a shaker

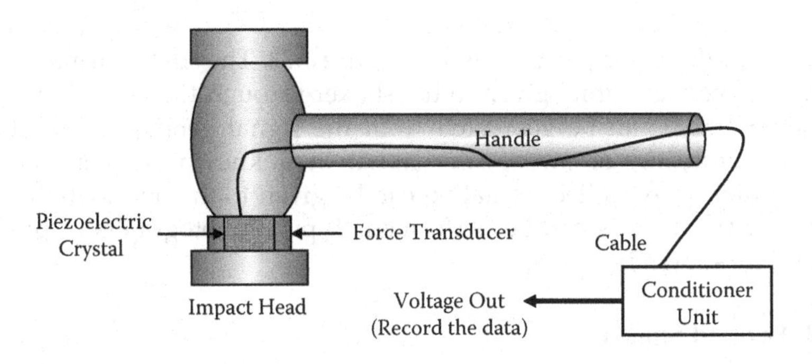

FIGURE 5.20
An instrumented hammer.

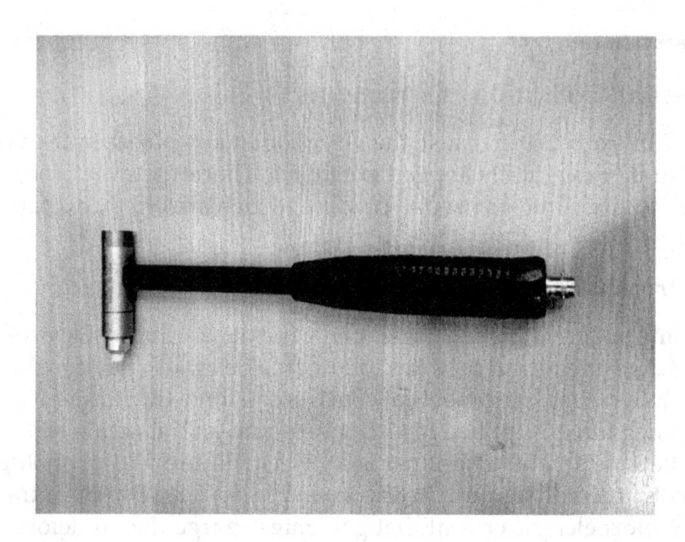

FIGURE 5.21
Photograph of a small instrumented hammer.

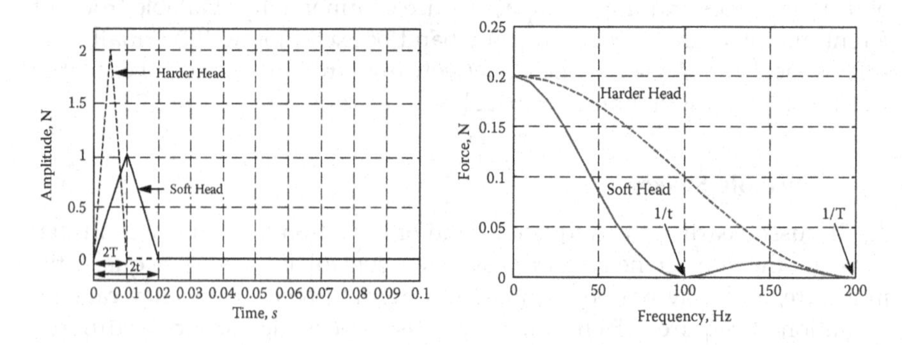

FIGURE 5.22
Effect of different impact heads of an instrumented hammer. (a) Time waveforms. (b) Spectra of the applied forces.

and its working principle is shown in Figure 5.23. The shaker armature is a follower of voltage signal given to the shaker through the power amplifier. A known signal can be generated from the signal generator (sinusoidal, random, multisine, or sweep-sine signal in a known frequency band); hence, a highly controlled excitation can be given to the object and a more deterministic analysis can be performed. Typical photographs of portable shakers are shown Figure 5.24.

5.6.3 Payload Shaker

The working principle of this class of shakers is similar to that of the portable shakers. The only difference is that the sizes and load capacities of these

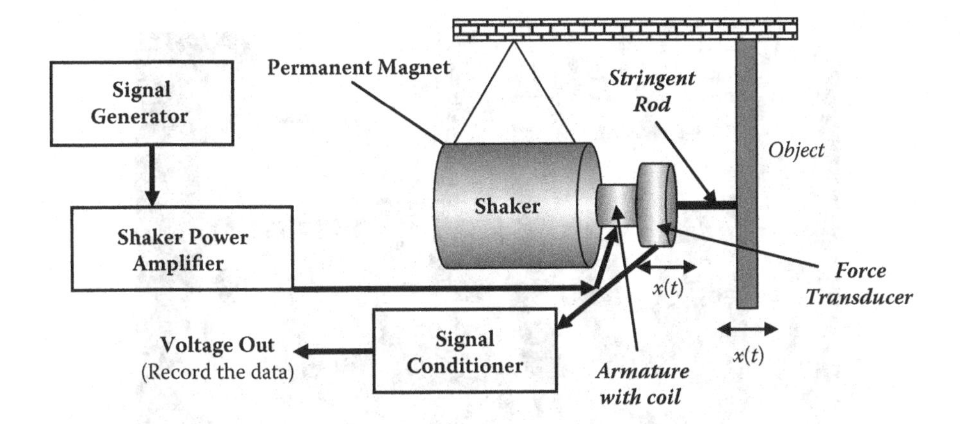

FIGURE 5.23
A schematic of a shaker and its working principle.

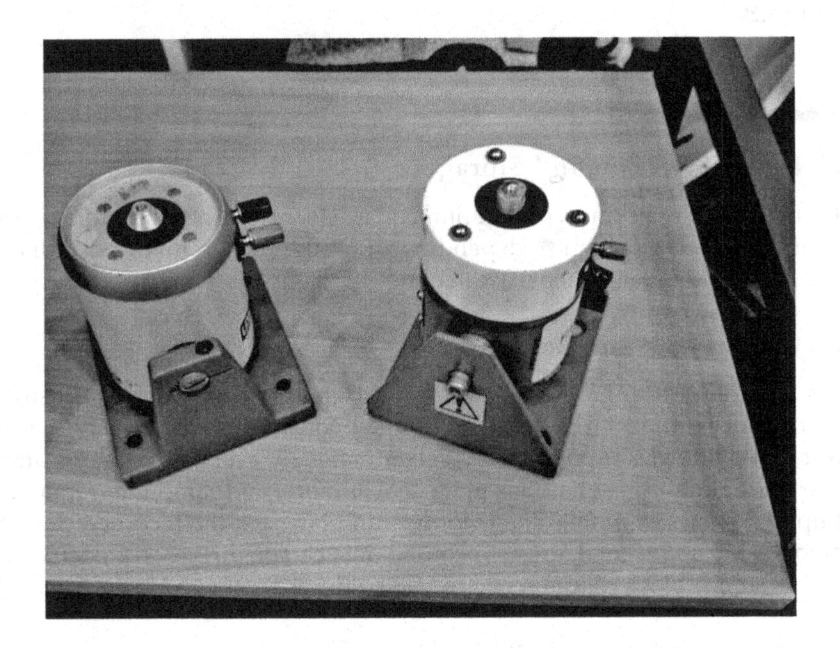

FIGURE 5.24
Photograph of small portable modal shakers.

shakers are high enough for the test object to be directly mounted on the shaker. Such shakers are useful for simulating earthquake motions. Several structural components, machinery, and instruments can be directly qualified for seismic loads, and any other types of loads expected during service condition. A typical small payload shaker is shown in Figure 5.25.

FIGURE 5.25
Photograph of typical medium-sized payload shakers.

5.7 Data Collection and Storage

The following two kinds of simultaneous data recording and storage are commonly used in practice, depending upon the person involved in measurements or the instruments available.

5.7.1 Digital Data Tape Recorder

This is a compact unit where data from different sensors during measurement at the plant site can be recorded simultaneously into a digital tape so that offline data analysis is possible. Most commercially available recorder units have input and output channels up to 256 (or more) and sampling frequency of up to 25 kHz (or more). The recorder can play back the recorded data in analogue form if needed, or the recorded data can be directly transferred to a PC. A simple block diagram of a tape recorder is shown in Figure 5.26.

5.7.2 PC-Based Data Acquisition and Storage

These days, PC-based data acquisitions are more popular. There are several types of data acquisition (DAQ) cards (see Figure 5.27) already available with their driver software, which are quite easy to use. It is required to plug in to the PC, and the driver software simply does the data acquisition. The number of inputs per DAQ card generally ranges from 2 input channels to as high as 32 channels or higher, with different sampling rates. These DAQ cards accept voltage input signals.

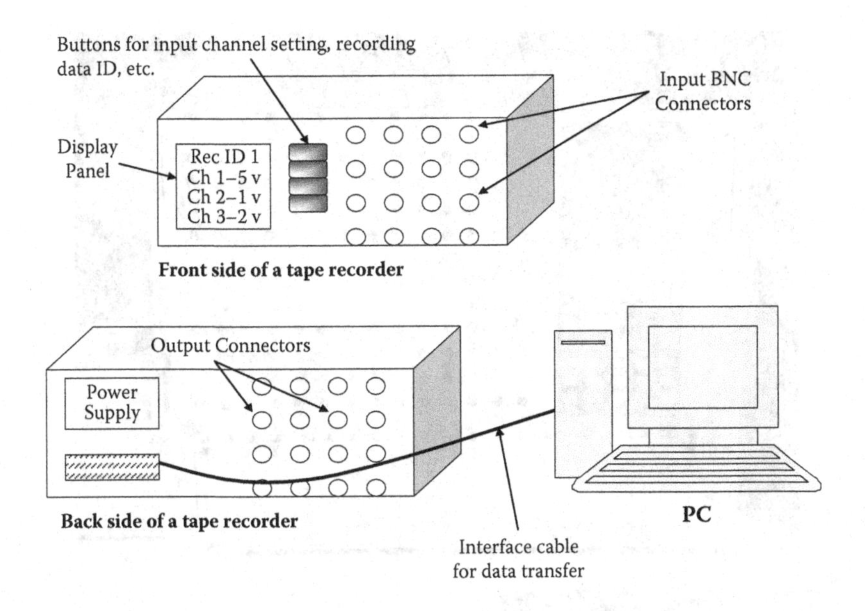

FIGURE 5.26
Typical view of the digital tape recorder.

5.7.3 Precautions during Data Recording/Collection

The following are some important aspects to consider while collecting data on a PC or tape recorder through the aid of a DAQ device:

- Must be aware of the technical specifications of the recording system, i.e., DAQ card
  - Maximum Sampling Frequency for each input channel
  - Bits of the DAQ card to understand ADC
  - Equipped with antialiasing filter or not
- Technical details of sensors
- Utilize maximum possible dynamic range for ADC to get meaningful recorded data/signals

5.8 Concept of Sampling Frequency, f_s

Figure 5.28 is the output voltage signal from a conditioner unit connected to a sensor. As shown in Figure 5.28, the output voltage signal is generally a continuous analogue signal. Generally, all the data in a continuous analogue

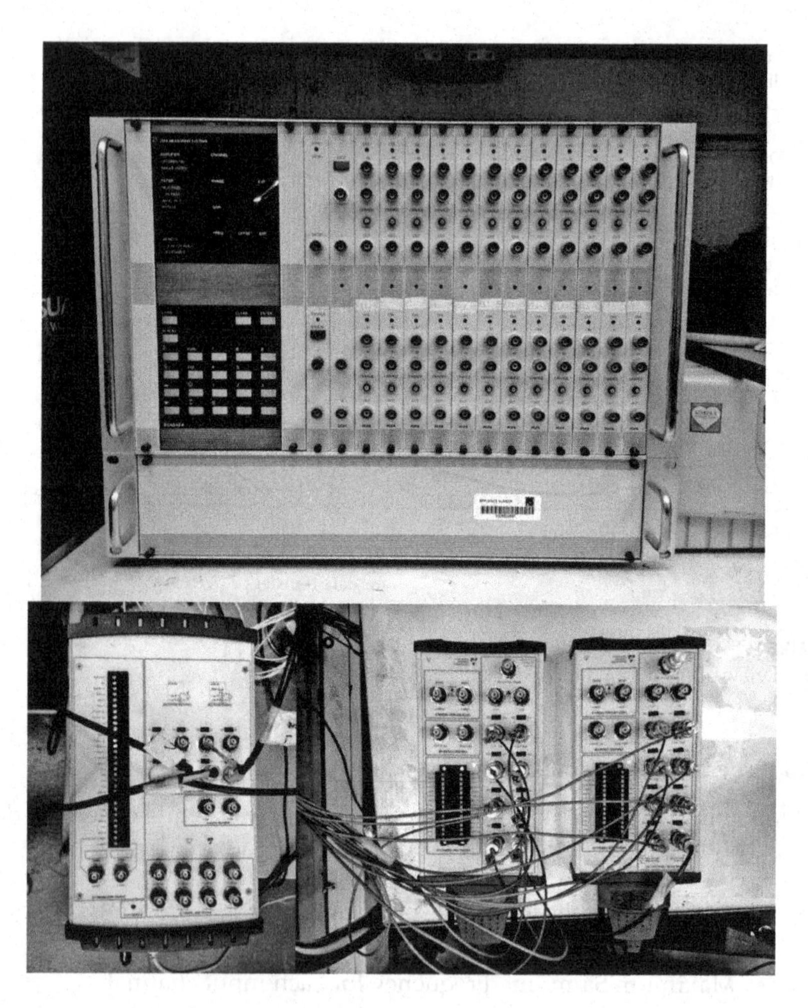

FIGURE 5.27
A few multi-input-channel DAQ devices.

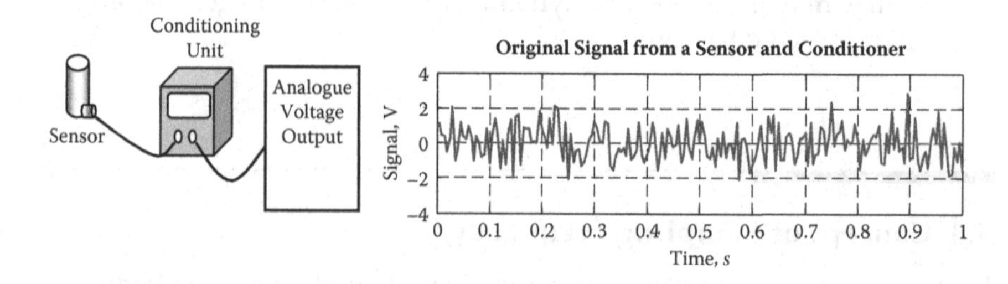

FIGURE 5.28
An original analogue signal.

signal are difficult to collect into the PC; hence, a definite rate of data collection needs to be defined, e.g., 20 samples/s, which is called a data sampling rate or sampling frequency, f_s. This means that 20 data samples from the signal will be collected in a second. Each datum is collected at equal intervals of 1/20 = 0.05 s. The time interval or time step (dt or Δt) between two consecutive samples of the data collection becomes equal to $1/f_s$. If the data length is 5 s, then a total of 101 samples (1 sample at 0 s + 20 samples × 5 s) will be collected into the PC. This is known as the analogue-to-digital conversion (ADC) process.

The data will then be collected in digital form in the following way for the original signal shown in Figure 5.28 using the sampling frequency f_s = 20 samples/s (or Hz).

Data	Time	Signal Amplitude
0	0	V0
1	dt = 0.05 s	V1
2	2 dt = 0.10 s	V2
3	3 dt = 0.15 s	V3
.	.	.
.	.	.
.	.	.

Figure 5.29 shows the reconstructed signals after data collection into the PC at the sampling rates f_s = 20 Hz and f_s = 40 Hz, respectively, from the original

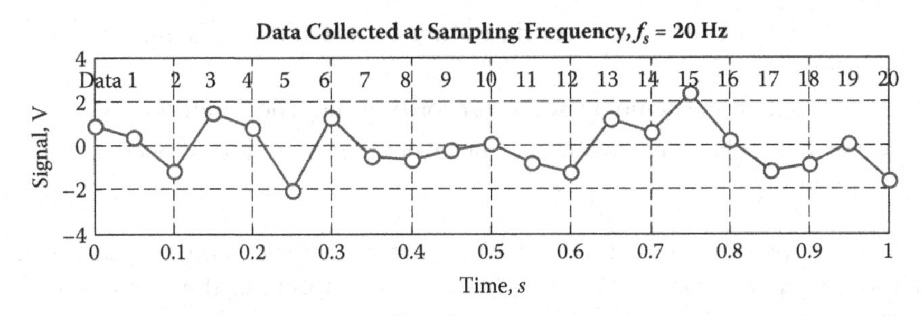

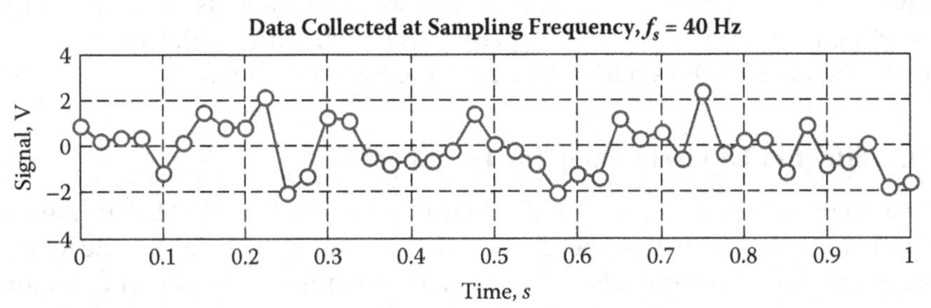

FIGURE 5.29
Collected (sampled) signals using different sampling frequencies.

analogue signal in Figure 5.28. It is obvious that the reconstructed signals are significantly different from the original signal. They can only be improved by increasing the data sampling rate. Hence, an appropriate selection of the sampling frequency, f_s, is important during the process of data collection.

5.9 Aliasing Effect and the Selection of Sampling Frequency, f_s

You may observe/notice the following often when you watch any video of a rotating wheel:

1. Wheel is always rotating in the forward direction.
2. Rotating speed is changing with time.
3. Sometimes the spokes in the wheel are observed to be moving forward, sometime they appear stationary, and other times they appear to be rotating backward.
4. Even when the spoke indicates a forward motion, the speed of the spokes is observed to be much lower than the actual forward speed of the wheel.

It is vital to note the following:

No change in the wheel configuration except for the change in the speed with time.

No change in the camera used for recording the wheel motion.

What could be the reason for these behaviors of the spokes? Any thought?

This happens because the camera may have a fixed scanning rate, e.g., 100 frames per second (FPS), for capturing the motion of the wheel, which may introduce such errors. The scanning rate of a camera is nothing but the sampling frequency. To further explain this effect, a sinusoidal signal of frequency 500 Hz, as shown in Figure 5.30, is considered here.

5.9.1 Effect of Different Sampling Frequencies

Here, the frequency of the signal remains constant, i.e., 500 Hz, unlike the wheel motion where the speed was always changing. Hence, the object frequency remains constant. Note that the scanning rate for the camera remained constant while recording the wheel motion, but for the signal considered here is recorded using a different sampling frequency, f_s. Hence, the sampling frequency, f_s, is changing.

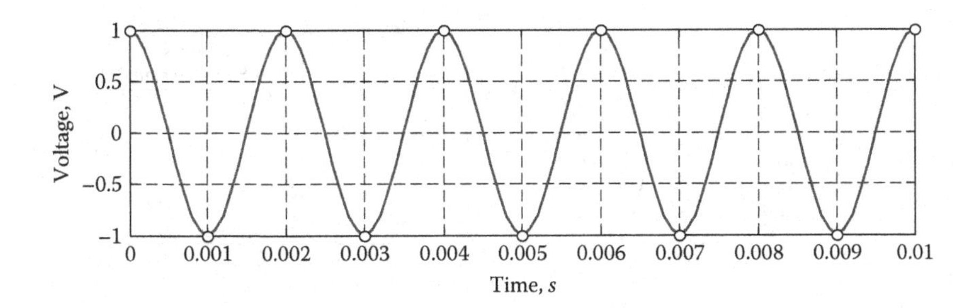

FIGURE 5.30
A typical sine wave of frequency of 500 Hz.

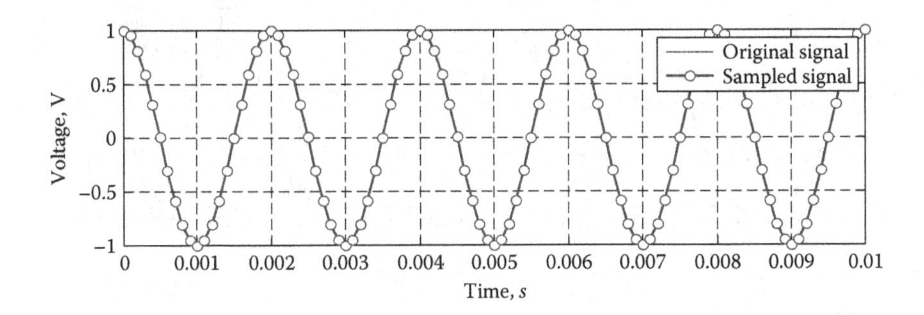

FIGURE 5.31
Original and sampled signals when $f_s = 10$ kHz.

The aim is to observe the effect of the sampling frequency on collection of data from the sinusoidal signal in Figure 5.30.

Case i: The sampling frequency $f_s = 10$ kHz, $dt = 0.1$ ms. The collected data (marked as a solid line with circles) shown in Figure 5.31 are the same as the original data in Figure 5.30. Hence, there is no error in the collected/recorded data.

Case ii: The sampling frequency $f_s = 5$ kHz, $dt = 0.2$ ms. The collected data (marked as solid line with circles) are compared with the original data (solid line) in Figure 5.32. The collected/recorded signal is nearly the same as the original signal.

Case iii: The sampling frequency $f_s = 2$ kHz, $dt = 0.5$ ms. The collected data (marked as solid line with circles) are compared with the original data (solid line) in Figure 5.33. The time period of 0.002 s of the recorded signal is the same as the original signal. The maximum and minimum voltages of the recorded signal are also the same as the original signal, but the number of samples per second were not enough to record a signal that is a replica of the original signal.

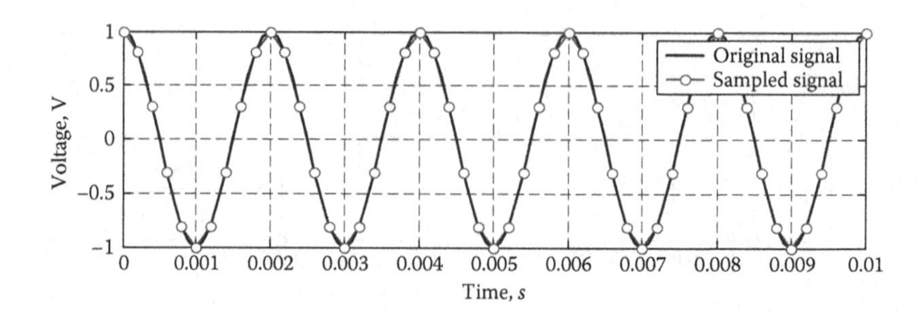

FIGURE 5.32
Original and sampled signals when $f_s = 5$ kHz.

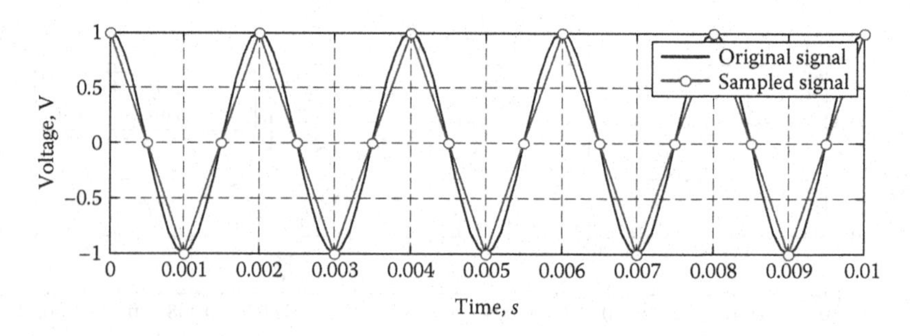

FIGURE 5.33
Original and sampled signals when $f_s = 2$ kHz.

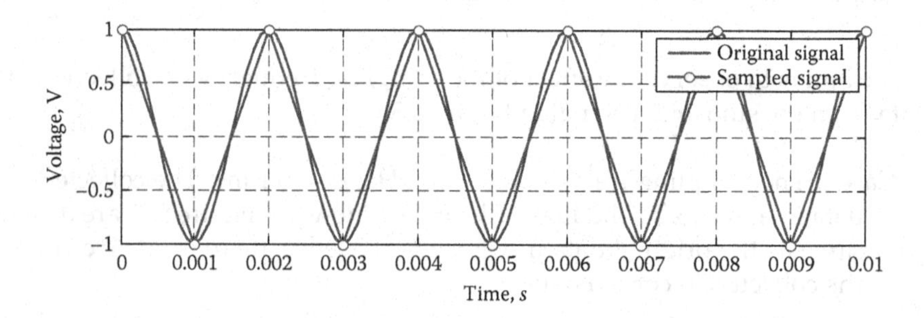

FIGURE 5.34
Original and sampled signals when $f_s = 1$ kHz.

Case iv: The sampling frequency $f_s = 1$ kHz, $dt = 1$ ms. The observation shown in Figure 5.34 is not much different from Case iii.

Case v: The sampling frequency $f_s = 750$ Hz, $dt = 0.00133$ s. The recorded signal (marked as solid line with circles) is compared with the original signal (solid line) in Figure 5.35. The time period of the recorded signal is 0.004 s, which is twice that of the original signal, 0.002 s. Significant error in amplitude is also observed in the recorded signal.

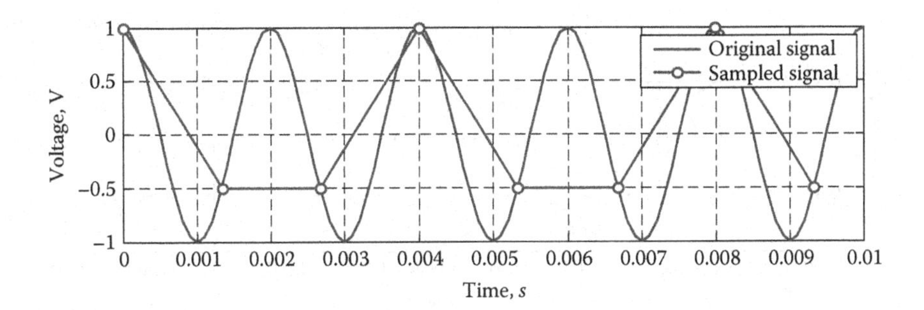

FIGURE 5.35
Original and sampled signals when $f_s = 750$ Hz.

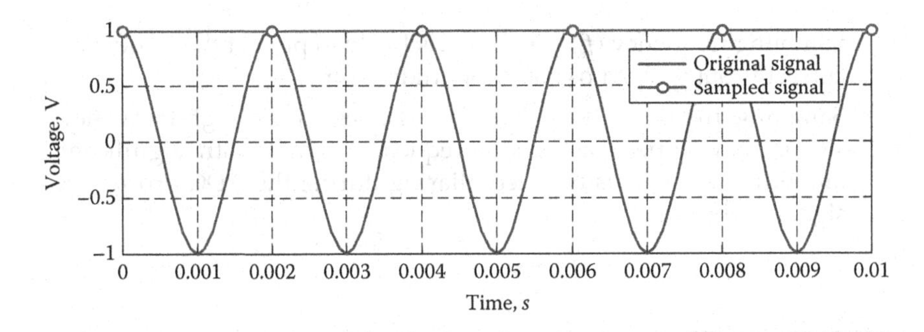

FIGURE 5.36
Original and sampled signals when $f_s = 500$ Hz.

Case vi: The sampling frequency $f_s = 500$ Hz, $dt = 0.002$ s. The recorded signal (marked as solid line with circles) is just a straight line compared to the original sinusoidal signal (solid line) in Figure 5.36. This means that the time period for the recorded signal is infinity. This is a typical case of synchronization of the signal frequency with the sampling rate, causing the frequency of the recorded signal to be zero.

Case vii: The sampling frequency $f_s = 400$ Hz, $dt = 0.0025$ s. The recorded signal (marked as a solid line with circles) is compared with the original signal (solid line) in Figure 5.37. The time period of the recorded signal is 0.01 s, which is much higher than the original signal, 0.002 s. Significant error in amplitude is also observed for the recorded signal.

5.9.2 Observations

- Original signal frequency, $f = 500$ Hz.
- Sampling frequency (f_s) much higher than $2f$ indicates no error.

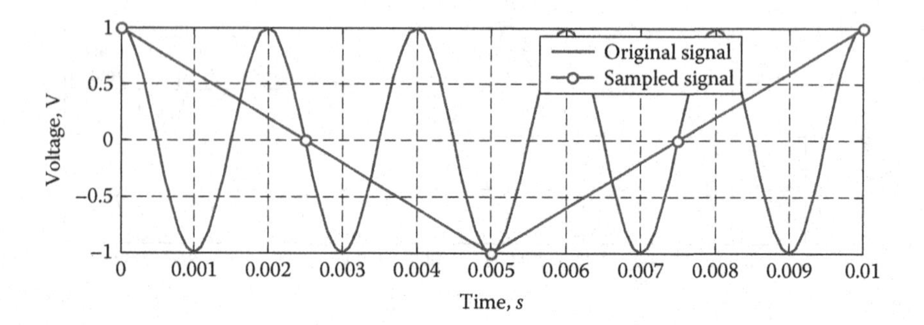

FIGURE 5.37
Original and sampled signals when $f_s = 400$ Hz.

- Sampling frequency (f_s) equal to $2f$ is likely to predict the frequency correctly, but with a possible amplitude error.
- Sampling frequency less than $2f$ indicates that a high-frequency signal may appear as a low-frequency signal with significant amplitude error. This is called aliasing during the ADC process in signal collection.

5.10 DAQ Device Bit for ADC

Once the sampling frequency is selected, then an appropriate DAQ card is required. The ideal DAQ card should be capable of supporting the required number of input channels and the sampling frequency for the simultaneous collection of data from all sensors/conditioner units.

DAQ is basically an electronic device and the analogue-to-digital conversion (ADC) is generally done using a binary representation for encoding the signal. This means that the smallest value of a signal using a 12-bit binary representation would be represented as

$$000000000000 = 0 \text{ and the largest value: } 111111111111 = 12$$

Let's assume that 0 represents 0 V and 12 represents 10 V; then the sensitivity of this device will be written as $S_d = (10 - 0)\text{V}/12 = 0.8333$ V/binary bin, where $(10 - 0)$V is the full-scale input voltage (FSIV) and the total number of binary bins, $BT = 12$. The DAQ device works on the principle of converting the input voltage into the number of the nearest binary bins (B), depending on the DAQ device sensitivity, S_d. The bit (b) of the DAQ device/card indicates the total number of binary bins (BT) for the card, which is calculated as $BT = 2^b$. This means that $BT = 2^4 = 16$ binary bins for a 4-bit card.

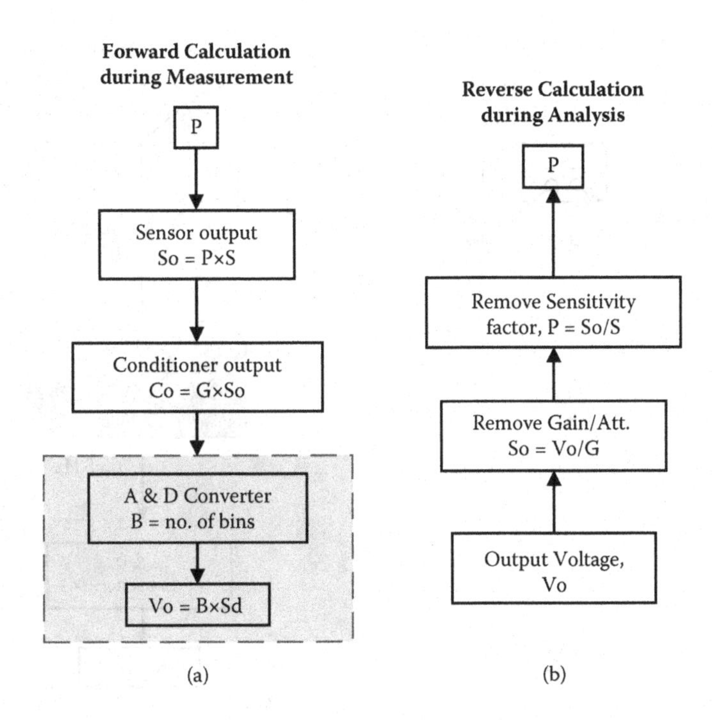

FIGURE 5.38
Forward and reverse calculation procedure.

The block with the dash line in Figure 5.38(a) shows the concept of encoding a signal into the PC through a DAQ device and then reconstruction of the measured data in Figure 5.38(b). The concept of ADC is further discussed in the following three cases for better understanding.

5.10.1 Case 1: 2-Bit DAQ Device

Consider an analogue voltage (C_o) sinusoidal signal of frequency 1 Hz shown in Figure 5.39 that needs to be collected into a PC through a DAQ device. As shown in Figure 5.39, a DAQ device with the following specifications is used to collect the signal (Figure 5.39) into a PC.

1. Four input channels.
2. Two-bit device (number of bits is generally an even number).
3. Maximum input voltage per channel is ±5 V; however, the user may select any input voltage range for individual input channels, which should be within ±5 V.
4. Maximum sampling frequency per input channel is 20 kHz, but any sampling frequency up to 20 kHz can be selected.

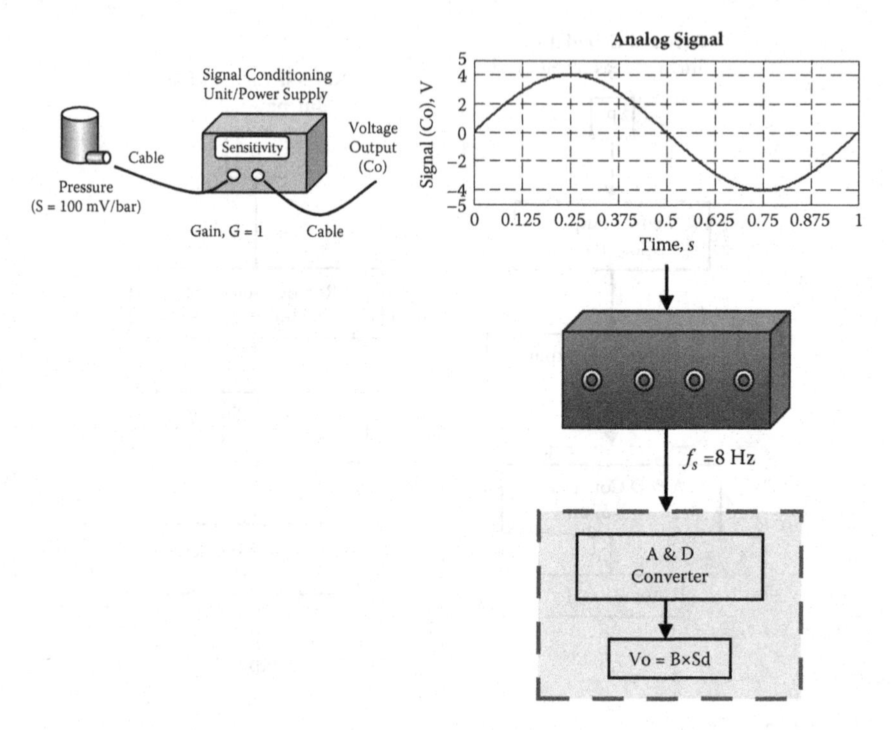

FIGURE 5.39
Demonstration of data collection during experiment.

For the present case, there is only one input, so the output voltage (C_o) of the signal is connected to one input channel. Since the maximum/minimum voltage of the signal is ±4 V, a ±5 V input voltage for input channel 1 on the DAQ device is selected. Now let's assume that the sampling frequency, f_s, is 8 Hz, i.e., the time step $dt = 0.125$ s.

Since the DAQ card is 2 bits, it will create $BT = 2^{DAQ\ bit} = 2^{b} = 2^2 = 4$ numbers of total binary bins. This means two bins for 0 to +5 V and two bins for 0 to −5 V, as shown in Figure 5.40. This is a simple mathematical representation of the actual process. Note that the complex electronic circuit is not discussed here so as to simplify the ADC process and for better understanding. The sensitivity of the DAQ card is defined as the ratio of the full-scale input voltage (FSIV) to the total number of bins (BT), mathematically expressed as

$$S_d = \frac{FSIV}{BT} \tag{5.13}$$

For the present DAQ card, selected FSIV = max voltage − min voltage = (+5) − (−5) = 10 V, BT = 4 bins. So, DAQ device sensitivity $S_d = \frac{FSIV}{BT} = \frac{10}{4} = 2.5$ V/bin.

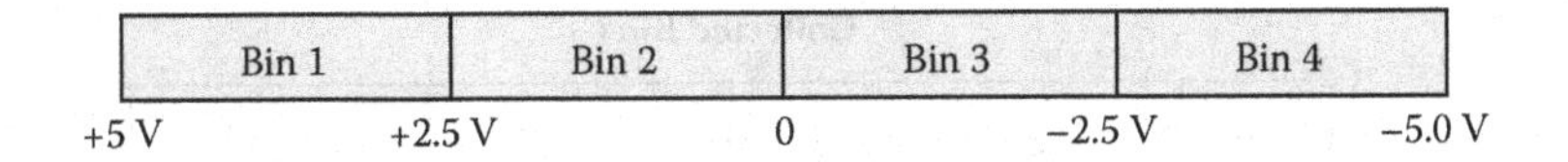

FIGURE 5.40
Voltage range for 2-bit DAQ to convert into binary bins.

Original Signal	A to D Conversion				Recorded Signal
t = 0, Co = 0	0	0	0	0	B = 0, Vo = 0
t = 0.125 s, Co = 2.828 V	+1	+1	0	0	B = +2, Vo = +5 V
t = 0.25 s, Co = 4.0 V	+1	+1	0	0	B = +2, Vo = +5 V
t = 0.375 s, Co = 2.828 V	+1	+1	0	0	B = +2, Vo = +5 V
t = 0.50 s, Co = 0	0	0	0	0	B = 0, Vo = 0
t = 0.625 s, Co = −2.828 V	0	0	−1	−1	B = −2, Vo = −5 V
t = 0.75 s, Co = −4.0 V	0	0	−1	−1	B = −2, Vo = −5 V
t = 0.875 s, Co = −2.828 V	0	0	−1	−1	B = −2, Vo = −5 V
t = 1.0 s, Co = 0	0	0	0	0	B = 0, Vo = 0

FIGURE 5.41
Step-by-step illustration of ADC for a typical signal in Figure 5.39.

The distribution is shown in Figure 5.40. Hence, the input voltage is then converted into the number of bins.

RULE: If a signal has 0 V, then all bins are empty and the data will be recorded as 0.

If the signal voltage is +4.5 V, then +2.5 V will go in bin 2 and the remaining in bin 1. The data will then be recorded as $B = +1 + 1 = +2$ and $V_o = +2 \times 2.5 = +5$ V.

Similarly, if the signal voltage is −2.6 V, then −2.5 V will go in bin 3 and the remaining −0.1 V in bin 4. Hence, the data will be recorded as $B = -1 - 1 = -2$ bins and $V_o = -2 \times 2.5 = -5$ V.

Now consider the signal shown in Figure 5.39 to be collected using this DAQ device into a PC as per the following:

1. Sampling frequency, $f_s = 8$ Hz ($dt = 0.128$ s)
2. The above rule for converting the data into bins (B) and then to voltage V_o

The stepwise collection is illustrated in Figure 5.41. Hence, the recorded signal in terms of the collected bins and the recorded voltage are shown

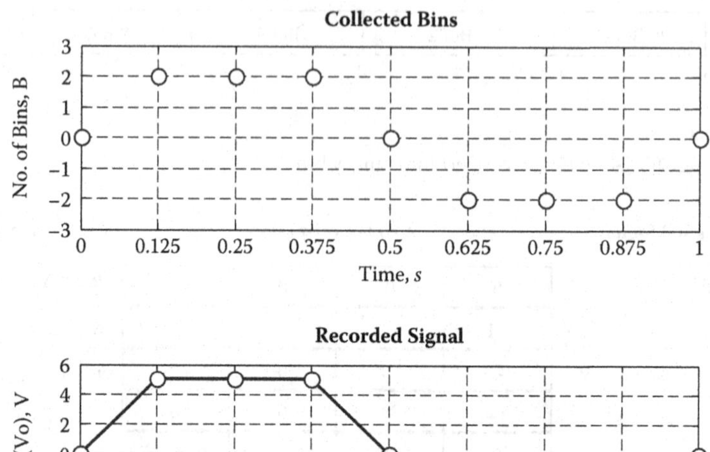

FIGURE 5.42
Collected bins and the recorded signal for the signal in Figure 5.39.

in Figure 5.42. The recorded/collected voltage signal is definitely different from the original signal.

To improve the quality of the recorded signal, the following two points are vital:

1. High-bit DAQ device to improve its sensitivity, S_d
2. Sufficiently high sampling frequency, f_s

5.10.2 Case 2: 16-Bit DAQ Device

Let us consider a 16-bit DAQ card with two input channels. The input voltage level for each channel is assumed to be ±5 V. Hence, the sensitivity (S_d) of the DAQ card is $\frac{FSIV}{2^b} = \frac{10}{2^{16}} = 0.15258$ mV/bin. It is also assumed that the data are collected at the sampling frequency, $f_s = 200$ Hz (i.e., $dt = 0.05$ s).

ADC process for this DAQ device is listed in Table 5.1. Figures 5.43 to 5.45 show the ADC of three different cases of pure sine waves of frequency 10 Hz. Tables 5.2 and 5.3 give the list of the actual voltage of the signals, the recorded number of bins, and voltages using the rule in Table 5.1.

TABLE 5.1
Rule for Analogue-to-Digital Conversion Using 16-Bit DAQ Device

Actual Signal Voltage Range	Digital Conversion (no. of bins, B)	Reconstruction of Signal from Bins ($B \times S_d$)
–5 V to less than –(5-1.5259e-4) V	–32,768	–5.00 V
–(5-1.5259e-4) V to less than –(5-2x1.5259e-4) V	–32,767	–(5-1.5259e-4) V
–(5-2x1.5259e-4)V to less than –(5-3x1.5259e-4)V	–32,766	–(5-2x1.5259e-4) V
⋮	⋮	⋮
–2x1.5259e-4 V to less than –1.5259e-4 V	–2	–2x1.5259e-4 V
–1.5259e-4 V to less than 0	–1	–1.5259e-4 V
0	0	0.00 V
Greater than 0 to +1.5259e-4 V	+1	+1.5259e-4 V
Greater than +1.5259e-4 V to +2x1.5259e-4 V	+2	+2x1.5259e-4 V
⋮	⋮	⋮
Greater than +(5-3x1.5259e-4) V to +(5-2x1.5259e-4) V	+32,766	+(5-2x1.5259e-4) V
Greater than +(5-2x1.5259e-4) V to +(5-1.5259e-4) V	+32,767	+(5-1.5259e-4) V
Greater than +(5-1.5259e-4) V to +5 V	+32,768	+5.00 V

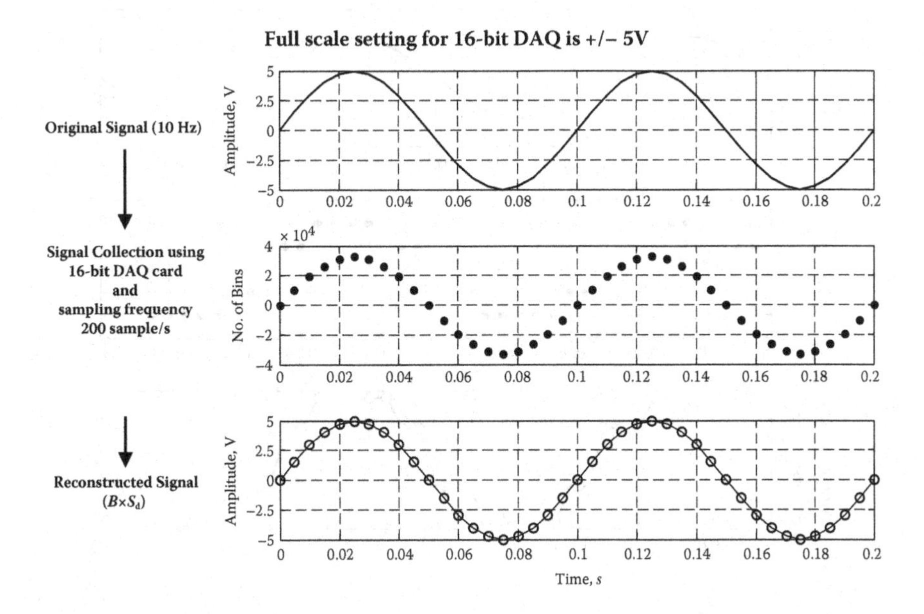

FIGURE 5.43
Original signal, collected bins, and the recorded (reconstructed) signal.

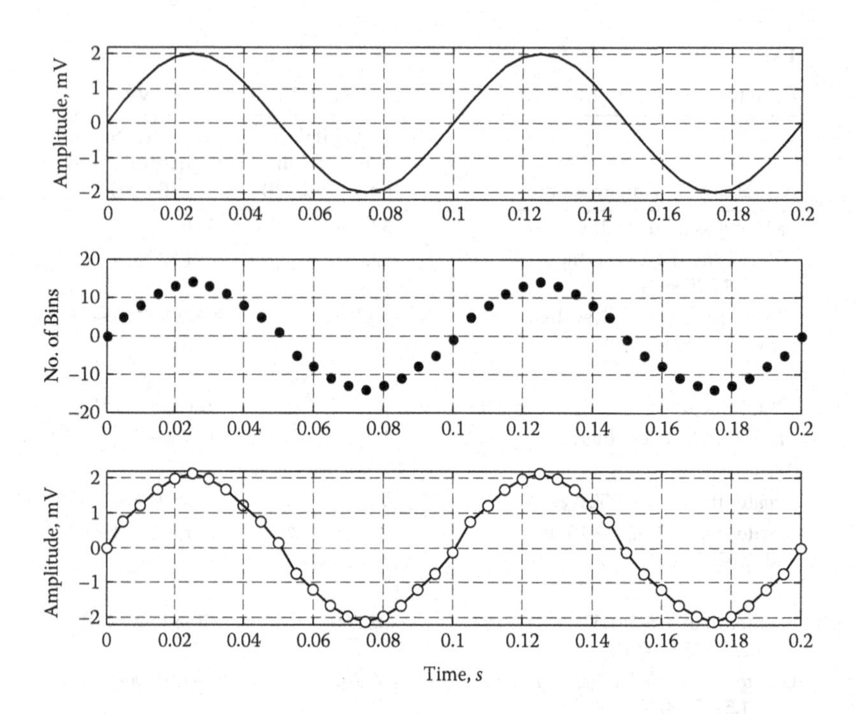

FIGURE 5.44
Original signal, collected bins, and the recorded (reconstructed) signal.

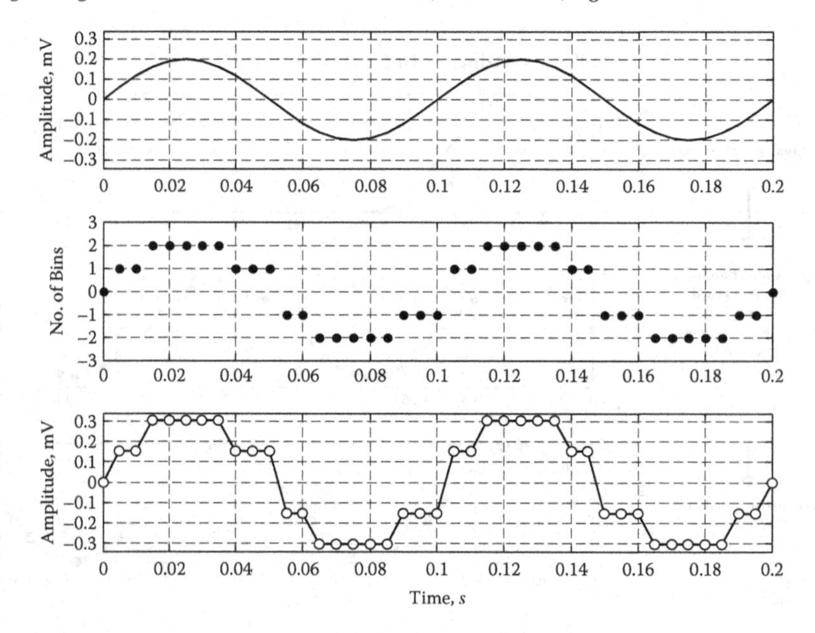

FIGURE 5.45
Original signal, collected bins, and the recorded (reconstructed) signal.

TABLE 5.2

ADC process for the signal in Figure 5.43

	±5 V Signal		
Time, s	**Actual Data, V**	**Digital Data, Bins (B)**	**Redata, V**
0	0	0	0
0.0050	1.5451	10,126	1.5451
0.0100	2.9389	19,261	2.9390
0.0150	4.0451	26,510	4.0451
0.0200	4.7553	31,165	4.7554
0.0250	5.0000	32,768	5.0000
0.0300	4.7553	31,165	4.7554
0.0350	4.0451	26,510	4.0451
0.0400	2.9389	19,261	2.9390
0.0450	1.5451	10,126	1.5451
0.0500	0.0000	1	0.0002
0.0550	−1.5451	−10,126	−1.5451
0.0600	−2.9389	−19,261	−2.9390
0.0650	−4.0451	−26,510	−4.0451
0.0700	−4.7553	−31,165	−4.7554
0.0750	−5.0000	−32,768	−5.0000
0.0800	−4.7553	−31,165	−4.7554
0.0850	−4.0451	−26,510	−4.0451
0.0900	−2.9389	−19,261	−2.9390
0.0950	−1.5451	−10,126	−1.5451
0.1000	−0.0000	−1	−0.0002
0.1050	1.5451	10,126	1.5451
0.1100	2.9389	19,261	2.9390
0.1150	4.0451	26,510	4.0451
0.1200	4.7553	31,165	4.7554
0.1250	5.0000	32,768	5.0000
0.1300	4.7553	31,165	4.7554
0.1350	4.0451	26,510	4.0451
0.1400	2.9389	19,261	2.9390
0.1450	1.5451	10,126	1.5451
0.1500	−0.0000	−1	−0.0002
0.1550	−1.5451	−10,126	−1.5451
0.1600	−2.9389	−19,261	−2.9390
0.1650	−4.0451	−26,510	−4.0451
0.1700	−4.7553	−31,165	−4.7554
0.1750	−5.0000	−32,768	−5.0000
0.1800	−4.7553	−31,165	−4.7554
0.1850	−4.0451	−26,510	−4.0451
0.1900	−2.9389	−19,261	−2.9390
0.1950	−1.5451	−10,126	−1.5451
0.2000	0	0	0

TABLE 5.3

ADC process for the signals in Figures 5.44 and 5.45

	±2 mV signal			±0.2 mV signal		
Time, s	Actual Data, mV	Digital Data, bins (B)	Redata, mV	Actual Data, mV	Digital Data, bins (B)	Redata, mV
0	0	0	0	0	0	0
0.0050	0.6180	5	0.7629	0.0618	1	0.1526
0.0100	1.1756	8	1.2207	0.1176	1	0.1526
0.0150	1.6180	11	1.6785	0.1618	2	0.3052
0.0200	1.9021	13	1.9836	0.1902	2	0.3052
0.0250	2.0000	14	2.1362	0.2000	2	0.3052
0.0300	1.9021	13	1.9836	0.1902	2	0.3052
0.0350	1.6180	11	1.6785	0.1618	2	0.3052
0.0400	1.1756	8	1.2207	0.1176	1	0.1526
0.0450	0.6180	5	0.7629	0.0618	1	0.1526
0.0500	0.0000	1	0.1526	0.0000	1	0.1526
0.0550	–0.6180	–5	–0.7629	–0.0618	–1	–0.1526
0.0600	–1.1756	–8	–1.2207	–0.1176	–1	–0.1526
0.0650	–1.6180	–11	–1.6785	–0.1618	–2	–0.3052
0.0700	–1.9021	–13	–1.9836	–0.1902	–2	–0.3052
0.0750	–2.0000	–14	–2.1362	–0.2000	–2	–0.3052
0.0800	–1.9021	–13	–1.9836	–0.1902	–2	–0.3052
0.0850	–1.6180	–11	–1.6785	–0.1618	–2	–0.3052
0.0900	–1.1756	–8	–1.2207	–0.1176	–1	–0.1526
0.0950	–0.6180	–5	–0.7629	–0.0618	–1	–0.1526
0.1000	–0.0000	–1	–0.1526	–0.0000	–1	–0.1526
0.1050	0.6180	5	0.7629	0.0618	1	0.1526
0.1100	1.1756	8	1.2207	0.1176	1	0.1526
0.1150	1.6180	11	1.6785	0.1618	2	0.3052
0.1200	1.9021	13	1.9836	0.1902	2	0.3052
0.1250	2.0000	14	2.1362	0.2000	2	0.3052
0.1300	1.9021	13	1.9836	0.1902	2	0.3052
0.1350	1.6180	11	1.6785	0.1618	2	0.3052
0.1400	1.1756	8	1.2207	0.1176	1	0.1526
0.1450	0.6180	5	0.7629	0.0618	1	0.1526
0.1500	–0.0000	–1	–0.1526	–0.0000	–1	–0.1526
0.1550	–0.6180	–5	–0.7629	–0.0618	–1	–0.1526
0.1600	–1.1756	–8	–1.2207	–0.1176	–1	–0.1526
0.1650	–1.6180	–11	–1.6785	–0.1618	–2	–0.3052
0.1700	–1.9021	–13	–1.9836	–0.1902	–2	–0.3052
0.1750	–2.0000	–14	–2.1362	–0.2000	–2	–0.3052
0.1800	–1.9021	–13	–1.9836	–0.1902	–2	–0.3052

TABLE 5.3 (*Continued*)

ADC process for the signals in Figures 5.44 and 5.45

	±2 mV signal			±0.2 mV signal		
Time, s	Actual Data, mV	Digital Data, bins (B)	Redata, mV	Actual Data, mV	Digital Data, bins (B)	Redata, mV
0.1850	−1.6180	−11	−1.6785	−0.1618	−2	−0.3052
0.1900	−1.1756	−8	−1.2207	−0.1176	−1	−0.1526
0.1950	−0.6180	−5	−0.7629	−0.0618	−1	−0.1526
0.2000	0	0	0	0	0	0

5.10.3 Case 3: DAQ Card Bit, b = 8, FSIV = 10 V, $S_d = 10/2^8 = 0.0391$ V/bin

Example i: Data1 contains accurate voltage output (C_o) from a conditioning unit of a pressure sensor (sensitivity, S = 1 V/bar) with a gain, G = 1.

Assume that the sampling frequency f_s = 1000 Hz, the sensor output voltage $S_o = P \times S$, and the conditioner output voltage $C_o = S_o \times G$. The original analogue voltage signal from the conditioner is typically shown in Figure 5.46. Now if Figure 5.46 is replotted in Figure 5.47 with the y-scale having divisions of 0.0391 V, so as to ease the understanding of the data conversion into the number of bins (B), the voltage resolution of 0.0391 V in the plot is equal to the DAQ device sensitivity. The number of bins (B) corresponding to the y-scale division is also shown on the left-hand side of Figure 5.47 for clarity. Hence, it is very much obvious from Figure 5.47 that most of the data points of the signal (shaded portion) are between the band of 0.0782 V ($2 \times S_d$) and 0.1173 V ($3 \times S_d$). Hence, most of the data points from the original signal will be recorded as B = 3, with a few points below and above this voltage band as B = 2 and 4, respectively. This is typically shown in Figure 5.48(b), while the recorded and reconstructed voltage signal is shown in Figure 5.48(c). Figure 5.48(d) shows a comparison between the original and recorded signal in voltage. Figure 5.48(e) shows the error in the recorded voltage, i.e., $V_{error}(t) = V_{original}(t) - V_{recorded}(t)$, which is significant when compared to the original voltage signal. The error in the pressure measurement (bar), i.e., $V_{error}(t)/(G \times S)$, is shown in Figure 5.48(f).

Example ii: Data2 is the consideration of Data1 again, but with G = 10 and exactly the same DAQ card and setting as in Example i. Data2 is typically shown in Figure 5.49. This is exactly the same as Figure 5.46, except that the voltage scale now is 10 times higher than earlier due to the gain, G = 10. Now observe the impact of

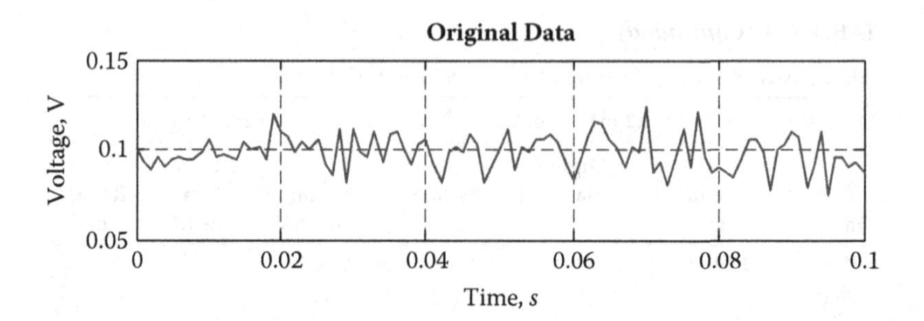

FIGURE 5.46
A typical original signal to be recorded.

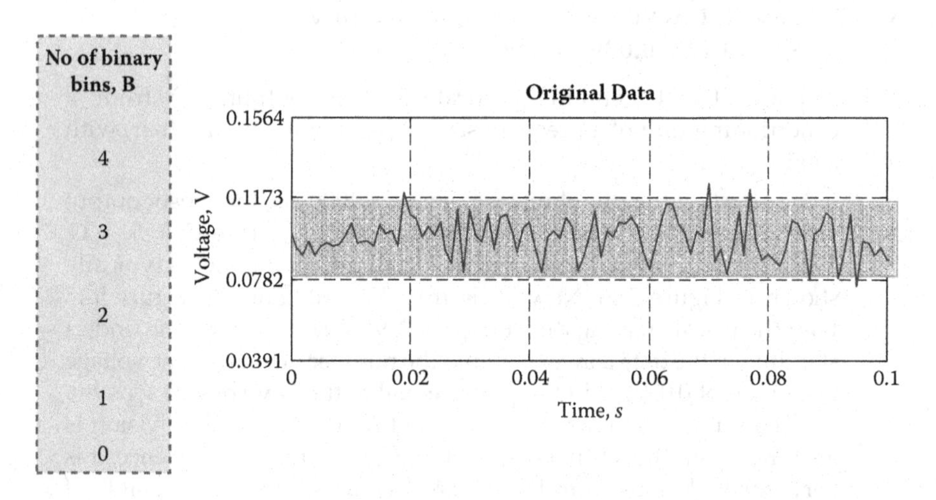

FIGURE 5.47
Replotting of the original signal with S_d = 0.0391 V division on the y-scale and likely number of binary bins.

G = 10 in the recorded signal in terms of the collected bins and the corresponding voltage in Figure 5.50. The error in the recorded signal is much smaller compared to Example i. The increased voltage range by the gain G = 10 allows the data collection in a much wider range of bins compared to just two to four bins in Example i.

This example highlights the fact that the amplification of signal voltage can improve the data collection even if a DAQ device with high ADC bit is not available. Another possibility is to adjust the setting of the FSIV

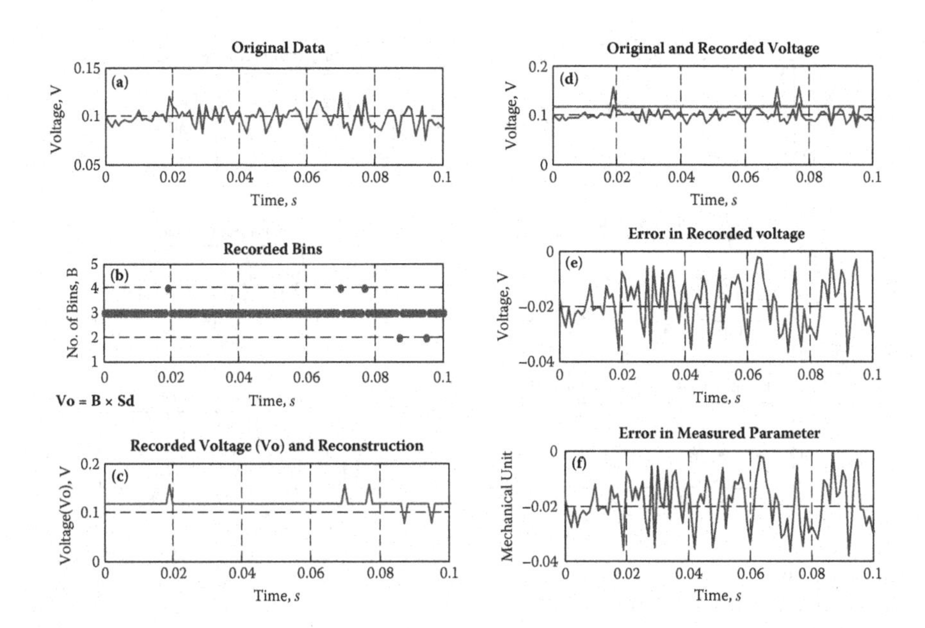

FIGURE 5.48
Original signal, collected bins, recorded signal, comparison, and errors in the recorded signal.

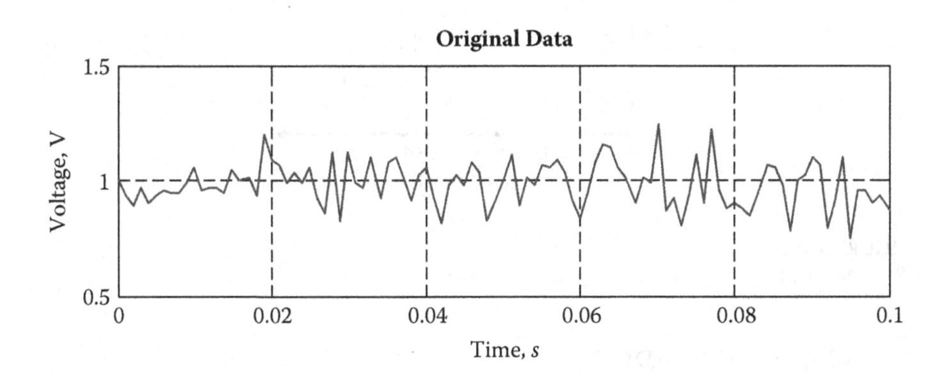

FIGURE 5.49
Original signal in Figure 5.46 but with gain $G = 10$.

depending upon the voltage of the signal to be recorded to enhance the DAQ device sensitivity, s_d. For example, FSIV = 2 – (–2) = 4 V will have $S_d = \frac{4}{2^8} =$ 15.625 mV/bin for the 8-bit DAQ card, and $S_d =$ 39.0625 mV/bin for FSIV = 5 – (–5) = 10 V. Hence, a smaller FSIV, depending upon the signal voltage, can also improve data collection and recording.

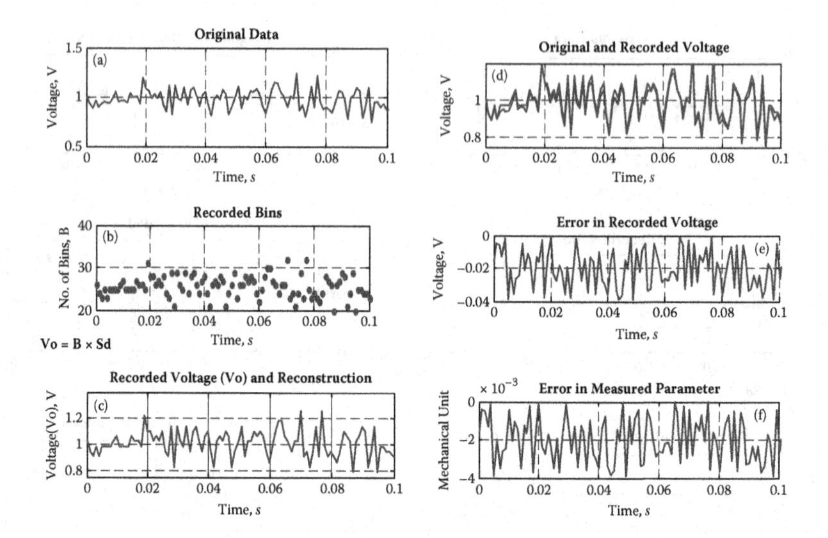

FIGURE 5.50
Improved recording with gain $G = 10$.

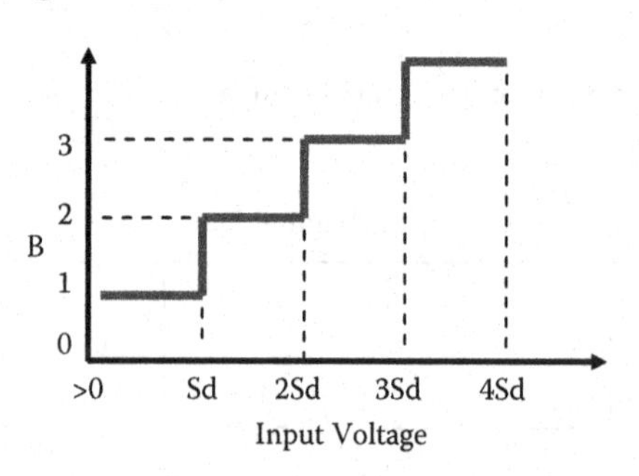

FIGURE 5.51
Rules for the voltage-to-the-binary bin conversion.

5.10.4 Summary of ADC

b = bits of DAQ card

BT = total number of binary bins = 2^b

FSIV = full-scale input voltage, $V_{upper} - V_{lower}$

$$\text{DAQ card sensitivity, } S_d = \frac{FSIV}{BT}$$

Recorded voltage, $V_o = B \times S_d$

Figure 5.51 basically illustrates the fact that the recorded number of bins will be the same for a certain range of input voltage of the signal to be recorded through a DAQ device. It depends upon the sensitivity of the device, S_d. Input voltage range from >0 to S_d will be recorded as 1, >S_d to $2S_d$ as 2, and so on. Hence, a lower value of the sensitivity S_d (means better DAQ sensitivity) is always preferred for measurement accuracy.

Aliasing effect:

1. Demonstrated this effect through the wheel motion.
2. Data collection using different f_s for a simple sinusoidal signal.
3. Introduced the concept of aliasing during data collection.
4. If a frequency or frequencies in a signal is greater than $f_s/2$, then these frequencies will appear as low frequencies in the recorded data.
5. $f_s \geq 2f_{max}$, where f_{max} is the highest frequency content in the signal to be recorded.

Accuracy depends on:

1. f_s: High f_s improves the accuracy in x-axis (time axis).
2. S_d: Better S_d improves the amplitude accuracy (y-axis) of the recorded/collected signals.
3. Removal of aliasing effect (unknown frequencies from the original signal).

6

Basics of Signal Processing

6.1 Introduction

Figure 6.1 illustrates various steps in the experimental setup, data collection, and signal processing required for vibration analysis experiments. The experimental setup and instruments are discussed in Chapter 5. This chapter deals with the signal data analysis commonly used in vibration analysis and diagnosis. This includes the concept of time domain signals, filtering, and data analysis in both the time and the frequency domain.

6.2 Nyquist Frequency

In signal processing, the signal sampling rate (e.g., 1 ksample/s, 5 ksample/s, etc.) is very vital (this is explained in Chapter 5). The sampling rate defines the maximum frequency limit of the recorded signal within which the signal can only be analyzed. It is noted in Section 5.9 that the maximum frequency in a signal that can be identified correctly is just half of the sampling frequency used for the analogue-to-digital conversion (ADC) during data collection through the data acquisition (DAQ) device. This means that, if the data are collected at the sampling rate of 1 ksample/s (f_s), then the signal will have a maximum frequency band up to 500 Hz (f_q) that can be seen correctly. This then implies that $f_{max} = f_q = f_s/2$, where f_q is called the Nyquist frequency (see Figure 6.1).

6.3 Time Domain Signals

The signals from various transducers are often time domain data. Three typical samples of the time domain signals measured from different sensors are shown in Figure 6.2. The first signal is purely sinusoidal at 10 Hz, the second

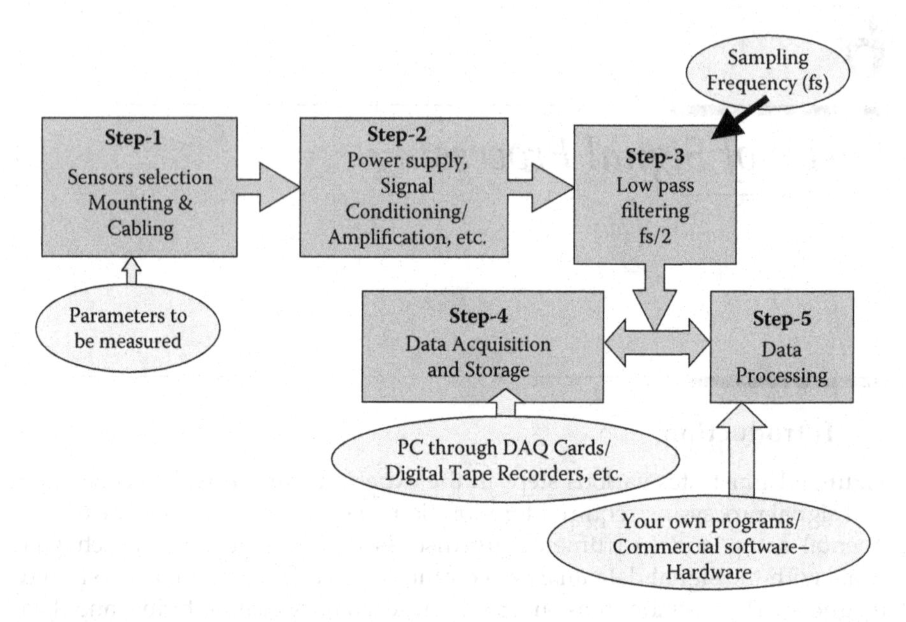

FIGURE 6.1
Measurement, data collection, and data analyses sequence.

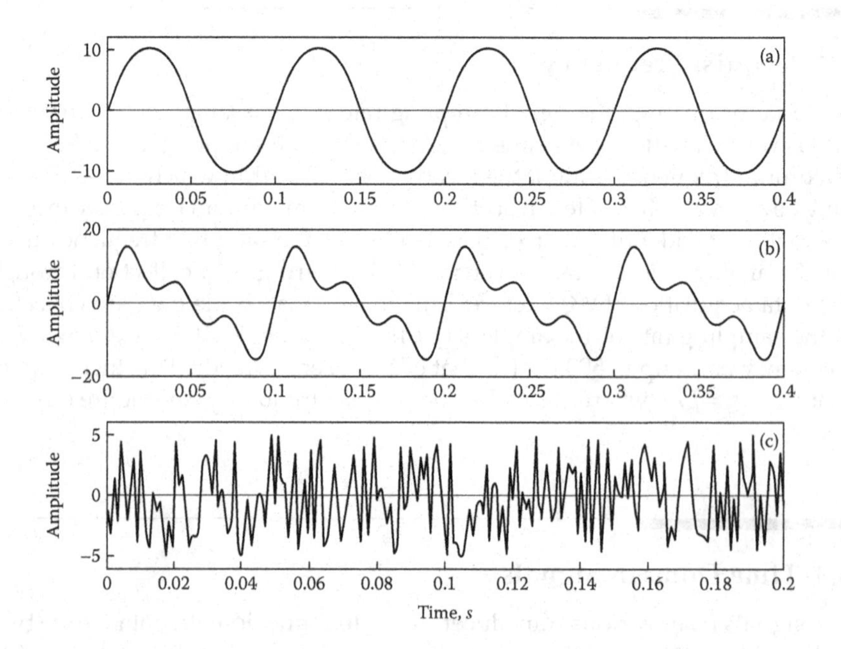

FIGURE 6.2
Time domain waveforms: (a) sinusoidal, (b) periodic, (c) random.

is periodic but contains signals of 10, 20, and 30 Hz, while the third is a random signal (neither sinusoidal nor periodic). The following processing is frequently used for time domain signals.

6.4 Filtering

The first step in the data acquisition process is the definition of a sampling rate, followed by the use of an antialiasing filter at the Nyquist frequency to filter out the frequency contents above the Nyquist frequency in the acquired signals, as shown in the step 3 of Figure 6.1. In the absence of the antialiasing filter, the high-frequency components may cause interference in the recorded signal, as discussed earlier in Section 5.9. This is often referred to as aliasing, which may generate some erroneous results during further signal processing. The effect of aliasing is further discussed in Section 6.8. The design of the antialiasing filter depends on the kind of system (software codes or hardware) used. Two typical filter designs used in practice are shown in Figure 6.3, with a signal acquired at a sampling rate (f_s). At this sampling rate, the useful signal information will span up to $f_q = f_s/2$. A typical low-pass digital filter at $f_q = f_s/2$ is generally desired, as shown in Figure 6.3. The software-based digital signal filter can possibly utilize up to the full frequency range, $f_q = f_s/2$; however, the hardware-based filter usually gives an amplitude linearity in the signal of up to frequency, $f_{Hardware\_filter} = f_s/2.56 = 400$ Hz. Hence, it is always preferable to select a sampling frequency, f_s, equal to or higher than 2.56 times the maximum frequency in the signals to be collected, and then use the antialiasing filter at $f_q = f_s/2$ during the data collection to avoid any possibility of aliasing.

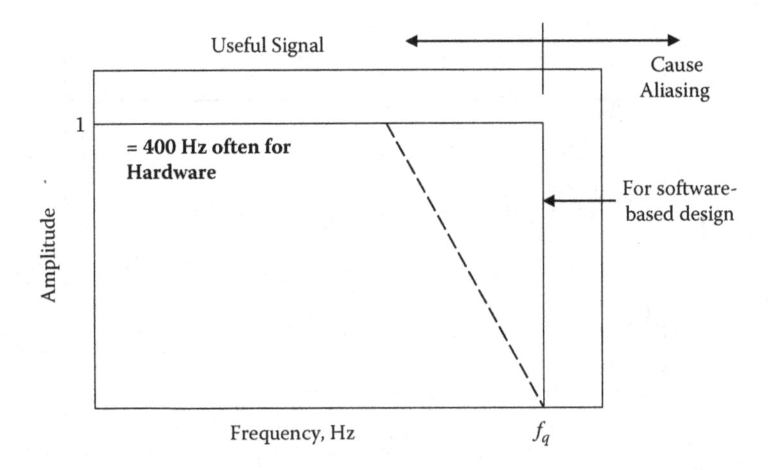

FIGURE 6.3
A typical low-pass filter characteristic.

Depending on the practical requirements of the signal processing, the three commonly used types of filters are discussed below.

6.4.1 Low-Pass (LP) Filter

Antialiasing filtering is also known as *low-pass filtering*, where the frequency components above the Nyquist frequency (f_q) in the signal are simply filtered out during data collection. However, depending on the requirements, signals can be low-pass filtered at any other frequency lower (f_{Low}) than the Nyquist frequency for the collected data during signal processing.

6.4.2 High-Pass (HP) Filter

A HP filter typically removes the frequency content from a signal at and below any particular frequency f_{High}. A typical high-pass filter at f_{High} = 10 Hz for a signal is shown in Figure 6.4. In many cases, it is required to filter out low-frequency noise, DC components, etc., from the recorded/measured signals.

6.4.3 Band-Pass (BP) Filter

This is a combination of both HP and LP filters. For example, a time signal with a sampling rate (f_s) of 1000 samples/s (Hz) can be band-pass filtered in a frequency band of f_{High}= 10 Hz to f_{Low} = 200 Hz, as shown in Figure 6.5. This simply means the frequency components below 10 Hz and above 200 Hz will be filtered out from the signal.

Typical examples of the above three filters are illustrated through measured data from a rotating machine in Figure 6.6.

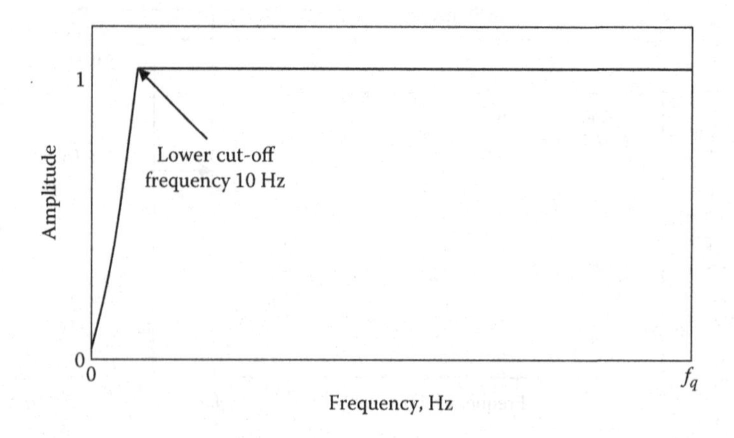

FIGURE 6.4
A typical high-pass filter characteristic.

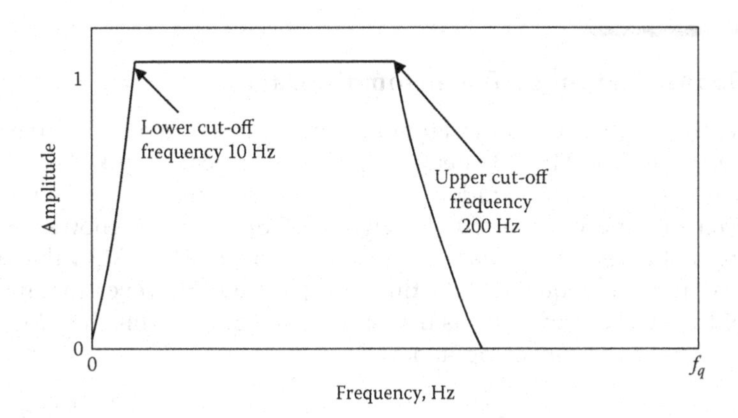

FIGURE 6.5
A typical band-pass filter characteristic.

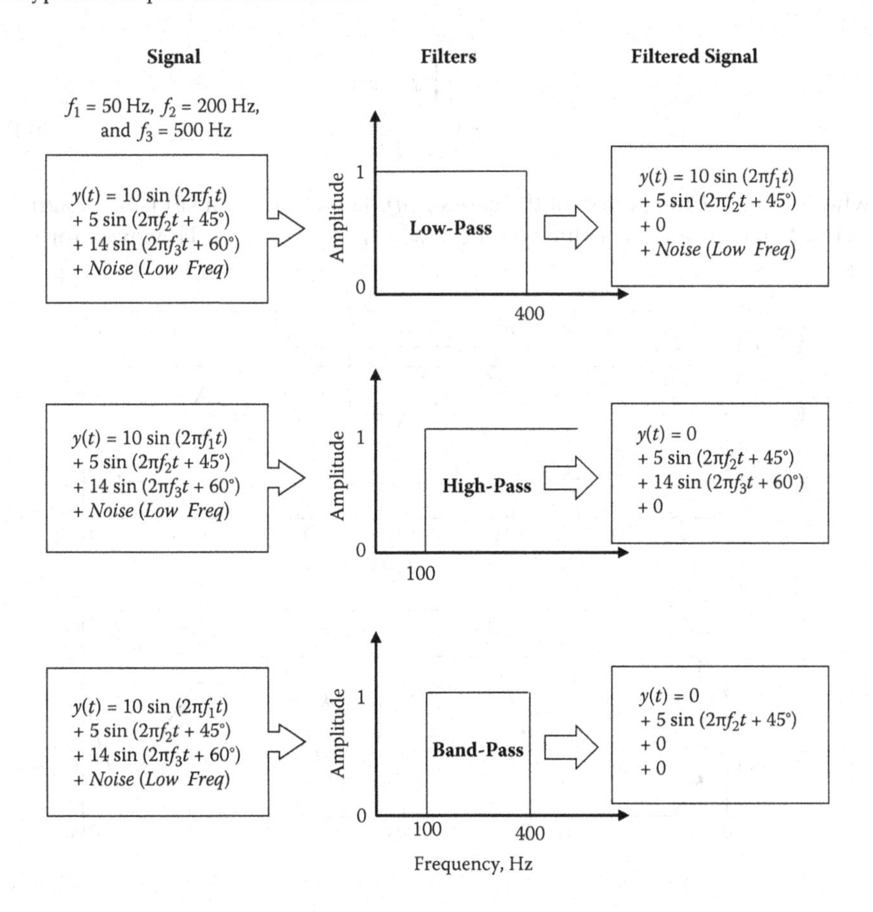

FIGURE 6.6
Working illustrations of different filters.

6.5 Quantification of Time Domain Data

The overall amplitude of any signal may be specified in any of three ways: root mean square (RMS), 0 to peak (pk), or peak-to-peak (pk-pk). Figure 6.7(a) represents a simple illustration of the RMS, pk, and pk-pk values of a displacement sine wave of 50 Hz. Figure 6.7(b), however, shows the time waveform of a periodic signal that is nonsinusoidal. The RMS value of any sine wave is always equal to 0.707 times the pk value. However, for any other kind of time waveform (such as that shown in Figure 6.7(b)), the RMS value needs to be numerically computed.

6.5.1 RMS Value

The RMS value of a signal $y(t)$ is defined as

$$y_{rms} = \sqrt{\frac{\int_0^T y^2(t)\,dt}{T}} \tag{6.1}$$

where T is the time period of the signal, $y(t)$. Therefore, for a sinusoidal signal, $y(t) = A\sin\omega t$, the RMS value will be $y_{rms} = A/\sqrt{2} = 0.707A$. However, for the

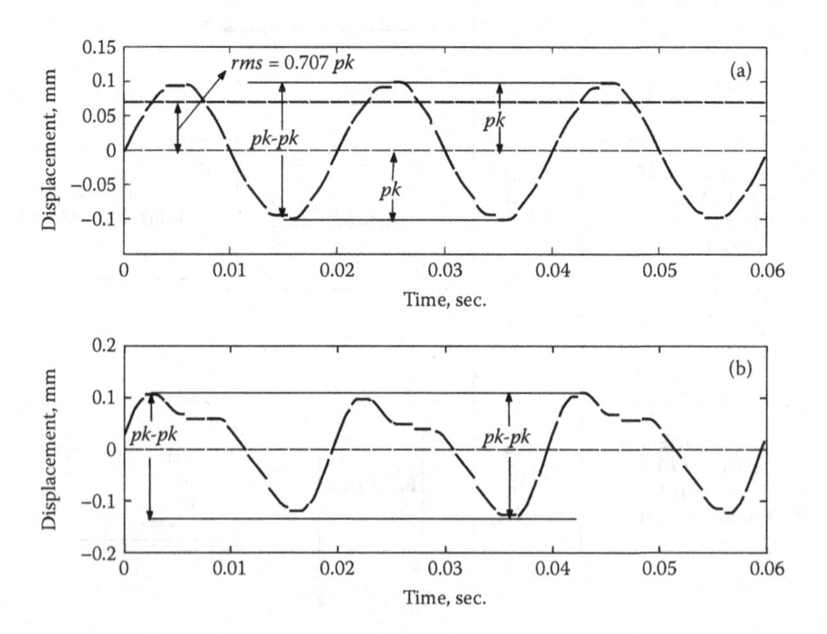

FIGURE 6.7
Time waveforms showing 0-peak, peak-peak, and RMS vibration amplitudes.

signal shown in Figure 6.7(b), the equation for the signal $y(t)$ with time (t) as a function needs to be defined for computing the RMS (y_{rms}) value using the above equation. It may not always be possible to define functional equations, particularly if the signal is totally random. Hence, it is good to estimate the RMS value numerically. The simplest approach is to consider a small but representative segment of the time domain signal, $y(t)$. Let us assume the selected segment contains p data points, represented as $y_i(t_i)$ at time, t_i where $t_i = (i - 1)dt$, $i = 1, 2, \ldots, p$. Then, the RMS can be calculated as

$$y_{rms} = \sqrt{\frac{\sum_{i=1}^{p} y_i^2}{p}} \tag{6.2}$$

6.5.2 Example 6.1: RMS Computation

Consider the signal shown in Figure 6.8. It is a 1 s ($= t_p$) datum for a signal defined in Equation (6.3) with the sampling rate of 10 samples/s, so the time step $dt = 0.1$ s.

Signal,

$$y(t) = 20\sin(3.5\pi t) + 25\cos(6\pi t) \tag{6.3}$$

where $t = 0, dt, 2dt, \ldots, t_p$. The signal data of $y(t) = 20\sin(3.5\pi t) + 25\cos(6\pi t)$ and its square values are listed in Table 6.1. The computed parameters—RMS,

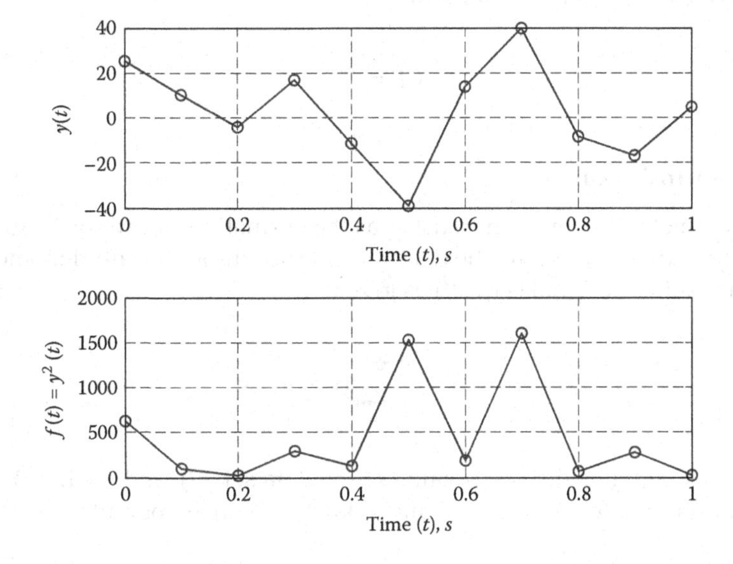

FIGURE 6.8
The signal $x(t)$ and the function $f(t) = y2(t)$.

TABLE 6.1

Data of Example 6.1

Time (t), s	Signal, $y(t)$	$y^2(t)$	Calculation Process	Parameters
0	25.0000	6.2500e+002	Mean = 0	RMS (Equation (6.2)) = 21.031
0.1000	10.0947	1.0190e+002		
0.2000	−4.0451	1.6363e+001	$p = 11$	0 to peak (pk) = 39.9792
0.3000	17.0967	2.9230e+002		
0.4000	−11.2957	1.2759e+002		
0.5000	−39.1421	1.5321e+003	$\sum_{i=1}^{q} y_i^2 = 4866.10$	Peak-to-peak (pk-pk) = 79.121
0.6000	13.9058	1.9337e+002		
0.7000	39.9792	1.5983e+003		
0.8000	−8.4697	7.1736e+001		Ratio (peak to RMS) = 1.8761
0.9000	−16.8052	2.8242e+002		
1.0000	5.0000	2.5000e+001		

0 to peak (pk), or peak-to-peak (pk-pk), are also listed in Table 6.1 to aid the understanding of the quantification process for a signal.

6.5.3 Crest Factor (CF)

This is a nondimensional parameter for any signal, which is defined as the ratio of the *peak value to the RMS value.* Table 6.1 shows the CF value of the signal of Example 6.1 equals 1.8761.

$$CF = \frac{y_{peak}}{y_{rms}} \tag{6.4}$$

6.5.4 Kurtosis (Ku)

This parameter is defined in statistics as the normalized fourth-order moment of the time domain signal. The definition of the kth-order central moment for any time domain signal (say, $y(t)$) is given by

$$M_k = \frac{1}{p}\sum_{i=1}^{p}(y_i - \bar{y})^k \tag{6.5}$$

where $y_i = y(t_i)$, $\bar{y}$ is the mean value of the data set $y(t)$, and $i = 1, 2, 3, \ldots, p$. The normalized fourth-order moment, kurtosis (ku), is computed as

$$ku = \frac{M_4}{(M_2)^2} \tag{6.6}$$

6.5.5 Example 6.2: Comparison between CF and Kurtosis

For illustration, three types of signals shown in Figure 6.9 are considered here. The signal $a_1(t)$ is a pure sine wave at 5 Hz, the signal $a_2(t)$ is a normally distributed random data, and the signal $a_3(t)$ is the transient vibration data due to the impulsive loading. The peak, RMS, CF, and Ku have been computed for these data and listed in Table 6.2 for comparison. The peak value is kept the same for all three signals so that easy comparison of different estimated parameters for the three signals can be done. It can be seen that the CF and Ku both increased significantly for both the random and transient data compared to the sine wave data. The kurtosis (Ku) shows a sharp increase in value for the impulsive signal compared to the other two signals. This indicates that kurtosis may be a more effective and useful parameter to detect the impulsive nature in a signal.

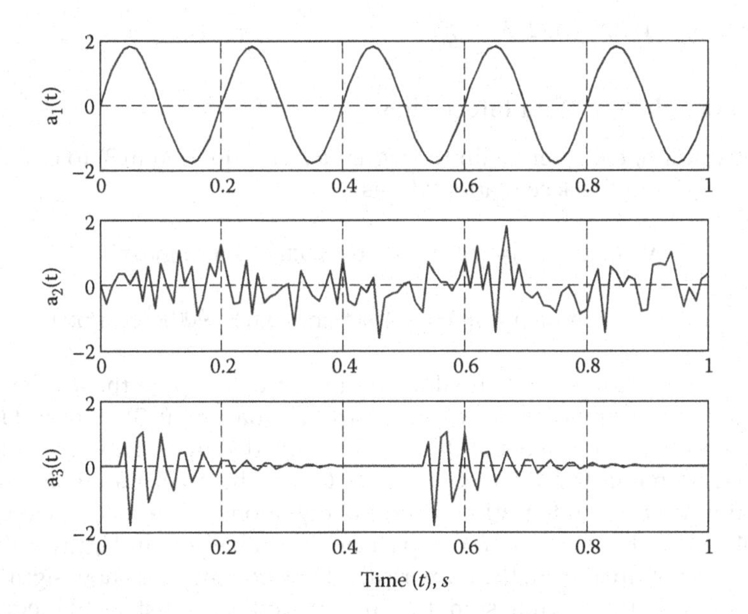

FIGURE 6.9
Three forms of the signals: sine, random, and transient.

TABLE 6.2
Comparison of Different Estimated Parameters for the Three Signals—Sine, Random, and Impulsive (Transient)

Signals	Mean	Peak	RMS	CF	Ku
Sine, $a_1(t)$	1.892e–16	1.80	1.2728	1.4142	1.515
Random, $a_2(t)$	–3.851e–2	1.80	0.6108	2.9469	3.2563
Impulsive, $a_3(t)$	2.737e–5	1.80	0.4400	4.0905	8.4486

6.6 Integration of Time Domain Signals

It is often required to convert the measured acceleration signal from an accelerometer into velocity and displacement. Let us assume that the acceleration signal is $\ddot{y}(t)$. The exact nature of the signal is difficult to express in sine and cosine terms, and so its analytical integration is often difficult. Therefore, it is better to compute numerically by assuming initial velocity, $\dot{y}_0(t = t_0) = 0$, and displacement, $y_0(t = t_0) = 0$. The simple integration process from the acceleration signal to the velocity and displacement is mathematically computed as

$$\text{Velocity at time } t_i,\ v_i = \dot{y}_i = \ddot{y}_i dt + \dot{y}_{i-1} \tag{6.7}$$

$$\text{Displacement at time } t_i,\ y_i = \dot{y}_i dt + y_{i-1} \tag{6.8}$$

where $t_i = (i - 1)dt$, $i = 1, 2, 3, \ldots, p$.

6.6.1 Example 6.3: Data Integration

For illustration, consider again the signal $y(t)$ in Equation (6.3) to derive the velocity $\dot{y}(t)$ and the acceleration $\ddot{y}(t)$ as

$$\text{Velocity, } v(t) = \dot{y}(t) = 70\pi\cos(3.5\pi t) - 150\pi\sin(6\pi t) \tag{6.9}$$

$$\text{Acceleration, } a(t) = \ddot{y}(t) = -245\pi^2\sin(3.5\pi t) - 900\pi^2\cos(6\pi t) \tag{6.10}$$

Now the time step $dt = 0.01$ s has been used to compute the acceleration, $a(t)$, velocity, $v(t)$, and displacement, $y(t)$, from Equations (6.3), (6.9), and (6.10). Figure 6.10 shows the plots of the data computed signals. Now the velocity and displacement data are also computed from the acceleration signal, $a(t)$, using the numerical integration scheme in Equations (6.7) and (6.8), which are also plotted in Figure 6.10 for comparison. It can be seen in Figure 6.10 that the signals obtained from the numerical integration method show significant deviation from the original signals. This only indicates that such integration introduces some error due to the fact that the initial conditions are generally not known, which results in DC being offset in the integrated signal due to the integration constant. But such DC component in the integrated signal can be considered as the very low frequency component in the signal. Hence, it is always recommended that the integrated signals must be high-pass filtered at a very low frequency. Here the integrated signals—velocity and displacement—have been high-pass filtered at 1 Hz. It can be seen that the filtered signals shown in Figure 6.11 are almost identical to the original signals after 0.5 s. Thus, it is also recommended to remove some initial data after integration to get the reliable data. For the present example, the data up to 0.5 s must be removed from the integrated data sets.

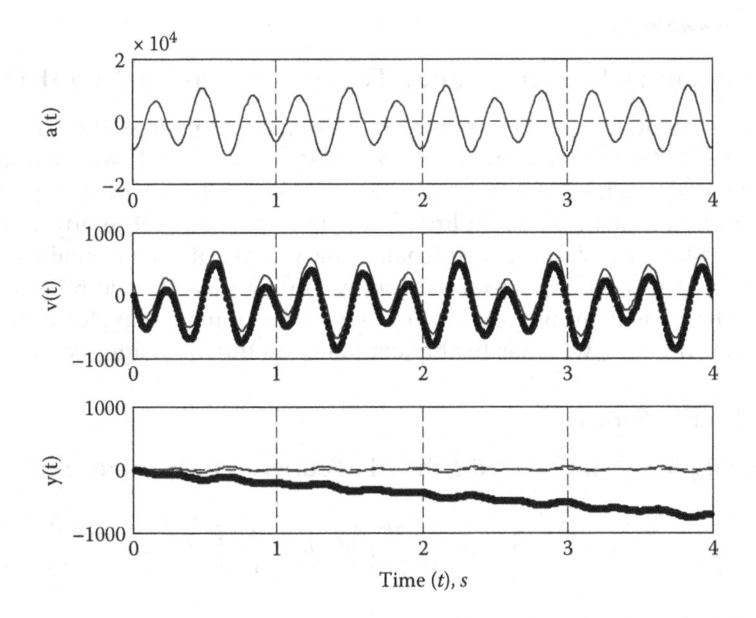

FIGURE 6.10
Original acceleration, velocity, and displacement signals (solid line) and comparison with velocity and displacement signals (solid line with dots) computed from the original acceleration signals using the numerical integration.

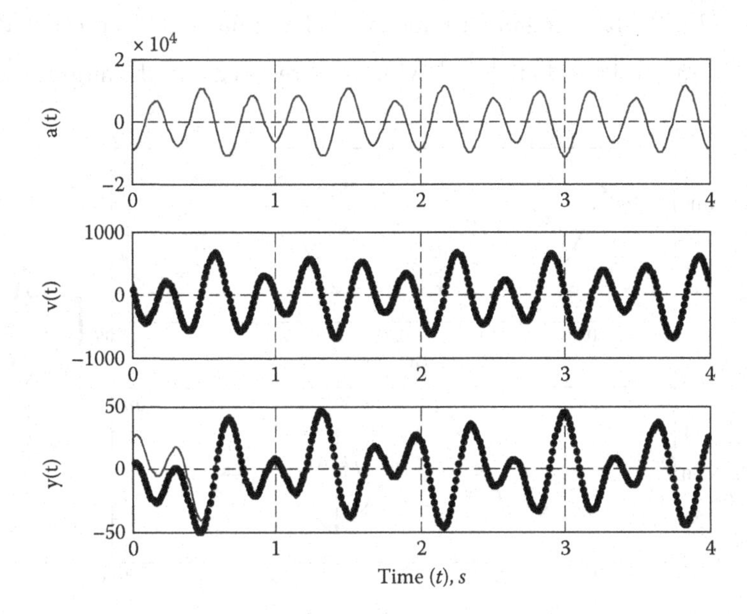

FIGURE 6.11
Original and integrated signals—velocity and displacement after high-pass filter at 1 Hz showing promising results.

6.7 Frequency Domain Signal: Fourier Transformation (FT)

It is important to convert a time domain waveform into the frequency domain in order to ascertain the different frequency contents within a signal. This helps to understand the physical behavior of the measuring parameters within a system. The amplitude versus frequency plot of any signal is known as the spectrum of the signal. Typical plots of time domain signals and their corresponding spectra are shown in Figures 6.12 and 6.13, respectively. The FT is therefore used to convert a time domain waveform into the frequency domain, which is further explained in the following sections.

6.7.1 Fourier Series

If a signal $y(t)$ has a time period T, then the Fourier coefficients are calculated as

$$a_0 = \frac{1}{T}\int_{-T/2}^{+T/2} y(t)dt \quad a_k = \frac{2}{T}\int_{-T/2}^{+T/2} y(t)\cos\left(\frac{2\pi k}{T}t\right)dt \quad b_k = \frac{2}{T}\int_{-T/2}^{+T/2} y(t)\sin\left(\frac{2\pi k}{T}t\right)dt \tag{6.11}$$

The Fourier series are, however, given by

$$y(t) = a_0 + \sum_{k=1}^{\infty} A_k \cos\left(\frac{2\pi k}{T}t + \phi_k\right) \tag{6.12}$$

where $1/T$ is the fundamental frequency of the signal and is represented by f. $A_k = \sqrt{a_k^2 + b_k^2}$ and $\phi_k = \tan^{-1}\left(\frac{b_k}{a_k}\right)$, which are, respectively, the amplitude and

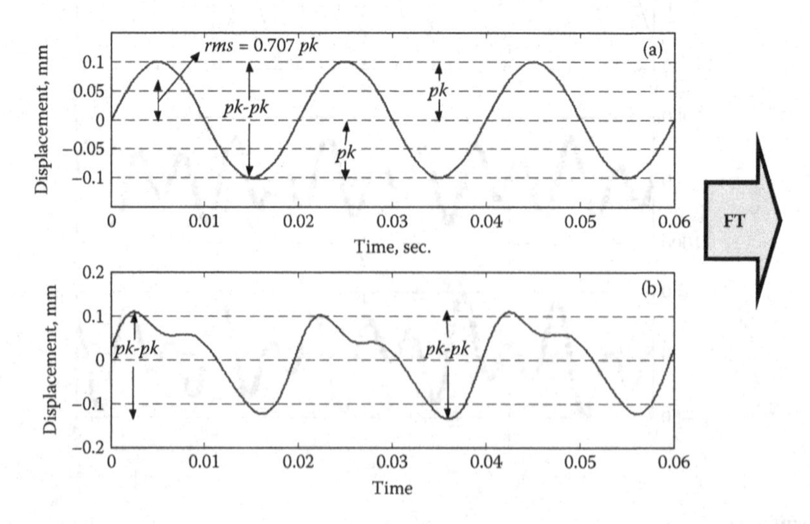

FIGURE 6.12
Time domain signals.

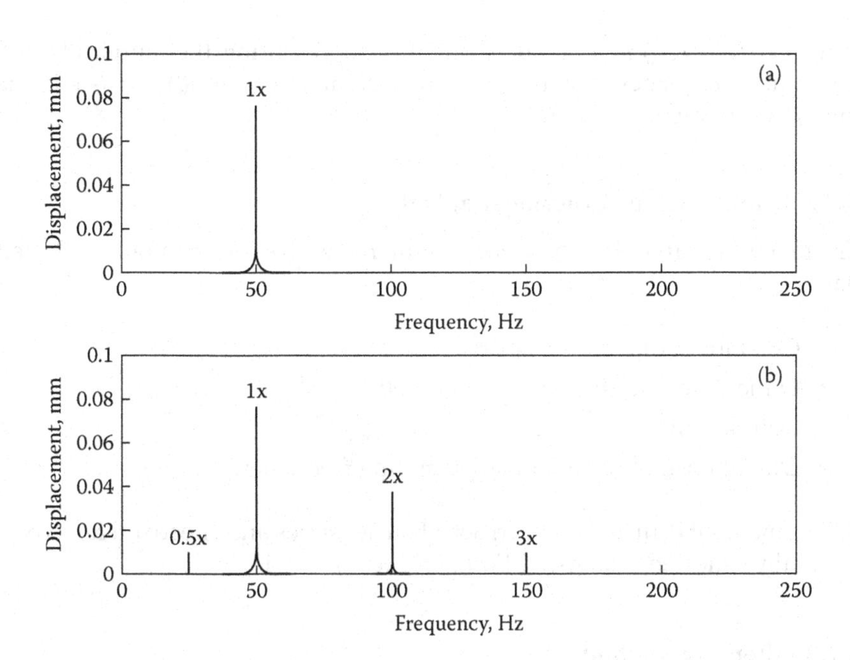

FIGURE 6.13
Frequency domain data.

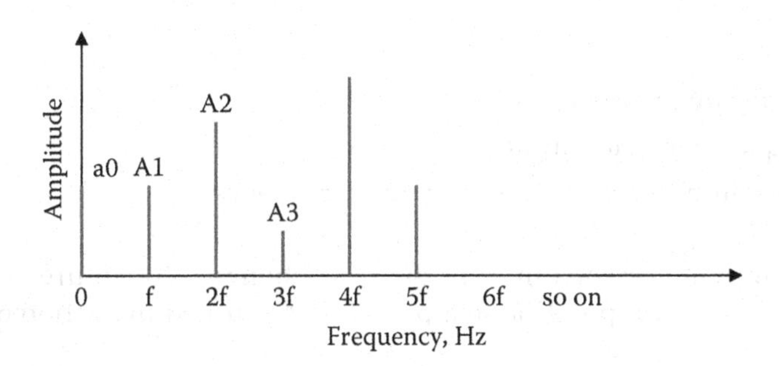

FIGURE 6.14
Amplitude spectrum: Representation of Fourier series amplitudes with their frequencies.

phase angle at the frequency of $kf = \frac{k}{T} = f_k$ within the time domain signal. The amplitude vs, frequency plot in Figure 6.14 is generally known as the *amplitude spectrum* of the time domain signal $y(t)$. In the frequency domain, the series for the time domain signal $y(t)$ can be represented as

$$Y\left(\frac{k}{T}\right) = Y(kf) = A_k e^{j\phi_k} \tag{6.13}$$

where $Y(kf) = Y(f_k)$ is a complex quantity representing the amplitude and phase at a frequency f_k within the time domain signal $y(t)$. $j = \sqrt{-1}$ is the imaginary number.

6.7.2 Limitations for Experimental Data

The following difficulties may be encountered with experimentally collected data:

- Contains actual data + noise
- Difficult to find the expression for $y(t)$
- Defining the time period (T) of the signal
- Digital form of the data with sampling frequency

This makes it difficult to construct a Fourier series and spectrum for experimentally collected data using Equations (6.11) to (6.13).

6.7.3 Alternate Method

Hence, the use of an alternative method for computing the FT of a time domain signal is required. The following considerations are generally needed:

- Artificially define T
- Frequency resolution, $df = 1/T$
- Maximum frequency up to Nyquist frequency, f_q

The computed spectrum can then be changed to Figure 6.15 from Figure 6.14. The computational approach is discussed in Section 6.7.4.

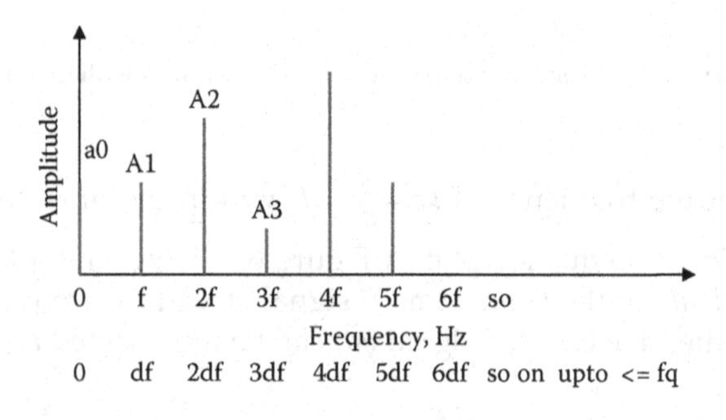

FIGURE 6.15
A typical computed amplitude spectrum.

6.7.4 Computation of Fourier Transform (FT)

As earlier mentioned, the Fourier transformation (FT) is a tool that is generally used for converting digital data in a time waveform into the frequency domain. It can be calculated using the following formula:

$$Y(kdf) = \frac{2}{N}\sum_{p=1}^{N} y(t_p)e^{-j\frac{2\pi(p-1)k}{N}} \tag{6.14}$$

where $Y(kdf) = Y(f_k)$ is the FT at the frequency $f_k = kdf$ for the time domain signal $y(t)$. $y_p = y(t_p)$ is the amplitude of the signal $y(t)$ at time $t_p = (p-1)dt$. N is the number of data points of the signal $y(t)$ used for the FT. The variables $k = 0, 1, 2, \ldots, (N/2-1)$ and $p = 1, 2, \ldots, N$. The frequency resolution $df = 1/T$, where T is the pseudo- (artificial) time period chosen for the signal for the FT and is defined as $T = Ndt$, where dt (= Δt) is the time between two consecutive data in the signal $y(t)$. The number of data points (N) used for the FT calculation should be equal to 2^r, where r is a positive integer number. The selection of number of data points N (or r) from the time domain signal data is totally the choice of the user, as this defines the artificial time period, T, and the frequency resolution, df, for the calculated spectrum. This approach is also known as fast Fourier transformation (FFT). Table 6.3 gives the values of different parameters used in the FFT for time signal data shown in Figure 6.16(a) collected at the sampling rate f_s = 100 samples/s for the following sinusoidal signal at the frequency f_e = 12.5 Hz.

TABLE 6.3

Parameters for FT Analysis

r	$N = 2^r$	$dt = \frac{1}{f_s}$	$T = Ndt$	$df = \frac{1}{T}$	$f_q = \frac{f_s}{2}\left(\frac{f_s}{2.56}\right)^{**}$	$n_f = \frac{N}{2}\left(\frac{N}{2.56}\right)^{**}$
1	2	0.01 s	0.02 s	50 Hz	50 Hz (39.0625 Hz)	1 (0)
2	4	0.01 s	0.04 s	25 Hz	50 Hz (39.0625 Hz)	2 (1)
3	8	0.01 s	0.08 s	12.5 Hz	50 Hz (39.0625 Hz)	4 (3)
4	16	0.01 s	0.16 s	6.25 Hz	50 Hz (39.0625 Hz)	8 (6)
5	32	0.01 s	0.32 s	3.125 Hz	50 Hz (39.0625 Hz)	16 (12)
6	64	0.01 s	0.64 s	1.5625 Hz	50 Hz (39.0625 Hz)	32 (25)
7	128	0.01 s	1.28 s	0.78125 Hz	50 Hz (39.0625 Hz)	64 (50)
8	256	0.01 s	2.56 s	0.390625 Hz	50 Hz (39.0625 Hz)	128 (100)
9	512	0.01 s	5.12 s	0.195313 Hz	50 Hz (39.0625 Hz)	256 (200)
10	1024	0.01 s	10.24 s	0.097656 Hz	50 Hz (39.0625 Hz)	512 (400)

** Values in bracket are for the hardware-based FFT analysis.

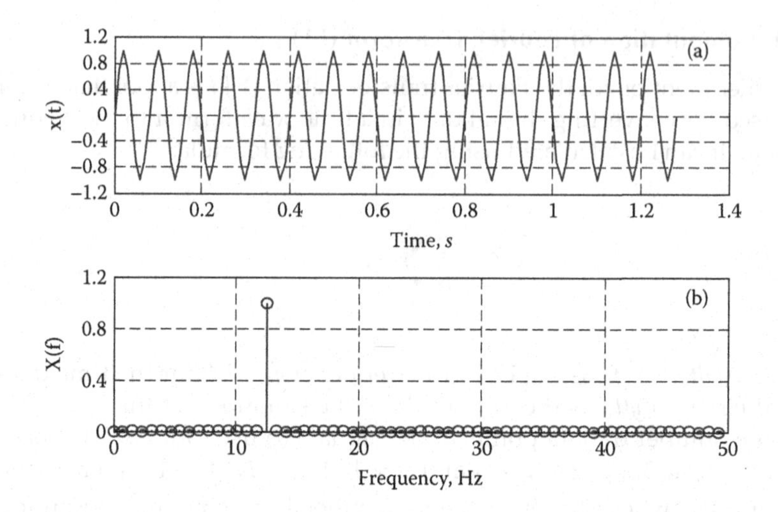

FIGURE 6.16
An example of the sinusoidal signal and its spectrum.

Signal,

$$y(t) = \sin(2\pi f_e t) \tag{6.15}$$

n_f is the number of frequency lines in the spectrum and is often known as the FFT points in the spectrum. The spectrum computed for the time domain signal in Figure 6.16(a), using the FT equation (6.14), is also shown in Figure 6.16(b). The only peak at 12.5 Hz in the spectrum confirms that the time domain data are a pure sine wave at frequency 12.5 Hz. The number of the data points $N = 128$ (for $r = 7$) is used to calculate the FT. The computed FT values at $n_f = N/2 = 64$ frequencies from 0 to 49.2188 Hz at the frequency resolution step of 0.78125 Hz are also listed in Table 6.4. The calculated FT values are complex quantities (i.e., $a_k + jb_k$) at the frequency f_k. This is due to the fact that the signals are generally combinations of a number of frequency components, having their respective amplitudes and phases. Hence, the amplitude and phase at a frequency f_k in a signal can be calculated as $\sqrt{a_k^2 + b_k^2}$ and $\tan^{-1}\left(\dfrac{b_k}{a_k}\right)$, respectively. The spectrum plot consists of only amplitudes at different frequencies up to the frequency f_q. The above computation is also known as the fast FT (FFT).

6.7.5 Example 6.4: Importance of Segment Size on FT Analysis

A sine wave, $y(t) = \sin(2\pi ft)$, where $f = 100$ Hz, collected at a sampling frequency of $f_s = 1024$ Hz for 2 s, is considered here for the FFT analysis. The collected signal is shown in Figure 6.17. The zoomed view of the signal in Figure 6.18 clearly indicates that the signal is a pure sine wave with time period equal to 0.01 s (i.e., 100 Hz).

TABLE 6.4
Frequency *f* and *FT Y*(*f*) Data

Frequency (*f*), Hz	*Y*(*f*)	$\|Y(f)\|$	Frequency (*f*), Hz	*Y*(*f*)	$\|Y(f)\|$
0	0.000	0.000	25.0000	–0.0000 + 0.0000*j*	0.000
0.7813	–0.0000 – 0.0000*j*	0.000	25.7813	–0.0000 – 0.0000*j*	0.000
1.5625	–0.0000 + 0.0000*j*	0.000	26.5625	0.0000 – 0.0000*j*	0.000
2.3438	–0.0000 – 0.0000*j*	0.000	27.3438	–0.0000 + 0.0000*j*	0.000
3.1250	–0.0000 + 0.0000*j*	0.000	28.1250	–0.0000 – 0.0000*j*	0.000
3.9063	–0.0000 – 0.0000*j*	0.000	28.9063	0.0000 – 0.0000*j*	0.000
4.6875	–0.0000 – 0.0000*j*	0.000	29.6875	0.0000 + 0.0000*j*	0.000
5.4688	0.0000 – 0.0000*j*	0.000	30.4688	–0.0000 + 0.0000*j*	0.000
6.2500	–0.0000 + 0.0000*j*	0.000	31.2500	–0.0000 – 0.0000*j*	0.000
7.0313	0.0000 – 0.0000*j*	0.000	32.0313	0.0000 + 0.0000*j*	0.000
7.8125	0.0000 + 0.0000*j*	0.000	32.8125	–0.0000 + 0.0000*j*	0.000
8.5938	–0.0000 + 0.0000*j*	0.000	33.5938	–0.0000 – 0.0000*j*	0.000
9.3750	–0.0000 – 0.0000*j*	0.000	34.3750	–0.0000 – 0.0000*j*	0.000
10.1563	–0.0000 – 0.0000*j*	0.000	35.1563	–0.0000 – 0.0000*j*	0.000
10.9375	–0.0000 – 0.0000*j*	0.000	35.9375	0.0000 – 0.0000*j*	0.000
11.7188	0.0000 – 0.0000*j*	0.000	36.7188	0.0000 – 0.0000*j*	0.000
12.5000	1.0000 – 0.0000*j*	1.000	37.5000	0.0000 – 0.0000*j*	0.000
13.2813	0.0000 + 0.0000*j*	0.000	38.2813	0.0000 + 0.0000*j*	0.000
14.0625	–0.0000 + 0.0000*j*	0.000	39.0625	0.0000 + 0.0000*j*	0.000
14.8438	0.0000 + 0.0000*j*	0.000	39.8438	0.0000 – 0.0000*j*	0.000
15.6250	–0.0000 + 0.0000*j*	0.000	40.6250	0.0000 + 0.0000*j*	0.000
16.4063	–0.0000 – 0.0000*j*	0.000	41.4063	–0.0000 + 0.0000*j*	0.000
17.1875	0.0000 – 0.0000*j*	0.000	42.1875	–0.0000 + 0.0000*j*	0.000
17.9688	0.0000 + 0.0000*j*	0.000	42.9688	0.0000 – 0.0000*j*	0.000
18.7500	–0.0000 + 0.0000*j*	0.000	43.7500	–0.0000 + 0.0000*j*	0.000
19.5313	0.0000 + 0.0000*j*	0.000	44.5313	–0.0000 – 0.0000*j*	0.000
20.3125	0.0000 – 0.0000*j*	0.000	45.3125	0.0000 – 0.0000*j*	0.000
21.0938	–0.0000 + 0.0000*j*	0.000	46.0938	0.0000 + 0.0000*j*	0.000
21.8750	–0.0000 – 0.0000*j*	0.000	46.8750	0.0000 + 0.0000*j*	0.000
22.6563	0.0000 + 0.0000*j*	0.000	47.6563	0.0000 + 0.0000*j*	0.000
23.4375	–0.0000 + 0.0000*j*	0.000	48.4375	0.0000 + 0.0000*j*	0.000
24.2188	0.0000 + 0.0000*j*	0.000	49.2188	–0.0000 + 0.0000*j*	0.000

The signal shown in Figures 6.17 and 6.18 is a pure sine wave of frequency 100 Hz, with amplitude equal to 1. Ideally, the spectrum should show a peak at 100 Hz with amplitude equal to 1. However, the selection of the data points, *N*, can impact the spectrum calculation, and this is demonstrated through the computed spectra shown in Figures 6.19 to 6.24 for different values of *N*. This clearly shows that a higher value of *N* can produce a much better spectrum, which provides better representation of a time domain signal.

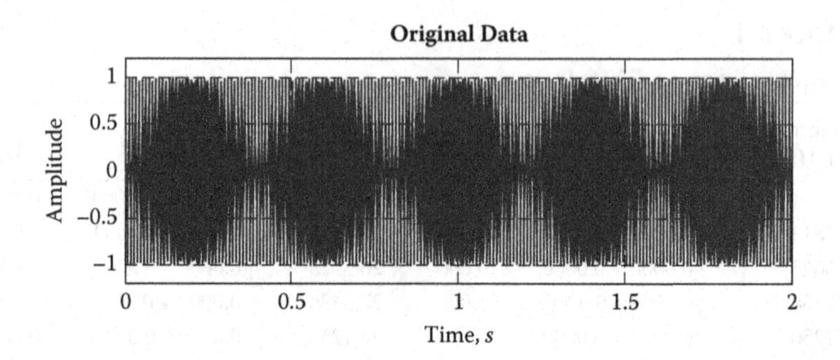

FIGURE 6.17
A typical 100 Hz sinusoidal signal for the FT.

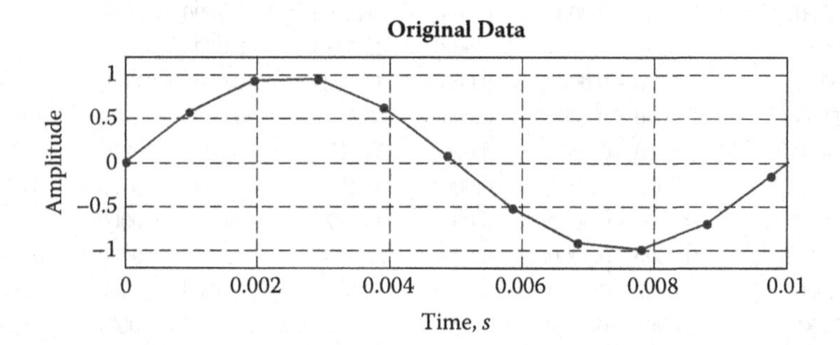

FIGURE 6.18
Zoom view of Figure 6.17 showing just a complete cycle.

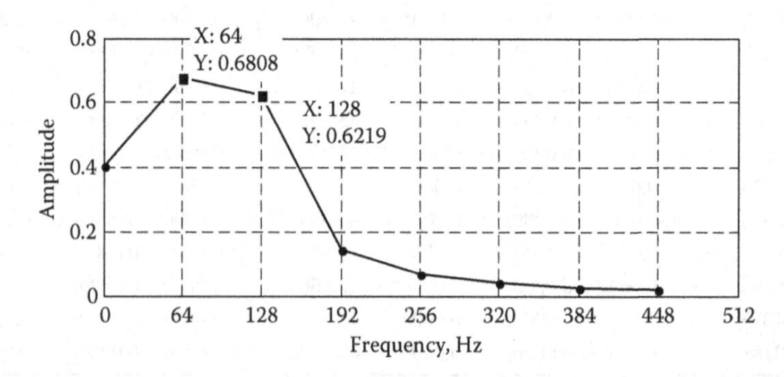

FIGURE 6.19
Amplitude spectrum for 100 Hz sine signal when $N = 2^4 = 16$, $T = Ndt = 0.0156$ s, $df = 1/T = 64$ Hz.

6.7.6 Leakage

The selection of a segment size (N) from a time domain signal for the FT may not match with the time period of the signal. A typical segment of size $N = 2^{10} = 1024$ and time $T = Ndt = 1.101$ s with the sampling frequency $f_s = 930$ Hz, shown in Figure 6.25, is used for the computation of the FT, and

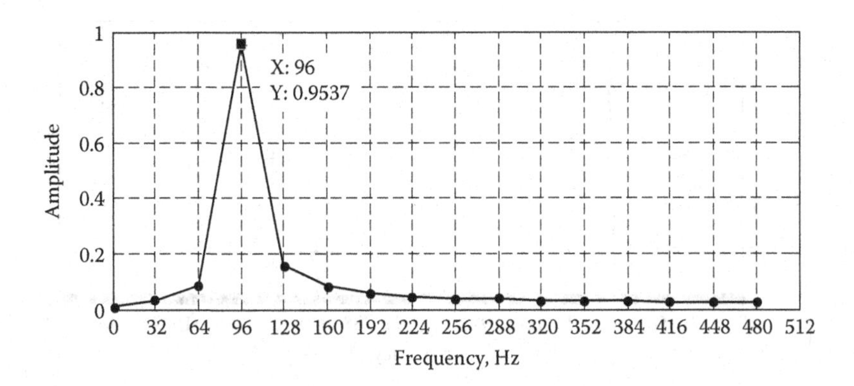

FIGURE 6.20
Amplitude spectrum for 100 Hz sine signal when $N = 2^5 = 32$, $T = Ndt = 0.0313$ s, $df = 1/T = 32$ Hz.

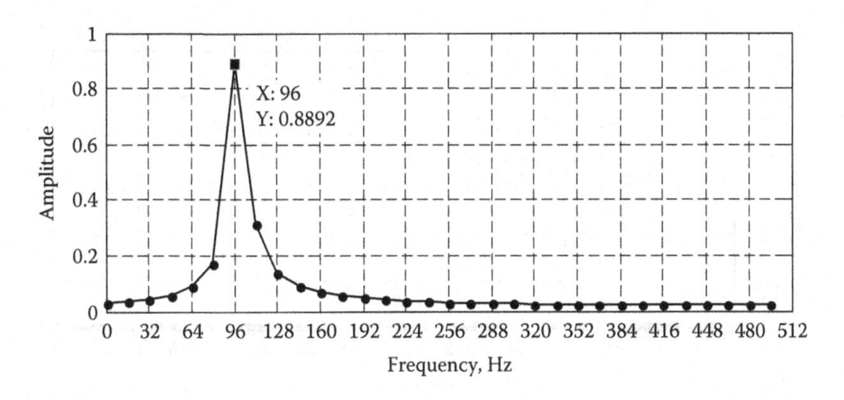

FIGURE 6.21
Amplitude spectrum for 100 Hz sine signal when $N = 2^6 = 64$, $T = Ndt = 0.0625$ s, $df = 1/T = 16$ Hz.

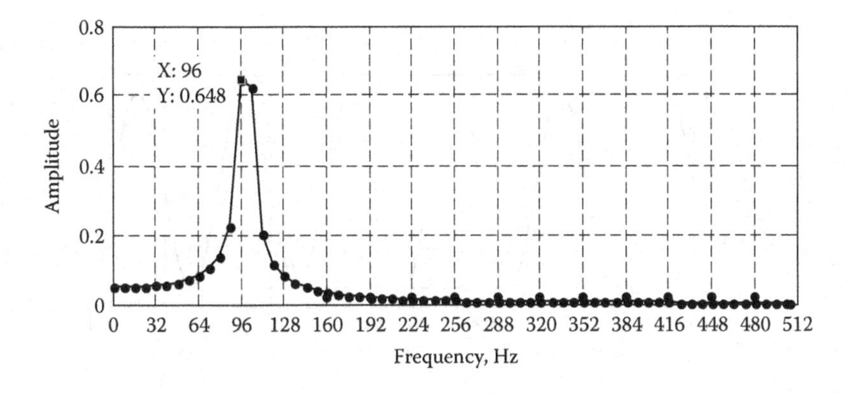

FIGURE 6.22
Amplitude spectrum for 100 Hz sine signal when $N = 2^7 = 128$. $T = Ndt = 0.125$ s, $df = 1/T = 8$ Hz.

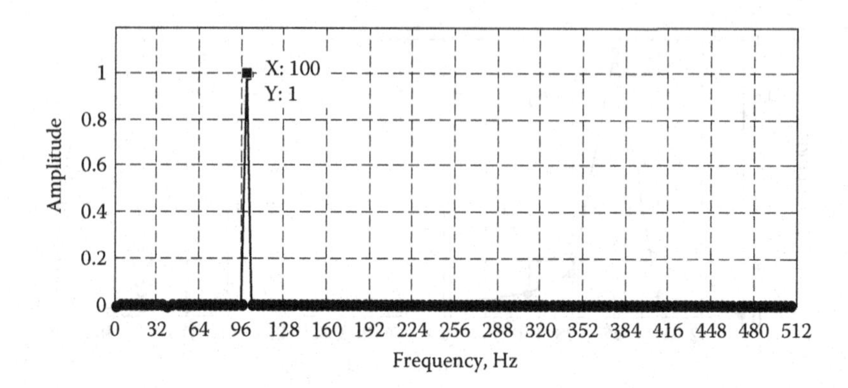

FIGURE 6.23
Amplitude spectrum for 100 Hz sine signal when $N = 2^8 = 256$, $T = Ndt = 0.25$ s, $df = 1/T = 4$ Hz.

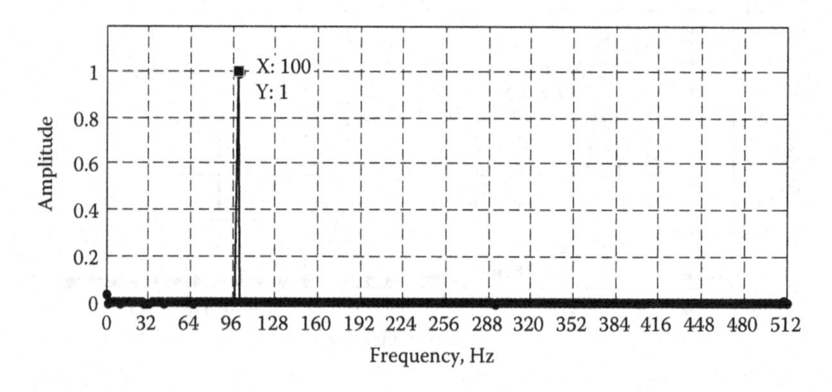

FIGURE 6.24
Amplitude spectrum for 100 Hz sine signal when $N = 2^9 = 512$, $T = Ndt = 0.5$ s, $df = 1/T = 2$ Hz.

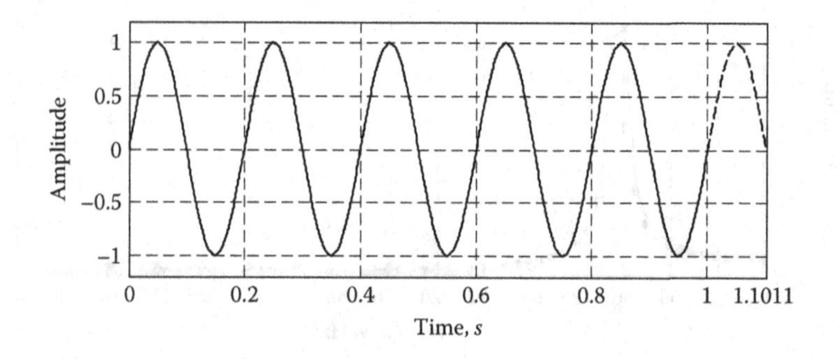

FIGURE 6.25
A typical segment size $T = Ndt$ of a sine waveform.

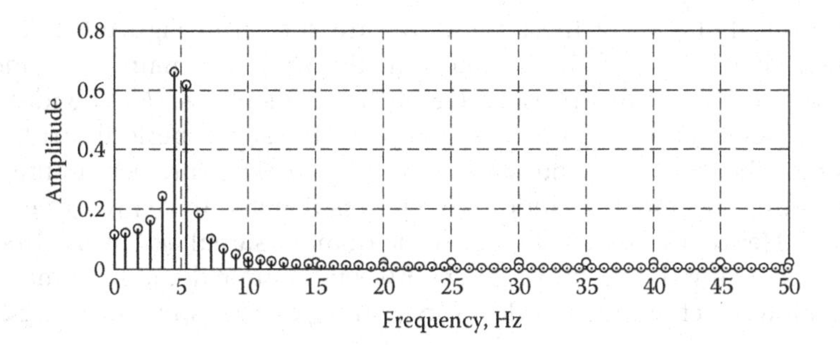

FIGURE 6.26
Frequency peak with side peaks indicates the leakage in the FT.

hence the spectrum. The time period of the signal (T_s) is equal to 0.2 s for the 5 Hz sinusoidal signal. The selected segment size (N) and the segment time length (T) are not integral multiples of the time period, T_s. The extra signal in the selected segment size $T = Ndt$ is deliberately shown as the dash line between 1 and 1.1011 s in Figure 6.25, which may impact the spectrum feature using Equation (6.14). Ideally, the spectrum should have a frequency peak at 5 Hz with amplitude equal to 1 for the time domain signal shown in Figure 6.25; however, the artificial selection of time period T for the segment of the signal for the FT may result in the spectrum shown in Figure 6.26. The spectrum is shown up to 50 Hz only to clearly observe the features around 5 Hz. The small side peaks on either side of 5 Hz are often due to the fact that the time length of the segment may not be equal to a multiple of the actual time period of the signal. This effect is called leakage in the spectrum computation.

6.7.7 Window Functions

Leakage in the spectrum is a serious problem and may result in misleading vibration analysis and conclusion. Hence, this error must be avoided during signal processing. It is well known that the selection of the segment size $T = Ndt$ is totally dependent on the number of data points, $N = 2^r$. This selection may not match with the time period of the signal, and in most cases, the time period of the measured signals is often difficult to know. The window function is suggested to overcome the leakage problem in signal processing. A number of window functions have been suggested and used in practice for this purpose. A few different types of windows are listed below:

1. Flat top window/rectangular window: This means there is no window used for the FT calculation and Equation (6.14) is directly used for the FT and spectrum calculation.
2. Gaussian window.
3. Hanning window.
4. Hamming window.

The typical shapes of these windows are shown in Figure 6.27 for the segment size $N = 2^{10} = 1024$. Note that the shapes remain the same for these window functions irrespective of the segment size, N. The idea is to make the selected segment of size $T = Ndt$ of the signal periodic artificially, by using either of these windows to reduce the leakage problem in the spectrum. Let us assume the window function is denoted by w_p, where $p = 1, 2, 3, \ldots, N$. Hence, the segment of the time domain signal, $y(t)$, is modified as $y_{w,p} = \vec{y}_p \vec{w}_p$ using the vector product between the time domain segment and the window function. The window function, w_p, can be estimated using equations $w_p = 0.5\left(1 - \cos\left(\frac{2\pi(p-1)}{(N-1)}\right)\right)$, $w_p = \alpha - \beta\cos\left(\frac{2\pi(p-1)}{(N-1)}\right)$ (where $\alpha = 0.54$ and $\beta = 1-\alpha$), and $w_p = e^{-\frac{1}{2}\left(\frac{(p-1)-\frac{(N-1)}{2}}{0.4\frac{(N-1)}{2}}\right)^2}$ for the Hanning, Hamming, and Gaussian windows, respectively (Bendat and Piersol, 1980; Oppenheim and Schafer, 1989). The modified time domain signal in Figure 6.25 using the Hanning window is shown in Figure 6.28, together with the Hanning window. The use of a window makes the amplitude nearly zero at the start and end of the chosen segment of the time domain signal. Now the signal looks more like a periodic signal, and hence the FT calculation on this modified signal helps in reducing the leakage effect in the spectrum. The use of

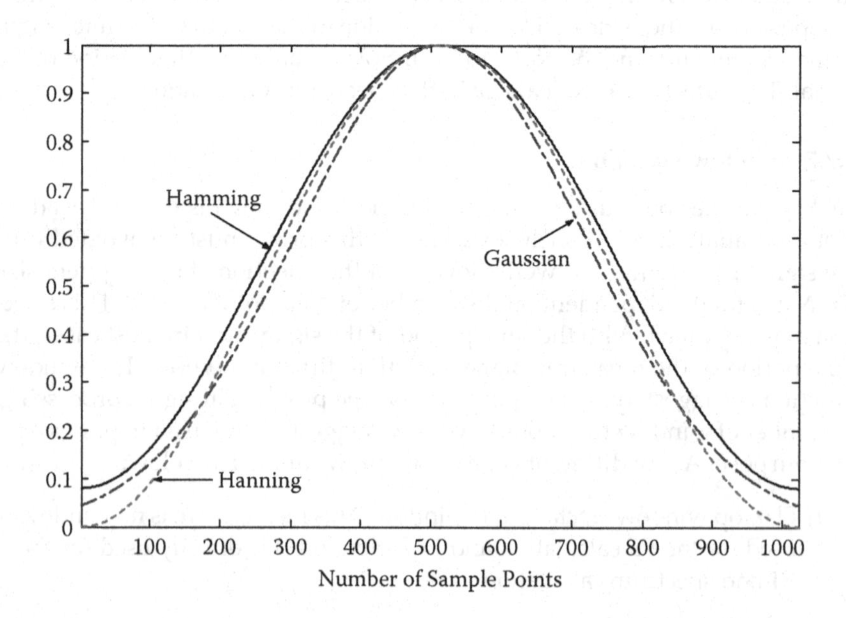

FIGURE 6.27
Typical view of a few different windows for the segment size $N = 1024$.

the window function affects the amplitude, and hence the FT calculation in Equation (6.14) is modified as

$$Y(kdf) = \frac{1}{C_w} \sum_{p=1}^{N} y_{w,p} e^{-j\frac{2\pi(p-1)k}{N}} \tag{6.16}$$

where C_w is a constant that compensates the effect of the window function on the amplitude of the computed spectrum. C_w is estimated as

$$C_w = \left(\frac{\sqrt{\sum_{p=1}^{N} w_p^2}}{\left(\sum_{p=1}^{N} w_p \Big/ N \right)} \right) \tag{6.17}$$

The spectrum estimated for the modified signal, shown in Figure 6.28 using Equation (6.16), is shown in Figure 6.29. It is clearly seen that the leakage

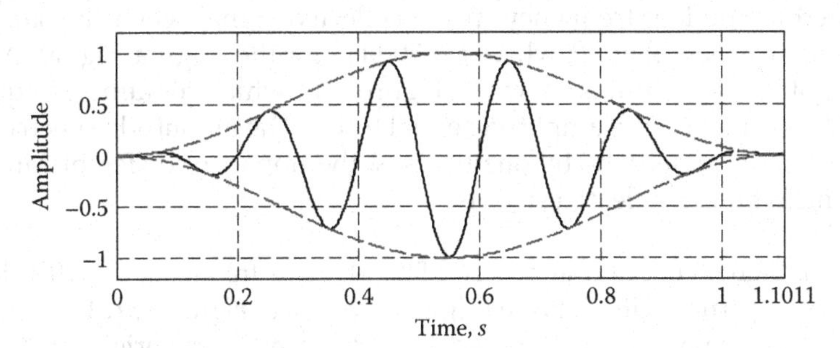

FIGURE 6.28
Sine waveform of Figure 6.25 with Hanning window.

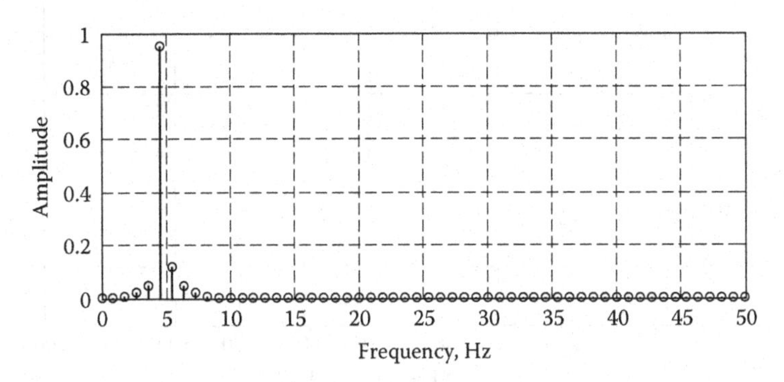

FIGURE 6.29
Spectrum with reduced leakage.

effect is reduced significantly, and a distinct peak at 5 Hz with amplitude of nearly 1 is observed.

6.8 Aliasing Effect

Aliasing can occur during data acquisition and is explained through a sinusoidal signal of frequency 500 Hz in Section 5.9. It is explained through the different sampling frequencies as Cases i to vii in that section. It is also shown that the period of the collected signal remains unchanged and correct when the sampling frequency is greater than or equal to more than twice the frequency of the signal. In this case, it is 1000 Hz and above for the sinusoidal signal of 500 Hz. When the sampling frequency is chosen below 1000 Hz, then the collected signal is observed to be significantly different, including the time period of the collected signals when compared to the original signal. It is observed that the high-frequency component in the signal appeared as the low frequency in the collected signal when the sampling frequency was less than $2f$, where f is 500 Hz for the original signal. A typical amplitude spectrum is shown in Figure 6.30, which contains a frequency peak at 500 Hz with the amplitude equal to 1 V. The amplitude spectrum for all cases is discussed one by one to show the impact of different sampling frequencies on the collected signals.

Case i: Signal collected at f_s = 10 kHz, Nyquist frequency f_q= 5000 Hz. The spectrum of the collected signal is shown in Figure 6.31. There is no error, as expected, and the spectrum is the same as the original signal.

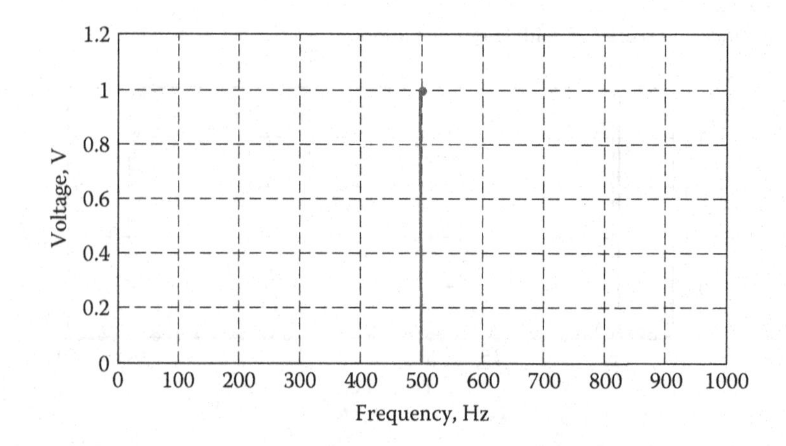

FIGURE 6.30
Expected ideal spectrum for the 500 Hz pure sine wave signal in Figure 5.30.

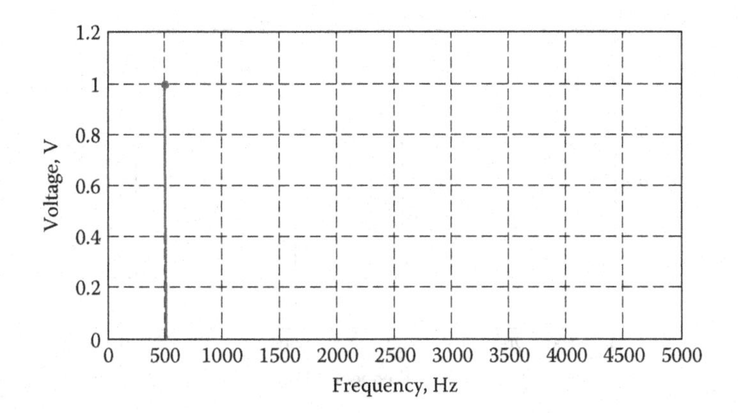

FIGURE 6.31
Spectrum without error when $f_s = 10$ kHz.

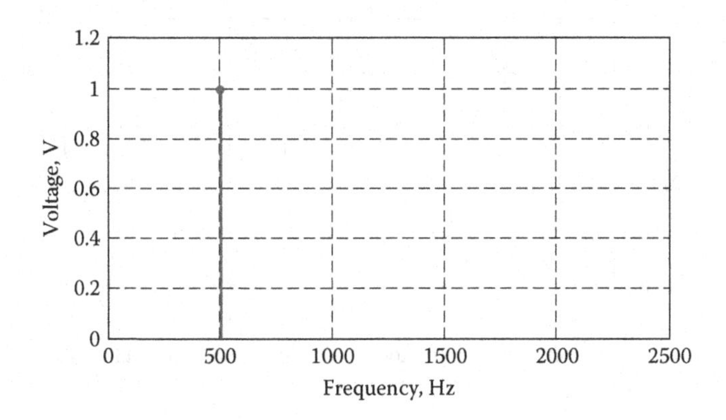

FIGURE 6.32
Spectrum without error when $f_s = 5$ kHz.

Case ii: Signal collected at $f_s = 5$ kHz, Nyquist frequency $f_q = 2500$ Hz. The time period of the collected signal is the same as the original signal (2 ms), and there is nearly negligible error in amplitude. Hence, the spectrum of the collected signal shown in Figure 6.32 is nearly the same as the original.

Case iii: Signal collected at $f_s = 2$ kHz, Nyquist frequency $f_q = 1000$ Hz. The time period of the collected signal is still the same as the original signal (2 ms), but there may be a very small error in the amplitude. Hence, the spectrum of the collected signal is shown in Figure 6.33.

Case iv: Signal collected at $f_s = 1$ kHz, Nyquist frequency, $f_q =$ 500 Hz. The time period of the collected signal is still the same as the original, but there is more error in the amplitude compared

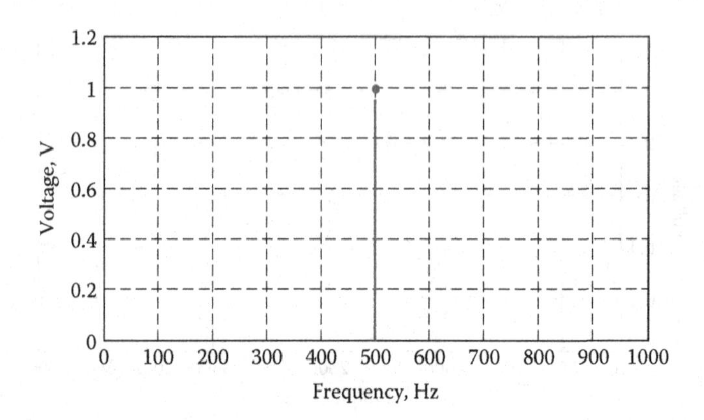

FIGURE 6.33
Spectrum without significant amplitude error when $f_s = 2$ kHz.

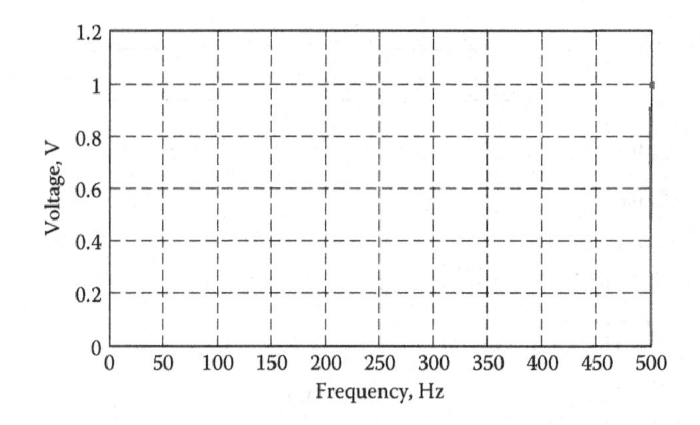

FIGURE 6.34
Spectrum expected to show 500 Hz when $f_s = 1$ kHz.

to Case iii. Hence, the spectrum of the collected signal is shown in Figure 6.34, which shows the peak at 500 Hz but the error in the peak amplitude.

Case v: Signal collected at $f_s = 750$ Hz, Nyquist frequency $f_q = 375$ Hz. Both amplitude and time period of the collected signal are changed significantly from the original signal. Now the time period $T = 0.004$ s, and hence the frequency $f = 1/T = 250$ Hz in the collected signal compared to 500 Hz in the original signal. The amplitude spectrum is shown in Figure 6.35. The peak at 250 Hz in the spectrum indicates a fold-back of 500 Hz in the original signal at the Nyquist frequency $f_q = 375$ Hz by a frequency amount equal to $(500 - 375) = 125$ Hz, and hence appeared at 250 Hz. This is called aliasing effect in the collected signal due to low sampling frequency.

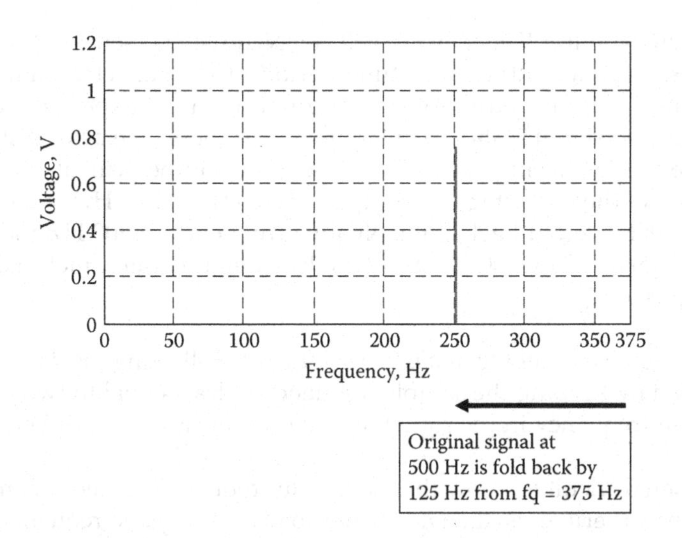

FIGURE 6.35
Spectrum showing aliasing effect when $f_s = 750$ Hz.

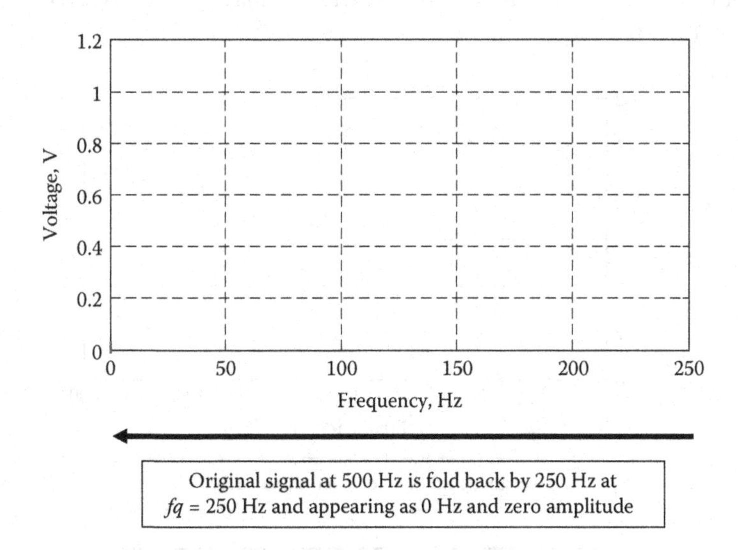

FIGURE 6.36
Spectrum showing no peak when $f_s = 500$ Hz (synchronized with signal frequency of 500 Hz).

Case vi: Signal collected at $f_s = 500$ Hz, Nyquist frequency $f_q = 250$ Hz. The collected signal is just a straight line compared to the original sinusoidal signal at 500 Hz, which means that the time period, T, tends to infinity and the frequency $f = 0$ Hz. Hence, the spectrum will have zero peak at 0 Hz, which is typically shown in Figure 6.36.

Case vii: Signal collected at f_s = 400 Hz, Nyquist frequency f_q= 200 Hz. Here both amplitude and time period of the collected signal are changed significantly from the original signal. The spectrum of the collected signal is shown in Figure 6.37. It shows the peak at 100 Hz due to the aliasing effect. The appearance of the peak at 100 Hz is due to the two-stage folding of (500 Hz – 200 Hz) = 300 Hz (i.e., 200 Hz fold-back at the Nyquist frequency of 200 Hz and then fold-forward by 100 Hz at 0 Hz). It is demonstrated pictorially in Figure 6.37.

Cases v to vii are known as *aliasing effect* while collecting the data. This can be avoided by keeping the sample frequency at least equal to two times the maximum frequency in the signal. In the present case, f_s should be 1000 Hz or more.

It is often difficult to know the maximum frequency content in any signal during the experiments; then it is better to use a low-pass frequency filter at Nyquist frequency, f_q, of the selected sampling frequency, f_s, to filter out high frequency above f_q. This process avoids the aliasing problem while collecting the data. This low-pass filter is called the *antialiasing filter,* which was also discussed in Section 6.4.

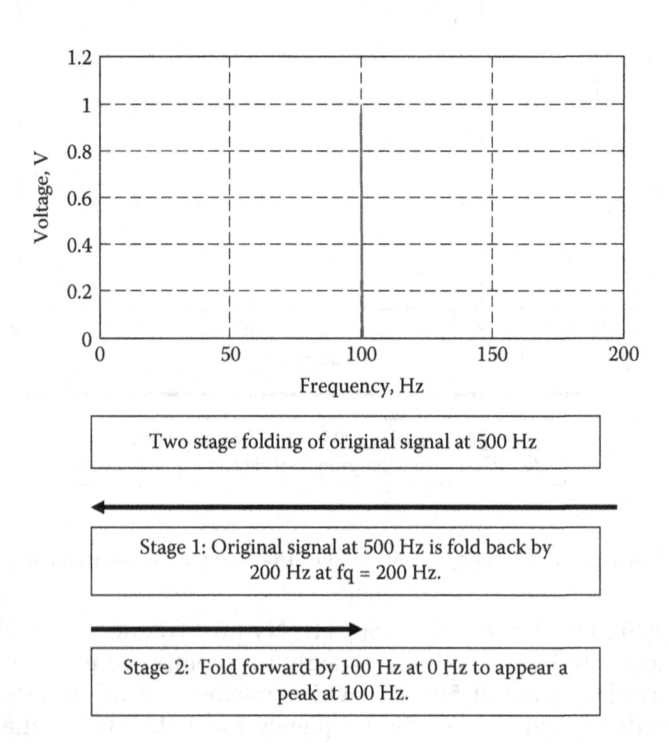

FIGURE 6.37
Spectrum showing aliasing effect when f_s = 400 Hz.

If the DAQ system is not equipped with the antialiasing filter, then select the correct sampling frequency by the trial-and-error approach. It is better to collect a few samples of the data at three to four different sampling frequencies and then compare the spectra of all sampling frequencies. If there is no aliasing problem at and above a typical sampling frequency, then select the typical sampling frequency for the experiments to avoid aliasing in the absence of the antialiasing filter.

6.9 Averaging Process for the Spectrum Computation

It is well known that the measured signal generally contains noise. It is also known that the FT computation is done through an approximated method, which heavily depends on the artificially selected time period $T = Ndt$. Hence, the computed spectrum may contain some erroneous frequency peaks. Such peaks can be removed by the averaging process in the spectrum computation.

6.9.1 Example 6.5: Averaged Spectrum

Consider a pure sine wave signal of frequency $f = 10.937$ Hz with a maximum amplitude of 1, which is sampled at 50 Hz. The sampled signal is shown in Figure 6.38, up to a time duration of 1.28 s.

Here, $f_s = 50$ Hz and $dt = 1/50 = 0.02$ s. For the spectrum analysis of this signal using the FT method, a selection of data points N is needed. Let us assume $N = 32$; hence, the time, $Ndt = 32 \times 0.02 = 0.64$ s can be assumed as the artificial time period (T) for the signal. Since the total data length is 1.28 s, the signal can be divided into two segments. Segment 1 is denoted as y_1 (from 0 to 0.64 s), and segment 2 is denoted as y_2 (from 0.64 to 1.28 s). The FT is then computed for these two segments. The calculated FTs for both segments, Y_1 and Y_2, are listed

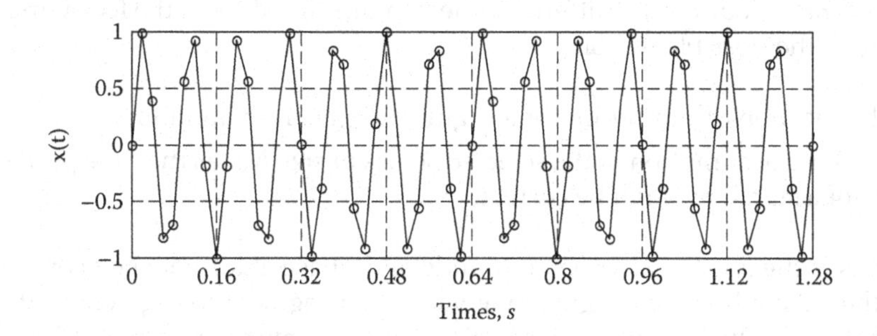

FIGURE 6.38
Typical time signal of Example 6.5.

TABLE 6.5

Computed FT Data for Two Data Segments for the Signal in Figure 6.38

	FFT	
Frequency, Hz	**(Y_1, first 0.64s data)**	**(Y_2, 2nd 0.64s data)**
0	0.0000	0.0000
1.5625	0.0000 + 0.0000j	−0.0000 − 0.0000j
3.1250	0.0000 + 0.0000j	0.0000 + 0.0000j
4.6875	0.0000 + 0.0000j	0.0000 − 0.0000j
6.2500	−0.0000 + 0.0000j	0.0000 + 0.0000j
7.8125	−0.0000 + 0.0000j	−0.0000 + 0.0000j
9.3750	0.0000 + 0.0000j	0.0000 − 0.0000j
10.9375	−0.0000 − 1.0000j	+0.0000 + 1.0000j
12.5000	0.0000 − 0.0000j	−0.0000 + 0.0000j
14.0625	0.0000 − 0.0000j	−0.0000 − 0.0000j
15.6250	−0.0000 − 0.0000j	0.0000 − 0.0000j
17.1875	0.0000 − 0.0000j	0.0000 + 0.0000j
18.7500	0.0000 − 0.0000j	0.0000 + 0.0000j
20.3125	0.0000 − 0.0000j	−0.0000 + 0.0000j
21.8750	0.0000 + 0.0000j	0.0000 − 0.0000j
23.4375	0.0000 + 0.0000j	0.0000 − 0.0000j

in Table 6.5. The complex quantities at the frequency equal to 10.9375 Hz for the two segments are deliberately shown as $-0 - 1j$ and $+0 + 1j$, respectively, to foster a clear and simple understanding of the averaging process.

6.9.2 Concept of Power Spectral Density (PSD)

The amplitude spectrum of either segment of the signal is shown in Figure 6.39. Both segments have identified the frequency and amplitude correctly, but the addition of Y_1 and Y_2 becomes 0 at all frequencies. Different errors may occur for the different time domain signals, and the following are some of the possible reasons:

1. Selection of start point for a segment is generally random.
2. Blind assumption of the selected segment size Ndt as the time period of a signal may not be correct.

This is the reason why the amplitude and phase at a frequency computed by the FT method for a signal may vary from segment to segment within a signal. It may become worst for a noisy complex time domain signal. There is always high probability that the FT for a segment may contain some spurious frequency peaks in addition to the real frequency peaks related to the signal.

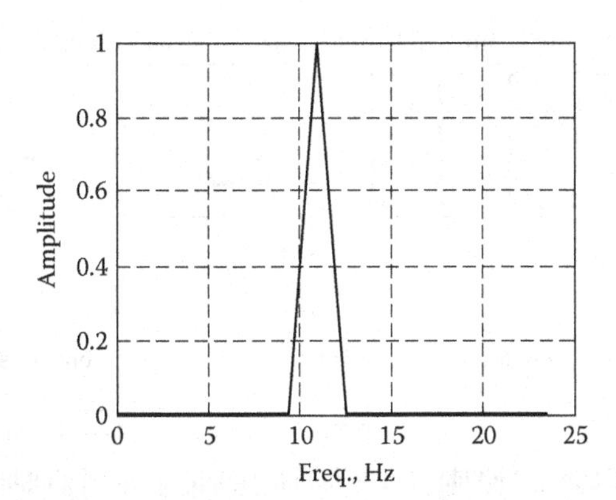

FIGURE 6.39
Identical computed spectra for both segments of the signal in Figure 6.38.

Hence, a large number of averaging is generally recommended to reduce the presence of spurious or random frequency peaks in a signal. But the averaging based on direct summation may not be correct. Hence, the following approach is suggested in practice for the averaging. This is known as the power spectral density (PSD) $S_{yy}(f_k)$ at the frequency f_k for the time domain signal $y(t)$.

$$S_{yy}(f_k) = E[Y(f_k)\, Y^*(f_k)] \tag{6.18}$$

$$\Rightarrow S_{yy}(f_k) = \frac{\sum_{i=1}^{n_s}\left[Y_i(f_k)Y_i^*(f_k)\right]}{n_s} \tag{6.19}$$

where $Y_i(f_k)$ and $Y_i^*(f_k)$ are the FT and its complex conjugate at frequency f_k for the ith segment of the signal $y(t)$. E indicates the mean (average) operator. n_s is the number of segments of the signal $y(t)$ used for averaging. The multiplication of the FT and its complex conjugate for any segment within a signal creates a real number, and hence addition is possible irrespective of conditions 1 and 2 above. The process of dividing the data into a number of segments is shown in Figure 6.40. Each segment is of size N.

This process of averaging is now applied to Example 6.5 in Figure 6.38 as shown below, which yields the expected correct results.

$$S_{yy}(f_k) = E(Y(f_k)Y^*(f_k)) = \frac{Y_1(f_k)Y_1^*(f_k) + Y_2(f_k)Y_2^*(f_k)}{2}\, Amplitude^2 \tag{6.20}$$

$$Amplitude\ of\ the\ spectrum = \sqrt{S_{yy}(f_k)} \tag{6.21}$$

For this simple example, the averaged spectrum based on the PSD concept will yield the amplitude spectrum exactly as shown in Figure 6.39.

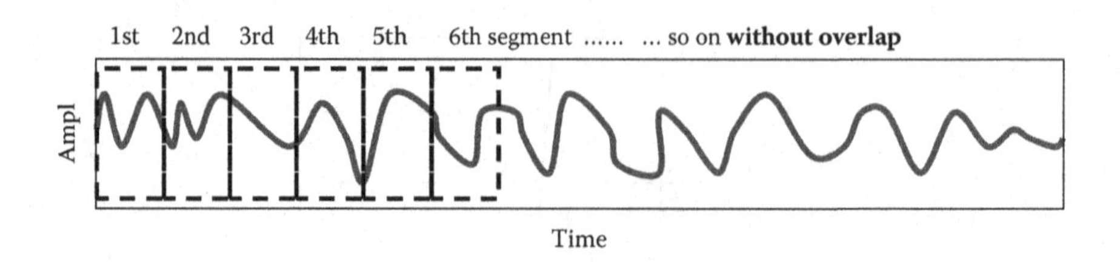

FIGURE 6.40
Process of dividing the time domain signal into a number of segments of size N.

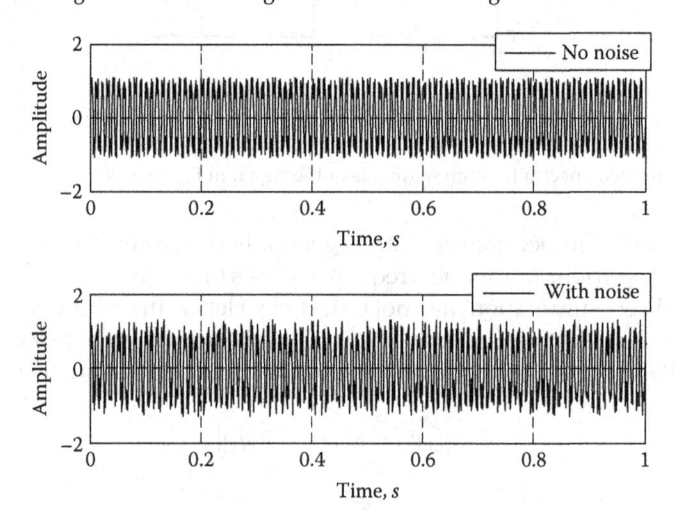

FIGURE 6.41
Example 6.6 time domain signals with and without noise.

6.9.3 Example 6.6: Comparison of the Averaged Spectra for the Signals with and without (i.e., No) Noise

Two signals (one with random noise in the signal) are considered here. They are shown in Figure 6.41, and their zoomed views are shown in Figure 6.42. The signals are represented by the following equations:

Signal without (no) noise:

$$y(t) = \sin(2\pi 100t) + 0.5\sin(2\pi 300t)$$

Signal with noise:

$$y(t) = \sin(2\pi 100t) + 0.5\sin(2\pi 300t) + Noise$$

The signal with noise has clearly affected the original signal, as can be seen in Figure 6.43. These signals were collected at f_s = 1024 Hz for 10 s.

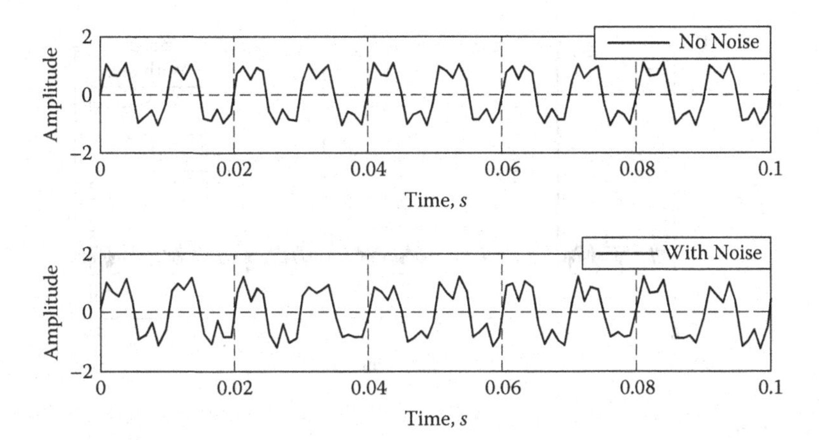

FIGURE 6.42
Zoomed view of signals in Figure 6.42.

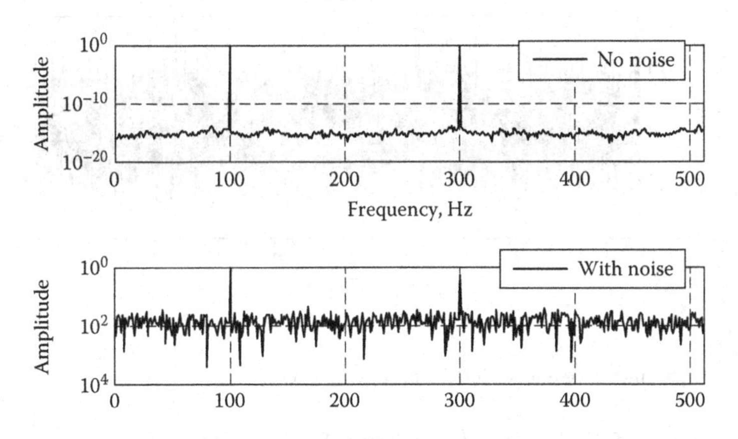

FIGURE 6.43
Spectra for the first data segment of the signals with and without noise in Figure 6.41.

They are then divided into 10 equal segments for the calculation of the averaged spectra. Each segment contains 1 s data, i.e., $N = 1024$ points.

Figure 6.43 compares the amplitude spectrum for the first segment for the signals with and without noise. The spectrum of the signal without noise just shows two frequency peaks without any background noise, even for the first segment of the data. However, the spectrum for the signal with noise shows background noise in addition to the two frequency peaks for the first segment. Hence, the averaging process is needed for the second signal with noise, so as to get a much clearer spectrum. The averaged spectrum of the second signal with noise using 10 averages is also shown in Figure 6.44, where the effect of the background noise has reduced to a much lower level. However, more averaging may further improve the spectrum.

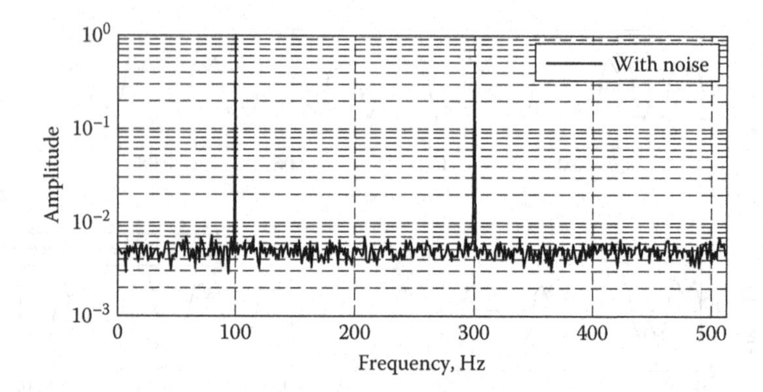

FIGURE 6.44
Averaged amplitude spectrum of the time domain signal with noise for Example 6.6.

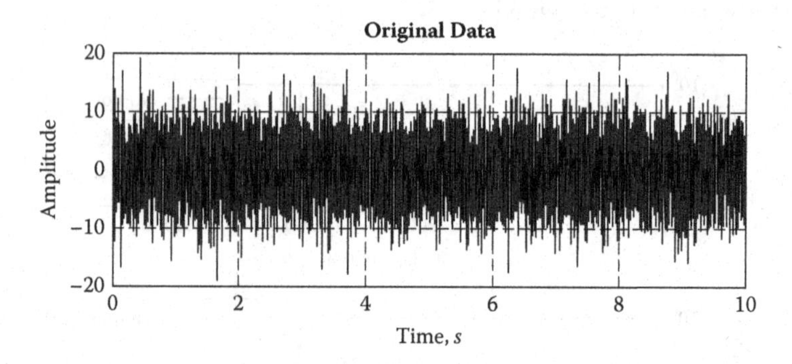

FIGURE 6.45
Collected noisy sine wave data of 100 Hz for Example 6.7.

6.9.4 Example 6.7: Averaged Spectrum for the Noisy Signal

This is a typical example of a noisy pure sine wave signal of frequency 100 Hz with amplitude equal to 1. The data are collected at f_s = 1024 Hz and are shown in Figure 6.45. The signal amplitude is higher than 1 due to the presence of noise in the signal. The zoomed view of the signal at the 0 to 0.01 s time span is shown in Figure 6.46, where 0.01 s is the time period for a 100 Hz sine wave, which does not show the presence of the pure sine wave in the collected signal. Now the 10 s data are divided into 20 equal segments of size N = 512 and $T = Ndt$ = 0.5 s for FT for the averaged spectrum computation. Figures 6.47 and 6.48 show the computed spectrum for a segment (without average) and the averaged spectrum of 20 segments, respectively. It is obvious from Figure 6.48 that the averaged spectrum almost accurately detects the presence of the sine wave at 100 Hz with amplitude of 1. This example illustrates the advantage of the averaging process in the spectrum computation.

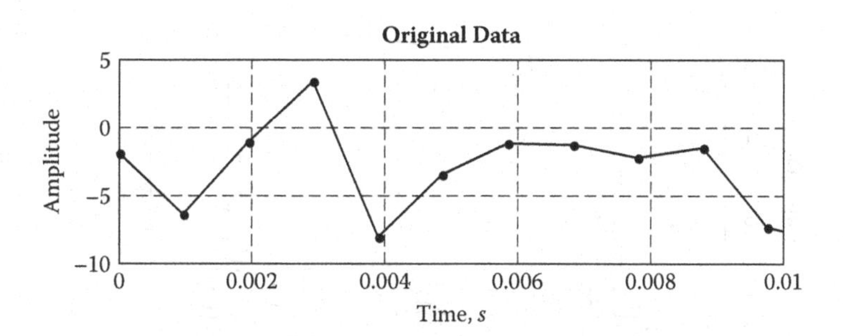

FIGURE 6.46
Zoomed view of the data in Figure 6.45.

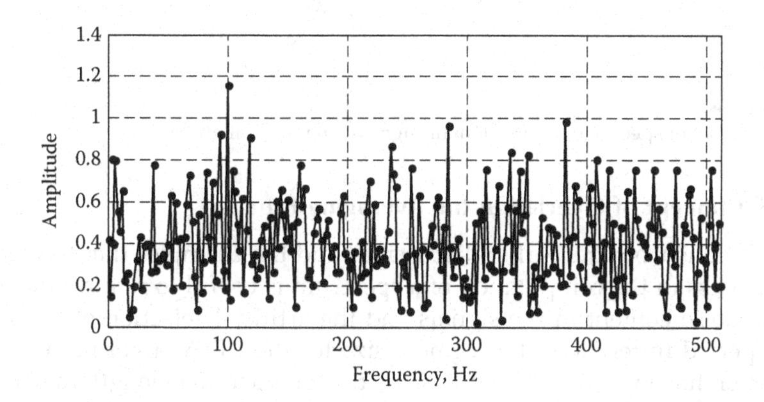

FIGURE 6.47
Spectrum without average showing no clear peak at 100 Hz.

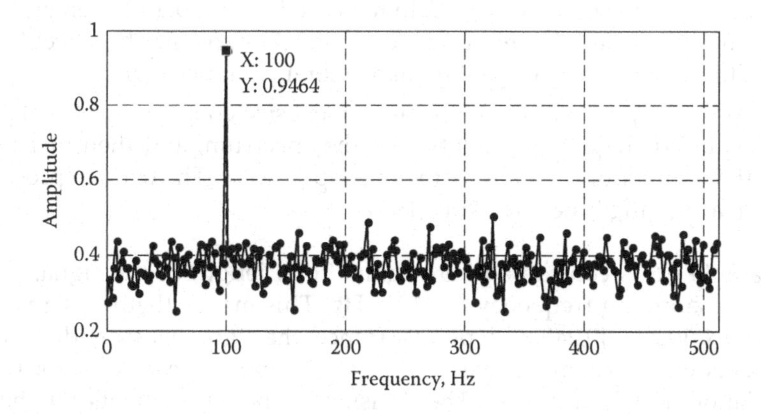

FIGURE 6.48
Spectrum with 20 averages showing a clear peak at 100 Hz.

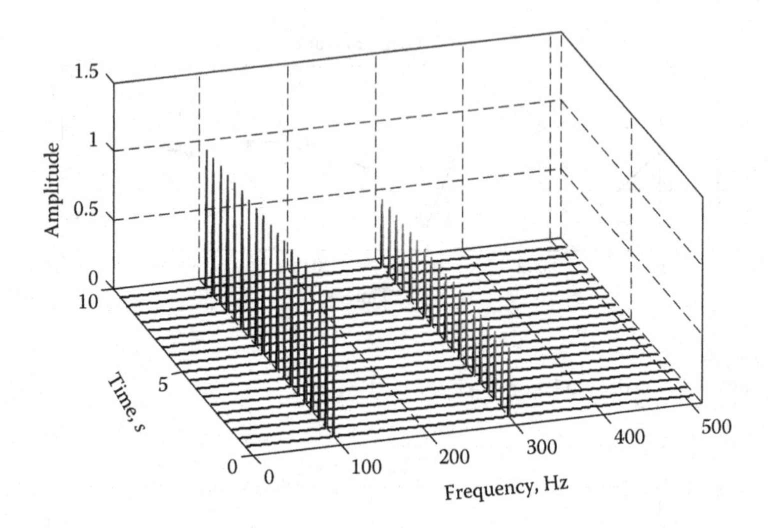

FIGURE 6.49
Waterfall plot of spectra for a time domain signal using STFT analysis.

6.9.5 Concept of Overlap in the Averaging Process

It is obvious from the earlier discussions and the examples that the averaging process in the computation of a spectrum is useful to remove the effect of the noise content in the signals and the artificial selection of the signal time period in terms of the segment size for the FT. A larger number may be better; however, on many occasions the recorded data length may not be long enough to have a large number of averages. Hence, a large number of averaging for the limited short-time domain data are possible in the following two ways:

1. Select small segment size, N, to increase the number of average, but this may impact the frequency resolution for the computed spectrum. This may not be preferred in many vibration applications.
2. Another approach is to keep the segment size good enough to get the required frequency resolution in the spectrum, and then increase the number of averages by the overlap process. The overlap process in averaging is now explained here.

Example 6.6 is considered here again. Total length of the signal is 10 s with the sampling frequency f_s = 1024 Hz. This means that the signal contains 10 × 1024 = 10,240 data points. Hence, the signal was divided into 10 segments with a segment size $N = 2^{10} = 1024$ and $T = Ndt = 1$ s for the FT calculation and 10 averages. The division of the 10 segments for the signal is illustrated in Table 6.6 for clear understanding. This process is called averaging without overlap; however, the overlap process can increase the number of averages. For example, 50% overlap can increase the number of averaging to 19. The 50% overlap means that the first segment contains the data from 0 s (first data point) to 1 s (1024th data point), and then the second segment starts at 0.5 s (513rd point) to 1.5 s (1536th point), and so on. This means that only 50% of the data points are dropped from the previous segment, and the new 50% data are added to the next segment. This is also demonstrated in Table 6.6. More of a percentage in the overlap can further increase the number of averages.

6.10 Short-Time Fourier Transformation (STFT)

The signal without noise in Example 6.6 is considered again to explain the STFT concept. The 50% overlap resulted in 19 segments and hence 19 spectra. Instead of using these 19 spectra for a single averaged spectrum, if these 19 spectra are plotted into a 3D waterfall plot with *x*-*y*-*z*-axes as frequency-time-amplitude, respectively, then it is known as the STFT. Let us assume the

TABLE 6.6
Key Elements in the Averaging Process

	Average without Overlap			Average with 50% Overlap		
Segment No.	Segment Size, *N*	Start Point	End Point	Segment Size, *N*	Start Point	End Point
1	1024	1	1024	1024	1	1024
2	1024	1025	2048	1024	513	1536
3	1024	2049	3072	1024	1025	2048
4	1024	3073	4096	1024	1537	2560
5	1024	4097	5120	1024	2049	3072
6	1024	5121	6144	1024	2561	3584
7	1024	6145	7168	1024	3073	4096
8	1024	7169	8192	1024	3585	4608
9	1024	8193	9216	1024	4097	5120
10	1024	9217	10,240	1024	4609	5632
11	—	—	—	1024	5121	6144
12	—	—	—	1024	5633	6656
13	—	—	—	1024	6145	7168
14	—	—	—	1024	6657	7680
15	—	—	—	1024	7169	8192
16	—	—	—	1024	7681	8704
17	—	—	—	1024	8193	9216
18	—	—	—	1024	8705	9728
19	—	—	—	1024	9217	10,240

1st spectrum for the 1st segment is related to the meantime $t_{s1} = \frac{0+1}{2} = 0.5$ s; similarly, 2nd, 3rd, 4th, … , 19th spectra are assumed to be associated with the mean time of each segment, $t_{s2} = \frac{0.5+1.5}{2} = 1.0$ s, $t_{s3} = \frac{1.0+2.0}{2} = 1.5$ s, $t_{s3} = \frac{1.5+2.5}{2} = 2.0$ s, … , $t_{s19} = \frac{9.0+10.0}{2} = 9.5$ s, respectively. The 3D waterfall is shown in Figure 6.49. The advantage of this STFT process is to show how the frequency content in a signal is changing with time. In this case, neither frequency nor the amplitude is changing with time. The segment size, *N*, and the overlap can be changed/adjusted to get the required information from the STFT analysis. The 3D waterfall plot in Figure 6.49 can also be represented as the 2D contour plot, as shown in Figure 6.50. Also, Figure 6.50 is sometimes referred to as the spectrogram.

A typical application of this STFT analysis is the tracking of the dynamic behavior of a rotating machine during machine run-up or run-down, i.e., changing rotating speed with time. The measured vibration acceleration from a rotating machine at bearing 1 during machine run-up is shown in Figure 6.51, and variation of the machine RPM with time in Figure 6.52. The 3D waterfall diagram in frequency-RPM-amplitude axes is shown in Figure 6.53. Note that the time axis is converted into the RPM axis using the relation between the time-RPM shown in Figure 6.52. It is observed that the vibration of the rig is mainly related to the machine RPM (frequency peak related to the speed, referred to as the 1X component) and its high

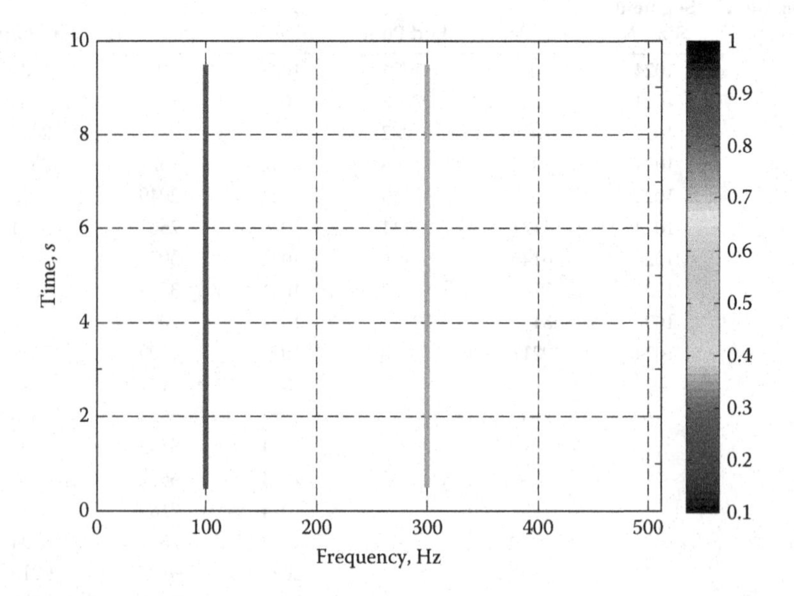

FIGURE 6.50
Two-dimensional contour plot of the waterfall plot of spectra in Figure 6.49.

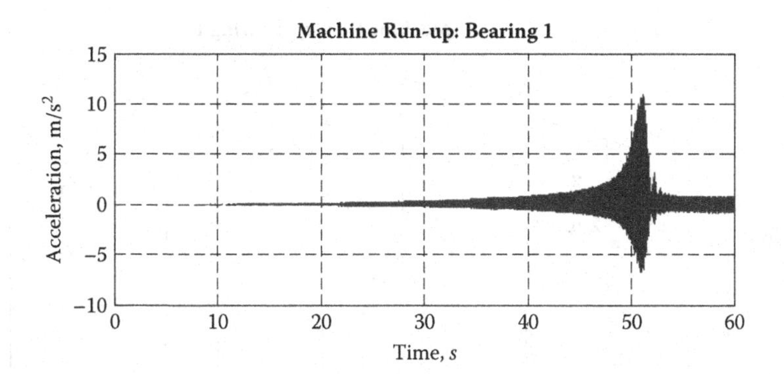

FIGURE 6.51
Measured acceleration response at bearing 1 of a machine.

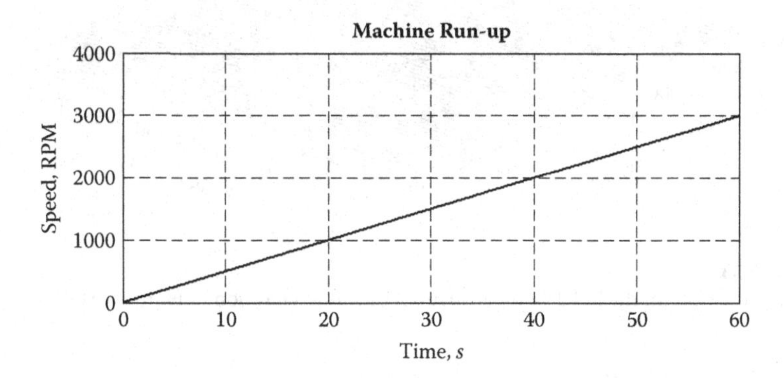

FIGURE 6.52
Machine run-up speed with time.

harmonic (2X, 3X, etc.) components. The high vibration (amplification) at around 42.7 2 Hz at 2571 RPM is likely to indicate the first natural frequency of the machine. It is also referred to as the first critical speed of the rotating machine. Hence, the vibration behavior of the complete machine can quickly be observed through the STFT analysis.

6.11 Correlation between Two Signals

6.11.1 Cross-Power Spectral Density (CSD)

Equation (6.18) explains the concept of the averaged auto-power spectral density (PSD), $S_{yy}(f)$, for a time domain signal $y(t)$. Similarly, the cross-power

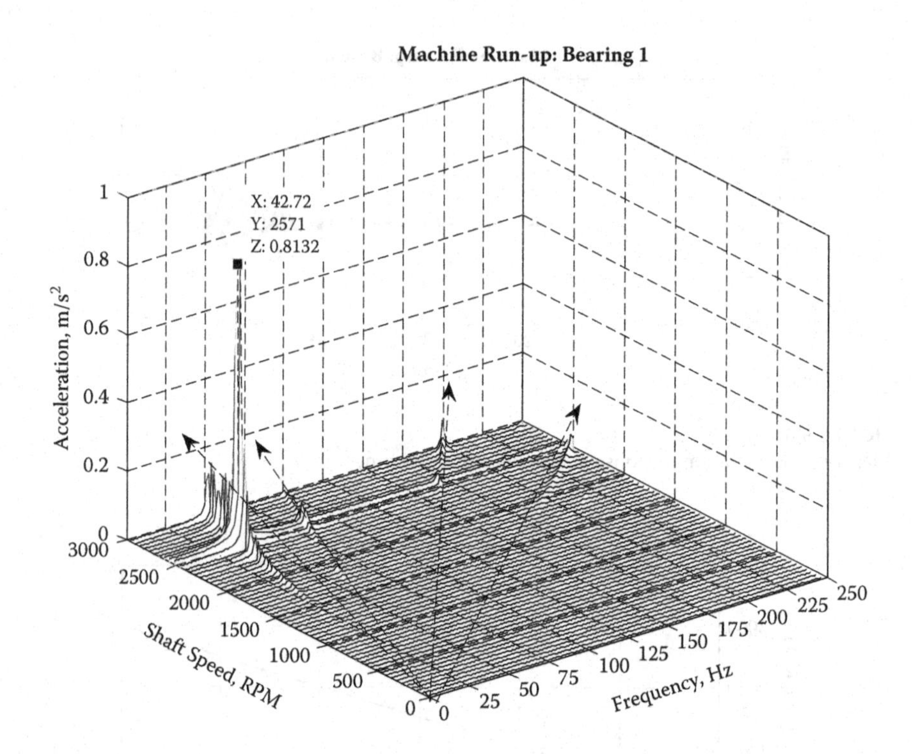

FIGURE 6.53
Waterfall diagram of the STFT of the measured machine vibration in Figure 6.51.

spectral density (CSD), $S_{xy}(f)$, at a frequency f between two time domain signals, $x(t)$ and $y(t)$, is given mathematically as

$$S_{xy}(f) = E(X(f)Y^*(f)) \tag{6.22}$$

$$\Rightarrow S_{xy}(f) = \frac{\sum_{i=1}^{n_s} (X_i(f)Y_i^*(f))}{n_s} \tag{6.23}$$

where $X_i(f)$ and $Y_i^*(f_k)$ are the FT and the complex conjugate of the FT at frequency f for the ith segment of the signals $x(t)$ and $y(t)$, respectively. E indicates the mean (average) operator. n_s is the number of equal segments of the signals $x(t)$ and $y(t)$, used for averaging.

6.11.2 Frequency Response Function (FRF)

The transfer function $x(t)/y(t)$ in the frequency domain, i.e., $X(f)/Y(f)$, is known as the FRF. This is a useful tool in signal processing for comparing the relative amplitude and the phase of the signal *x*(*t*) with respect to the signal *y*(*t*) at any frequency *f*. The averaged FRF is defined as

$$FRF(f) = \frac{S_{xy}(f)}{S_{yy}(f)} \tag{6.24}$$

$$\Rightarrow FRF(f) = \frac{\dfrac{\sum_{i=1}^{n_s}(X_i(f)Y_i^*(f))}{n_s}}{\dfrac{\sum_{i=1}^{n_s}(Y_i(f)Y_i^*(f))}{n_s}} \tag{6.25}$$

First, the averaged cross-power spectral density (CSD) between the two signals *x*(*t*) and *y*(*t*), and the averaged PSD of the signal *y*(*t*) are computed separately for the numerator and denominator of the FRF, respectively. Then the averaged CSD is divided by the averaged PSD to compute the FRF. The computed FRF contains a complex quantity at each frequency, which represents the amplitude ratio and phase of the signal *x*(*t*) with respect to the signal *y*(*t*) at that frequency.

6.11.3 Ordinary Coherence

The ordinary coherence between two vibration signals, *x*(*t*) and *y*(*t*), at a frequency is defined as (Bendat and Piersol, 1980)

$$\gamma^2(f) = \frac{|S_{xy}(f)|^2}{S_{xx}(f)S_{yy}(f)} \tag{6.26}$$

$$\Rightarrow \gamma^2(f) = \frac{\left|\dfrac{\sum_{i=1}^{n_s}(X_i(f)Y_i^*(f))}{n_s}\right|^2}{\left(\dfrac{\sum_{i=1}^{n_s}(X_i(f)X_i^*(f))}{n_s}\right)\left(\dfrac{\sum_{i=1}^{n_s}(Y_i(f)Y_i^*(f))}{n_s}\right)} \tag{6.27}$$

The coherence between two signals indicates the degree to which two signals are linearly correlated at a given frequency. The scale of the coherence function is between 0 and 1; 0 at a frequency indicates no relation between the two signals at that frequency, and 1 indicates perfect (100%) relation. The following can also be inferred about the coherence (Balla et al., 2006):

1. Coherence close to unity means the response is linearly correlated to the input excitation.
2. In general, measurement of structural responses and input are contaminated with noises (generated due to measuring instruments and structures), which may cause reduction in coherence.
3. Coherence reduces due to nonlinear relation between response and input.
4. Coherence reduces when the response is due to extraneous inputs.

6.11.4 Example 6.8: FRF and Coherence

The following two signals are considered to illustrate the concept of the FRF and coherence function:

$$x(t) = A_{x1}\sin(2\pi f_1 t + \theta_{x1}) + A_{x2}\sin(2\pi f_2 t + \theta_{x2}) + Noise \tag{6.28}$$

$$y(t) = A_{y1}\sin(2\pi f_1 t + \theta_{y1}) + A_{y2}\sin(2\pi f_2 t + \theta_{y2}) + Noise \tag{6.29}$$

The values of the parameters in Equations (6.28) and (6.29) are listed in Table 6.7. Both signals are assumed to be collected at the sampling frequency $f_s = 1000$ Hz, and the samples of the collected signals containing noises are shown in Figure 6.54. It is difficult to know the frequency contents and their respective amplitudes in both signals based on the visual observation of signals in Figure 6.54 and the relation between these two signals. The FRF and coherence functions help to understand these relations. Hence, the FRF of signal $x(t)$ with respect to $y(t)$ and their coherence are now computed as per Equations (6.25) and (6.27), respectively, which are shown

TABLE 6.7

Parameters of the Signals, $x(t)$ and $y(t)$

Frequency, Hz	Amplitude	Phase, degree
$f_1 = 100$	$A_{x1} = 10$	$\theta_{x1} = +90$
	$A_{y1} = 8$	$\theta_{y1} = +45$
$f_1 = 200$	$A_{x2} = 12$	$\theta_{x2} = -90$
	$A_{y2} = 6$	$\theta_{y2} = +45$

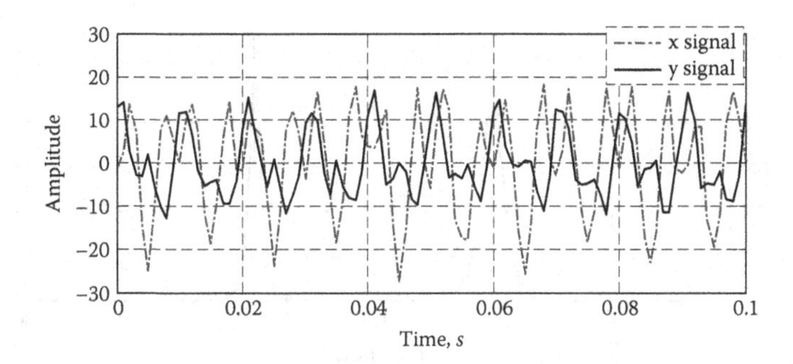

FIGURE 6.54
Typical sample of the signals x and y of Example 6.8.

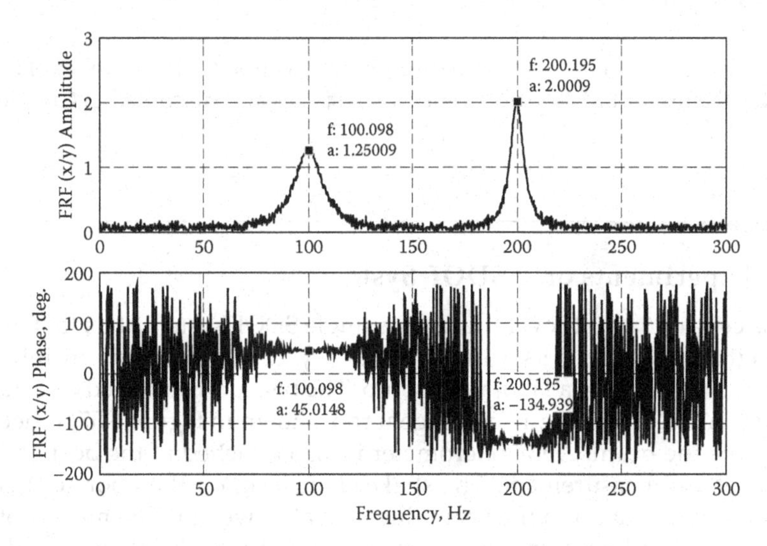

FIGURE 6.55
FRF amplitude and phase plot of Example 6.8.

in Figures 6.55 and 6.56. The following observations can be drawn from Figures 6.55 and 6.56:

1. FRF plot: Both signals just contain two frequencies—100 and 200 Hz.
2. FRF amplitude: Amplitude ratios between the signals (x/y) at the frequencies 100 and 200 Hz are 1.25 and 2, which are exactly the same as A_{x1}/A_{y1} and A_{x2}/A_{y2}, respectively.
3. FRF phase: Phase angles of the signals x with respect to y at the frequencies 100 and 200 Hz are +45° and –135°, which are again exactly the same as $(\theta_{x1} - \theta_{y1})$ and $(\theta_{x2} - \theta_{y2})$, respectively.
4. Coherence plot: An amplitude nearly equal to 1 at 100 and 200 Hz confirms that the both signals are well correlated at these two frequencies.

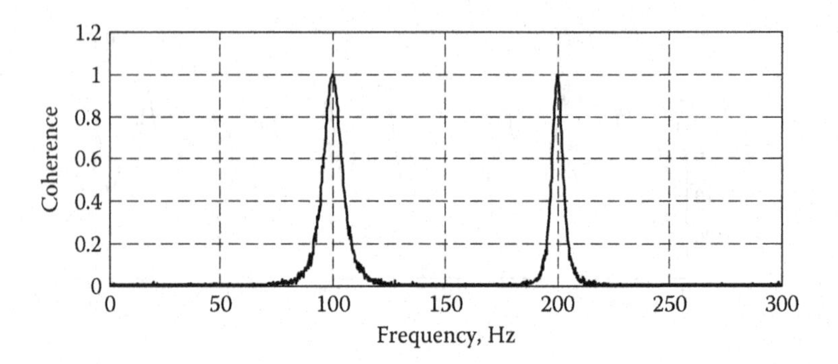

FIGURE 6.56
Coherence function plot of Example 6.8.

The use of FRF analysis is again discussed in Section 6.12. A number of examples for the use of FRF and coherence analysis are also discussed in Chapter 7.

6.12 Experiments on a SDOF System

Let us consider a cantilever beam (as a single SDOF system at its first mode) of length 1 m having its natural frequency of 100 Hz with an inherent damping (ζ = 10%) in a laboratory setup. The beam is connected to a shaker through a force sensor, and a stringent rod (shown in Figure 6.57) is needed to vibrate the beam. An accelerometer is also attached at the beam tip for the response measurement. The shaker is resting on the floor so there is no mass contribution to the beam due to shaker weight. The masses of the force sensor and accelerometer, and the stiffness of the stringer rod, are chosen to be very small so that they could not influence the dynamics of the beam significantly. Here, these influences are assumed to be zero for the demonstration.

A step-sine excitation was applied to the beam by the shaker starting from frequency 50 to 150 Hz at the steps of 1 Hz. This means the excitation frequency starts at 50 Hz and then shifts to 51 Hz, 52 Hz, and so on, to 150 Hz. The shaker controller was adjusted such that it gives 1 N excitation force at all frequencies, and the duration of the excitation at each frequency was kept to 10 s. The accelerometer and force sensor data were then recorded to a PC at the sampling frequency of 1024 Hz.

After the experiments, signal processing was carried out. The first 10 s of data are related to the exciting frequency of 50 Hz. It is also known that the initial acceleration data may contain some transient response; hence, the first 1 s of data were removed from both the force and accelerometer so that the remaining 9 s of data contain only the steady-state response.

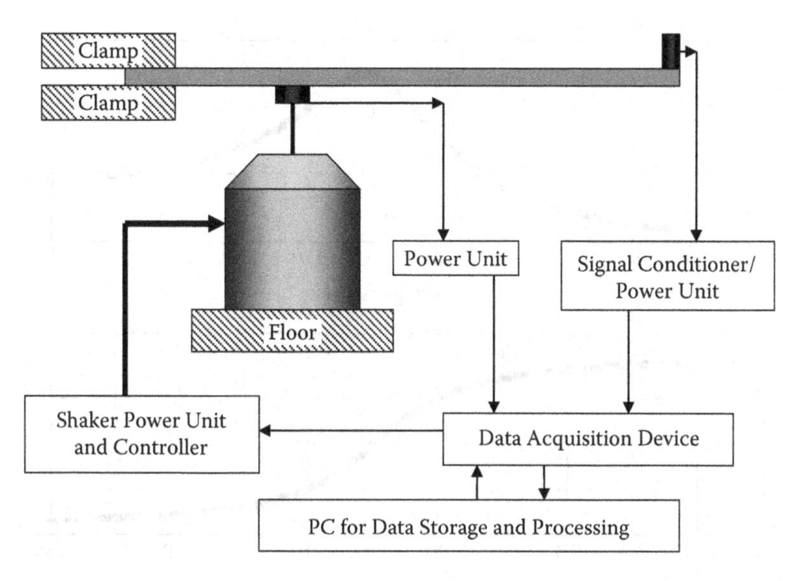

FIGURE 6.57
Experimental setup.

Now the FRF of the acceleration response with respect to the applied force was calculated as per Equation (6.25) using N = 1024 data points without overlap, thus resulting in a frequency resolution of 1 Hz with nine averages. The FRF data at the exciting frequency of 50 Hz was only then picked up from the complete FRF data points. Similarly, in the next step, the next 10 s of data for the exciting frequency of 51 Hz were considered and similar signal processing was performed, and again the FRF data point corresponding to the exciting frequency of 51 Hz was only stored. In a similar fashion, all the data related to all the exciting frequencies were processed, and then the FRF data at each exciting frequency plotted in Figure 6.58, which is in the frequency range of 50 to 150 Hz. The phase of the acceleration response to the force at the natural frequency of 100 Hz is found to be +90°, which is clearly seen from the time domain signal of the response in Figure 6.58.

The experimentally obtained FRF (A/F) is the ratio of the acceleration to the force, which is generally used in practice as the measurement is, in general, acceleration response using an accelerometer. However, to correlate the TF of the displacement to the force (Y/F) as discussed in Chapter 2 and Figure 2.13, the following computation was carried out to convert the measured $TF = FRF(A/F)$ to $TF = FRF(Y/F)$:

$$TF\left(\frac{Y(\omega_j)}{F(\omega_j)}\right) = -\frac{1}{(\omega_j^2)}TF\left(\frac{A(\omega_j)}{F(\omega_j)}\right) \tag{6.30}$$

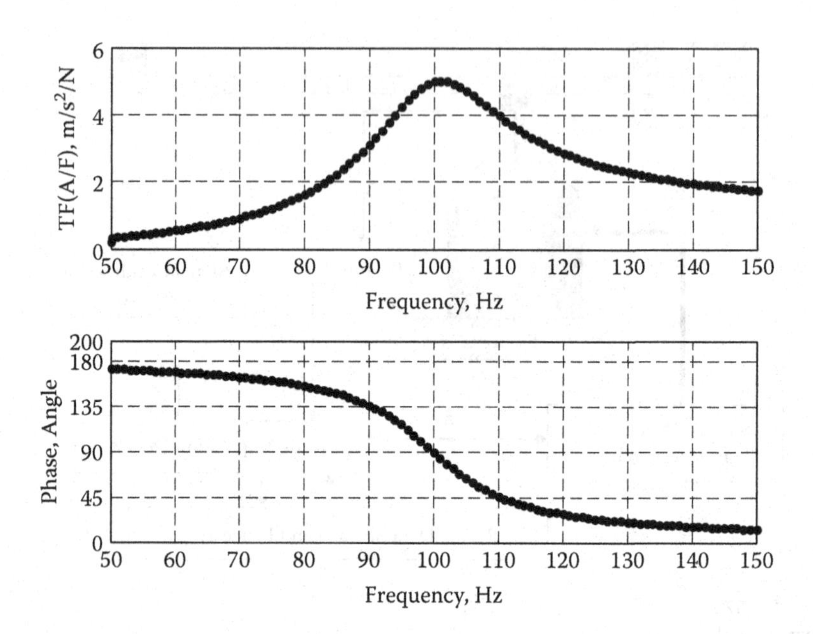

FIGURE 6.58
The TF steady-state acceleration response to the applied force.

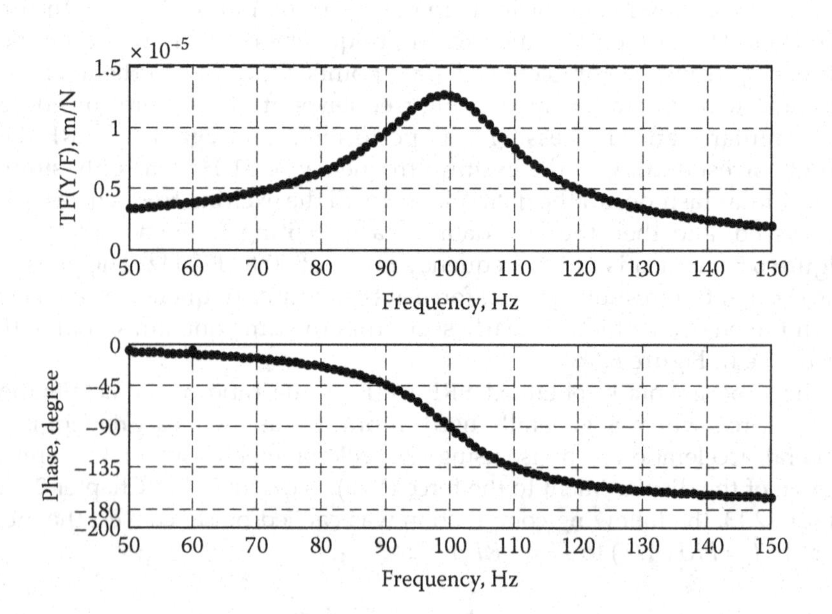

FIGURE 6.59
The TF steady-state displacement response to the applied force.

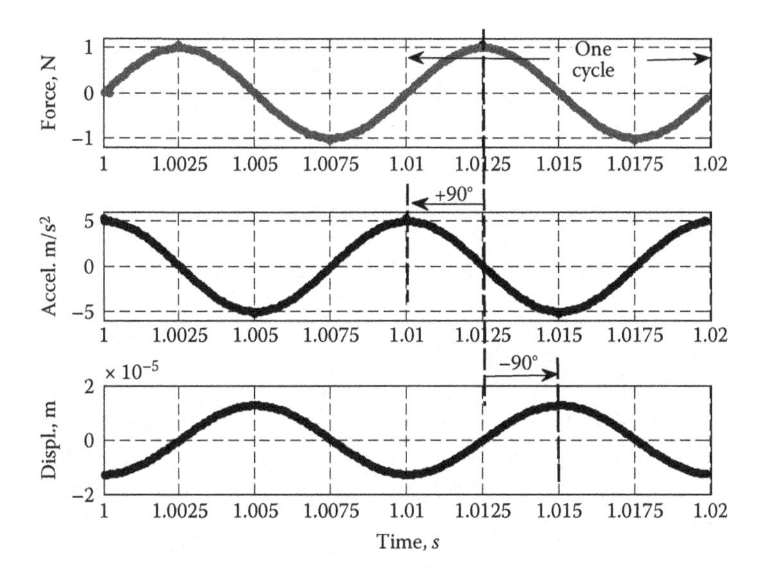

FIGURE 6.60
The steady-state responses to the applied load at the resonance at 100 Hz.

The estimated TF (Y/F) is also shown in Figure 6.59, which is similar to Figure 2.13. Figure 6.60 also gives the phase relation between the force and the displacement, which is –90° at 100 Hz, which is similar to the theory explained in Section 2.3.

References

Bendat, J.S., Piersol, A.G. 1980. *Engineering Applications of Correlation and Spectral Analysis*. Wiley, New York.

Balla, C.B.N.S., Meher, K.K., Sinha, J., Rao, A.R. 2006. Coherence Measurement for Early Contact Detection between Two Components. *Journal of Sound and Vibration* 290(1), 519–523.

Oppenheim, A.V., Schafer, R.W. 1989. *Discrete-Time Signal Processing*. Prentice-Hall, Englewood Cliffs, NJ.

[illegible]

[illegible]
[illegible] 2000 [illegible]
[illegible] 2000 [illegible]

7

Experimental Modal Analysis

7.1 Introduction

In Chapter 3, the finite element (FE) modeling of structures was discussed, and then the concept of the eigenvalues-eigenvectors was introduced to estimate the natural frequencies and mode shapes. This process needs detailed geometry, material properties, and boundary conditions for developing the mathematical FE model. Alternatively, the natural frequencies and mode shapes together with the modal damping at each mode can be extracted by experimental means without constructing the mathematical model of the structural system. This process is called experimental modal analysis.

In experimental modal analysis, an external dynamic force (excitation) to the structure is applied in a controlled frequency band, and simultaneously the vibration responses at a number of locations are picked up and then the collected vibration data are analyzed to extract the modal parameters, namely, natural frequencies, mode shapes, and modal damping. The step-by-step procedure for the modal testing, data analysis, and extraction of the modal parameters has been explained through a few typical examples ranging from a single degree of freedom (SDOF) systems to beam structure, tubes, etc.

7.2 Experimental Procedure

The schematic of a typical experimental setup for the modal testing of a cantilever beam is shown in Figure 7.1. As can been seen from the figure, six numbers of the accelerometers are mounted on the cantilever beam along its length and the external excitation at a point.

The concept of modal testing is to create the resonance condition at the natural frequencies for a small duration when excited by the external force so that the resonance peaks measured by the response accelerometers can be picked up. The following types of excitations are used in practice.

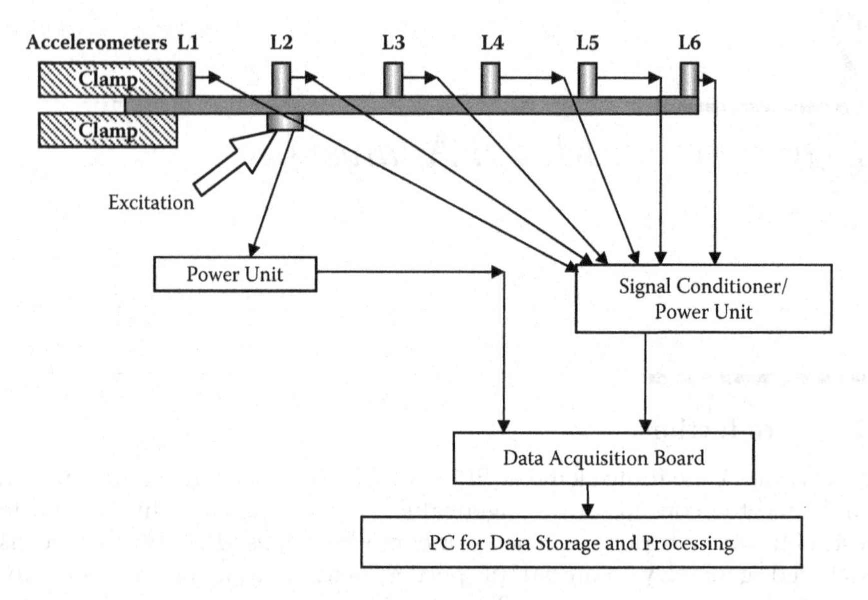

FIGURE 7.1
Schematic of the modal test setup.

7.2.1 Impulsive Load Input Using the Instrumented Hammer (see Section 5.6)

The impulsive force gives a broadband excitation to the structure for a short duration, which leads to resonance at all natural frequencies within the frequency band of excitation. The frequency band of excitation can be controlled by the use of different impact heads of the hammer. Low- to high-frequency ranges of excitation can be achieved by using a softer to harder impact tip, respectively, for hammering. This is the quickest/easiest way to do the modal testing to estimate the natural frequencies and mode shapes and, to an extent, the damping associated with each mode.

7.2.2 Shaker Excitation

The schematic of the shaker test is shown in Figure 5.23. The following types of excitation can be used:

1. **Band-limited random excitation (white noise):** It is not a preferred practice for modal testing, as the measured responses by the accelerometers may be noisy, which requires a large amount of averaging of the data in the frequency domain.
2. **Band-limited multisine excitation:** It is like the band-limited random excitation but consists of the sinusoidal waves for all frequencies in the frequency band of excitation at any time. It is by far a better option than the band-limited random excitation.

3. **Sweep-sine or chirp-sine excitation:** In this case, the frequencies of excitation vary from the lowest frequency to the highest frequency at the user-defined sweep (chirp) rate. It is a well-accepted approach and often used in modal testing.
4. **Step-sine excitation:** It is similar to the sweep-sine excitation, but excitation at each frequency in the excitation frequency band is considerably for a longer period so that full amplification at each frequency in the measured response can be seen. Refer to Section 6.12 for better understanding. It is an ideal approach for modal testing and particularly for accurate estimation of the damping ratio at each mode, but is very time-consuming.

7.3 Modal Test and Data Analysis

The measurement methods based on both external excitation and responses can be classified into the following two categories:

1. **SIMO analysis:** The schematic setup for the modal testing shown in Figure 7.1 is an example of SIMO analysis. It is *single input* (only one excitation force) and *multioutput* (a number of response locations on the structure). This is the simplest measurement scheme and generally useful for most practical purposes for modal testing.
2. **MIMO analysis:** In this case, a number of modal shakers at different locations are used simultaneously to excite the structure (*multi-inputs*) and the responses are measured at a number of locations (*multioutputs*). This scheme involves a very complex mathematical model for the signal processing to extract the modal parameters. Besides, the phase of excitation at different locations is very important for good results. This method is generally employed for very large structures when excitation from one exciter in not enough to excite the whole structure.

The SIMO analysis approach has been considered here for better understanding of the experimental modal analysis procedure. Now it has been assumed that the force and the response data for the setup shown in Figure 7.1 have been recorded into a PC for further processing. The recorded signals can be processed either in time domain or in frequency domain to extract the modal parameters. However, considering the fact that the frequency domain approach is more popular and well accepted and incorporated into a number of commercially available modal analysis software, only this approach has been discussed here. Following are the typical modal parameter estimation methods in the frequency domain.

1. **Peak pick method:** This is the simplest and quickest approach to modal parameter estimation and serves the purpose of modal analysis for most cases if the modes are well separated and the structure is lightly damped. The biggest limitation of the method is that it provides modal damping for each mode at all the measurement locations, which is clearly explained in Example 7.2.
2. **Least-squares SDOF circle fit method:** This is a much better approach than the peak pick method but is a little more involved. The advantage of this method is that it provides a global modal damping at each mode. The method also works well for structures with well-separated modes and light damping.
3. **MDOF methods:** This approach is a multidegree of freedom curve-fitting method for estimation of modal parameters. There are a number of estimation methods used for this approach. This method is useful if any structure has closely spaced modes and where there is a large difference in damping between the modes.

7.4 Example 7.1: Peak Pick Method

To explain this method, a simple example of a SDOF system consisting of a mass and a spring with an inherent damping, as shown in Figure 7.2 (Example 7.1), has been considered. Now it is assumed that the modal parameters—natural frequency (f_n), mode shape (φ), and modal damping (ζ)—have not been known and need to be estimated experimentally.

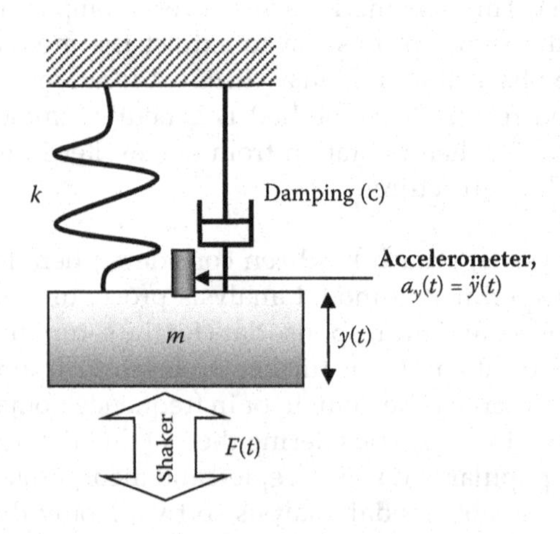

FIGURE 7.2
Example 6.2 of a SDOF system for modal testing.

7.4.1 Step 1: The Modal Testing Experiment

The modal testing experiment is conducted as per the measurement scheme shown in the schematic in Figure 7.1. The force and the response signal using the force sensor and the response accelerometers are collected and stored in a PC. In the presented case, the data have been collected at a sampling rate of f_s = 1024 samples/s. A sweep-sine excitation of constant amplitude of 1 N in a frequency band from 40 to 60 Hz has been given to the system through a shaker. For the case discussed here, the time period for a complete sweep-sine cycle from 40 to 60 Hz was kept for 16 s. A total of 10 cycles of the excitation was given, and both the force and acceleration responses were recorded to analyze the data in the frequency domain with 10 averages without any overlap. Typical recorded data for a cycle of excitation for both the shaker force and the acceleration response are shown in Figures 7.3 and 7.4. Figure 7.3(b) shows the linear change in frequency from 40 to 60 Hz with time during each cycle of the applied excitation. A few zoomed views of the measured force in Figure 7.3 are also shown in Figures 7.5 to 7.7 for clarity. The amplification in the measured acceleration response in Figure 7.4 in the region of 7 to 9 s indicates that the system may be passing through a resonance. The identification of natural frequency and the estimation of mode shape and damping have further been explained in the following steps.

7.4.2 Step 2: Computation of the Amplitude Spectrum, FRF, and Coherence

Since the sample frequency, f_s, is 1024 Hz and data have been collected for 160 s for both force and acceleration response (16 s for each cycle of the

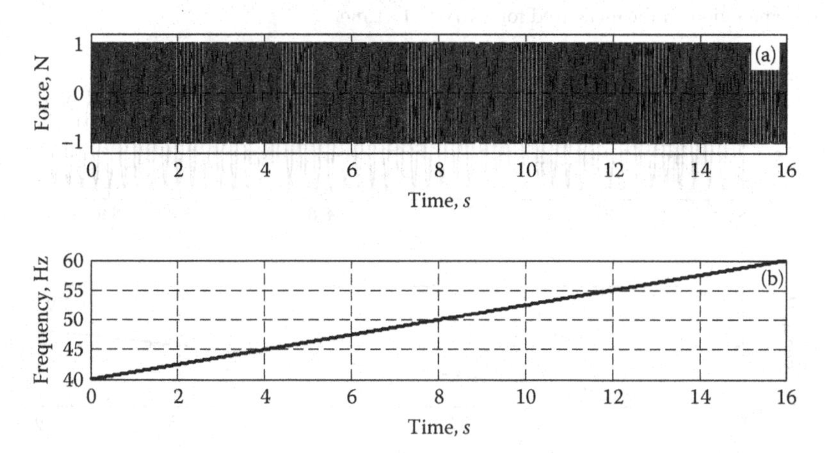

FIGURE 7.3
The measured force. (a) A complete cycle of the sweep-sine from 40 to 60 Hz. (b) Linear change in the exciting frequency.

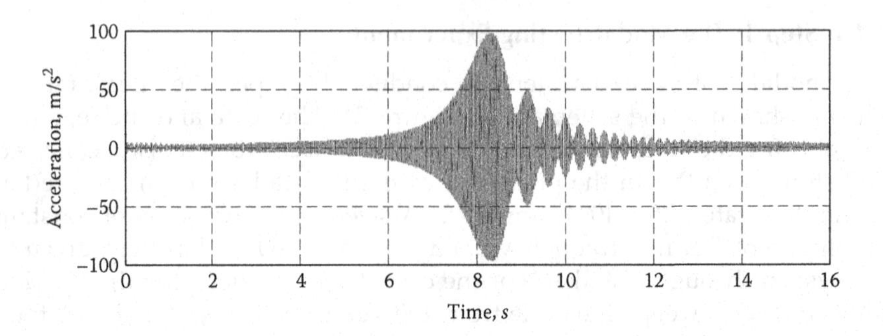

FIGURE 7.4
The measured acceleration response for the applied force in Figure 7.3.

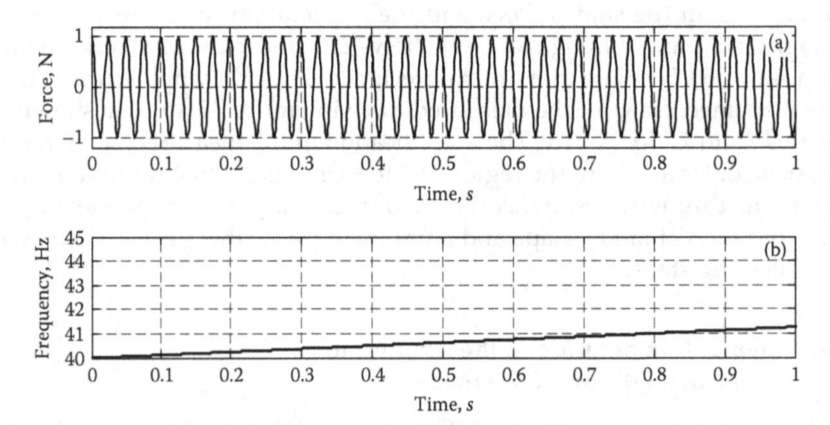

FIGURE 7.5
The zoomed view of the measured force in 0 to 1 s time.

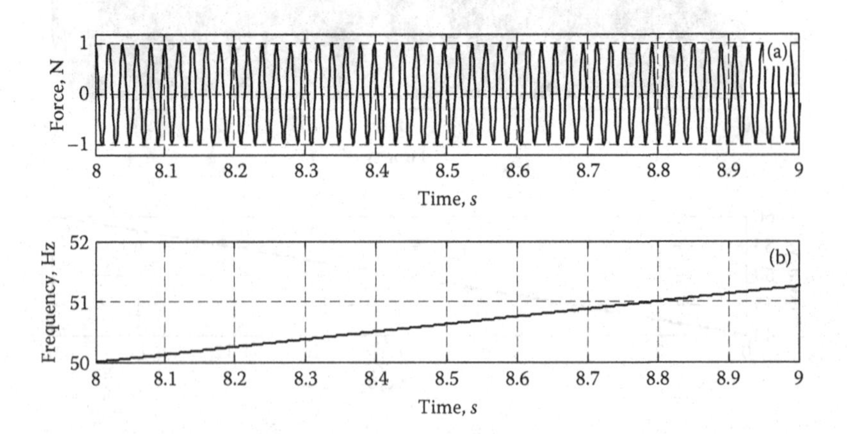

FIGURE 7.6
The zoomed view of the measured force in 8 to 9 s time.

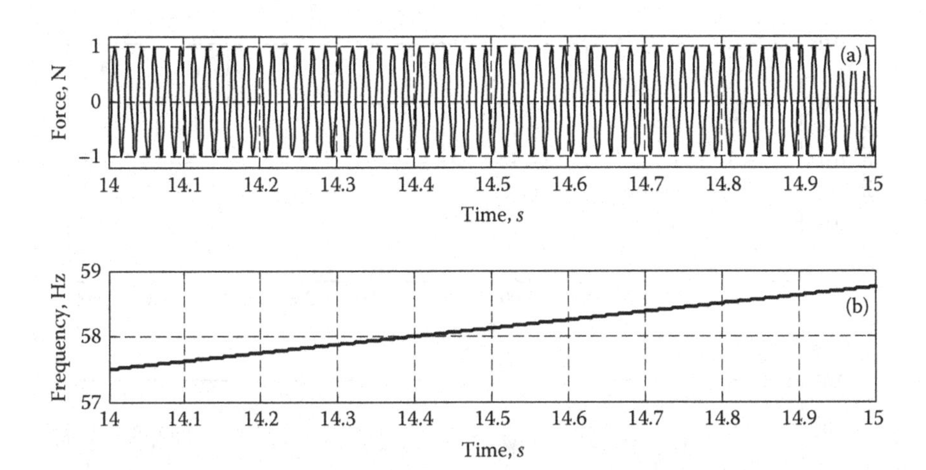

FIGURE 7.7
The zoomed view of the measured force in 14 to 15 s time.

sweep-sine excitation times 10 cycles), $N = 16 \times 1024 = 16{,}384$ data points for the FFT calculation were used, and then 10 averaging without any overlap was done. This leads to the frequency resolution $df = \frac{f_s}{N} = \frac{1024}{16384} = 0.0625$ Hz and the Nyquist frequency $f_q = \frac{f_s}{2} = 512$ Hz. The calculated averaged amplitude spectrum for the force data is shown in Figure 7.8(a), and the zoomed view in the frequency range 40 to 60 Hz in Figure 7.8(b). Similarly, the calculated amplitude spectrum for the acceleration response, and the frequency response function (FRF) and coherence for the input force and the output response are shown in Figures 7.9 to 7.11. Coherence is nearly 1 in the excitation frequency range of 40 to 60 Hz, confirming the fact that the force and response are linearly related and the quality of the measurements is very good. The only peak at 49.63 Hz seen in the acceleration spectrum and the FRF plot may be the natural frequency of the system.

7.4.3 Step 3: Identification of Natural Frequency from the FRF Plot

To identify whether the peak seen at 49.63 Hz is a natural frequency or not, the theory of the SDOF system discussed in Chapter 5 has been used here. The equation of displacement, Equation (2.37), in the frequency domain can be rearranged as

$$\frac{y(f)}{F(f)} = \frac{1}{\left((k - m\omega^2\right) + jc\omega} = \frac{(1/k)}{\left(1 - \left(\frac{f}{f_n}\right)^2\right) + j\left(2\zeta\frac{f}{f_n}\right)} \tag{7.1}$$

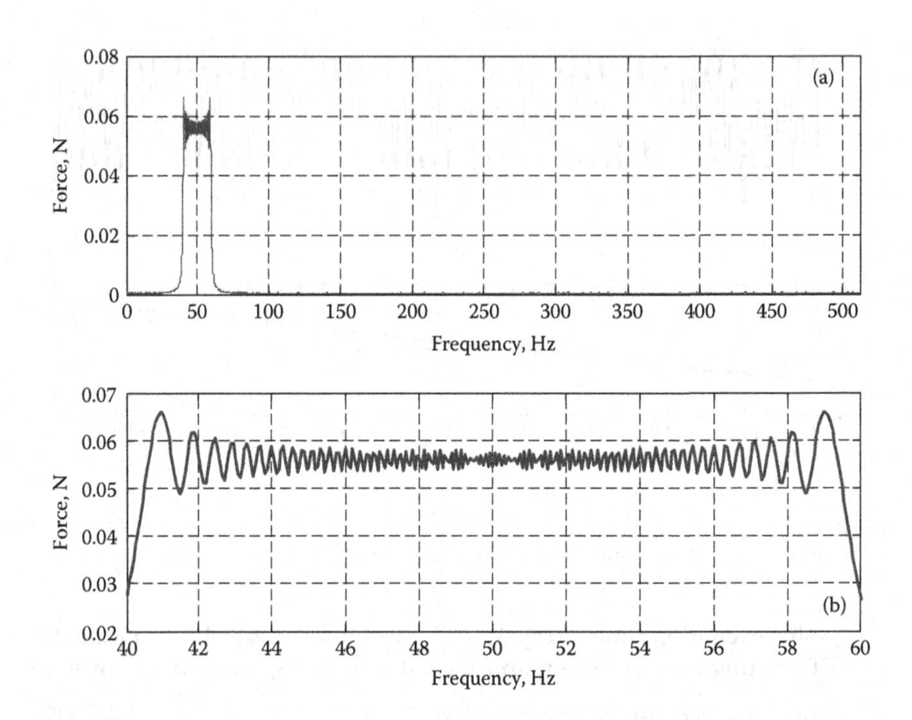

FIGURE 7.8
(a) Spectrum of the measured force. (b) Zoomed view in 40 to 60 Hz.

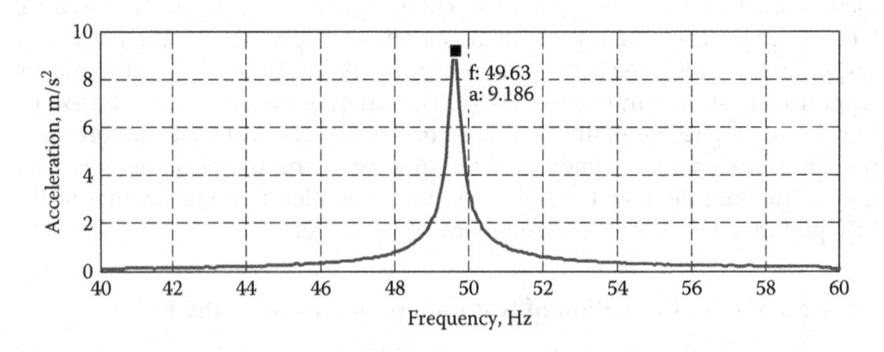

FIGURE 7.9
Spectrum of the measured acceleration response.

Equation (7.1) is nothing but the FRF of the system where the frequency, f, is the exciting frequency, and hence $F(f)$ and $y(f)$ are the applied force and the displacement response of the system at f. f_n is the natural frequency. The real part, $\left(k - m\omega^2\right)$ in Equation (7.1), is nothing but the resultant of the inertia and stiffness force in the system and the imaginary part; $c\omega$ is related

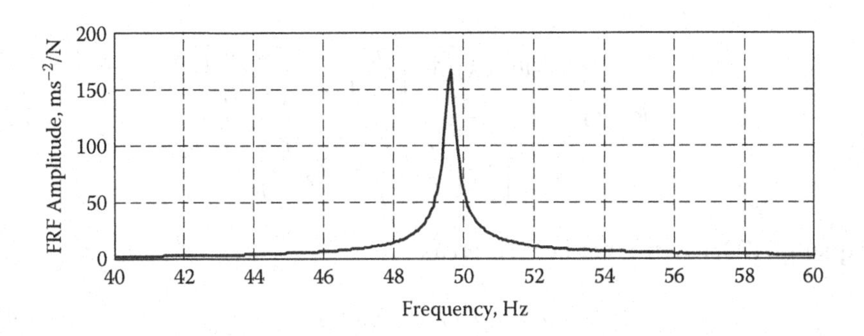

FIGURE 7.10
FRF plot.

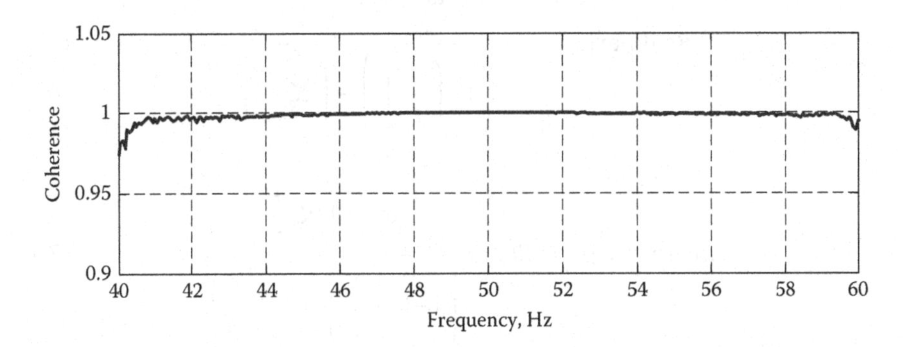

FIGURE 7.11
Coherence plot.

to the damping force. Since the measured response is the acceleration, the FRF in Equation (7.2) in terms of the acceleration can be written as

$$FRF(f) = \frac{a(f)}{F(f)} = \frac{-\omega^2/k}{\left(1-\left(\frac{f}{f_n}\right)^2\right)+j\left(2\zeta\frac{f}{f_n}\right)} \tag{7.2}$$

The FRF in Equation (7.2) is also known as inertance of the system. Equation (7.2) can also be written as

$$FRF(f) = \frac{-\omega^2/k}{\left(1-\left(\frac{f}{f_n}\right)^2\right)+j\left(2\zeta\frac{f}{f_n}\right)} \times \frac{\left(1-\left(\frac{f}{f_n}\right)^2\right)-j\left(2\zeta\frac{f}{f_n}\right)}{\left(1-\left(\frac{f}{f_n}\right)^2\right)-j\left(2\zeta\frac{f}{f_n}\right)}$$

$$FRF(f) = \frac{(-\omega^2/k)\left(1-\left(\frac{f}{f_n}\right)^2\right) - j\left(2\zeta\frac{f}{f_n}\right)}{\left(1-\left(\frac{f}{f_n}\right)^2\right)^2 + \left(2\zeta\frac{f}{f_n}\right)^2} \tag{7.3}$$

Equation (7.3) can further be written as

$$FRF(f) = A(f) + jB(f) \tag{7.4}$$

where

$$\text{Real part, } A(f) = \frac{(-\omega^2/k)\left(1-\left(\frac{f}{f_n}\right)^2\right)}{\left(1-\left(\frac{f}{f_n}\right)^2\right)^2 + \left(2\zeta\frac{f}{f_n}\right)^2} \tag{7.5}$$

$$\text{Imaginary part, } B(f) = \frac{(\omega^2/k)\left(2\zeta\frac{f}{f_n}\right)}{\left(1-\left(\frac{f}{f_n}\right)^2\right)^2 + \left(2\zeta\frac{f}{f_n}\right)^2} \tag{7.6}$$

and the phase angle between the response and the force at frequency, f, is

$$\varphi(f) = \tan^{-1}\left(\frac{B(f)}{A(f)}\right) = \tan^{-1}\left(\frac{\left(2\zeta\frac{f}{f_n}\right)}{\left(1-\left(\frac{f}{f_n}\right)^2\right)}\right) \tag{7.7}$$

Let's assume that the peak at the frequency 49.63 Hz in the FRF inertance plot in Figure 7.12 is a natural frequency, f_n. It is also assumed that the frequency of the applied force frequency f is now equal to f_n; then the real part, imaginary part, and phase of the FRF from Equations (7.5) to (7.7) are:

$$\text{Real part, } A(f_n) = 0$$

$$\text{Imaginary part, } B(f_n) = \frac{(\omega_n^2/k)}{2\zeta}$$

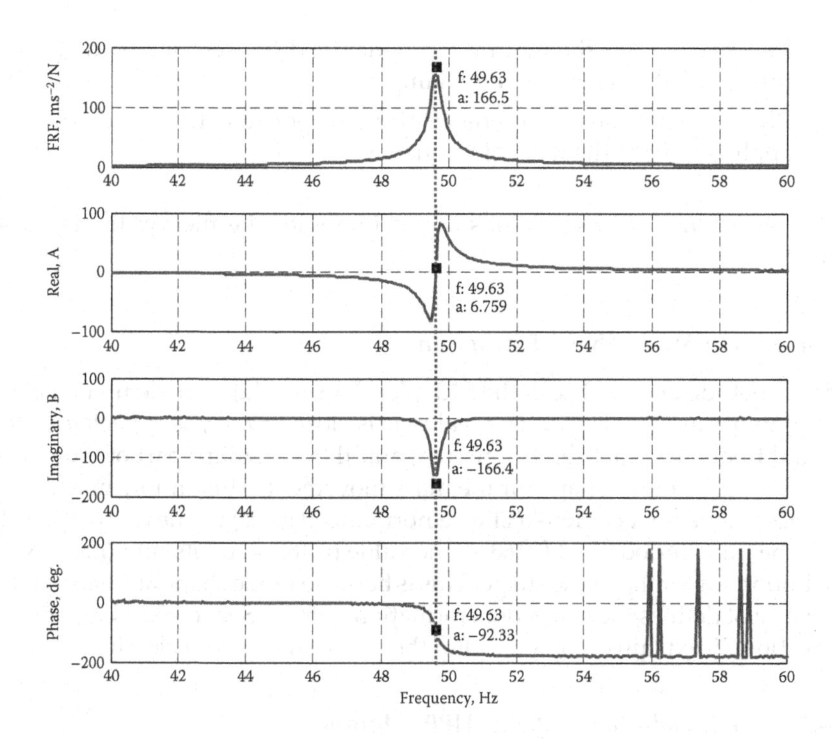

FIGURE 7.12
FRF plot showing amplitude, real, imaginary, and phase.

and the phase at frequency, f_n,

$$\varphi(f_n) = \tan^{-1}\left(\frac{B(f_n)}{A(f_n)}\right) = \tan^{-1}\left(\frac{(2\zeta)}{0}\right) = \pm 90^\circ \text{ (depending on force and response)}$$

Hence, it is clear from the above simple calculations that the FRF real part will be zero. The imaginary part may show either a positive or negative peak, and phase becomes equal to 90° at the resonance, i.e., when the exciting frequency equals the natural frequency. The peak at 49.63 Hz meets all these criteria, as clearly seen in Figure 7.12, and the frequency 49.63 Hz is the natural frequency of the SDOF system, as shown in Figure 7.2. The physical significance of the terminologies in the context of modal analysis can be summarized as follows:

1. **Real part:** It is basically the resultant of the stiffness force and the inertia force. A zero value indicates that the inertial force in the system is completely balanced by the stiffness force at the natural frequency. Refer to the theory section related to SDOF in Chapter 2 and Figure 2.14.
2. **Imaginary part:** The peak at the natural frequency shows that the applied force is completely balanced by the damping force. This means that the system completely loses its mechanical strength

(real part = 0) at the natural frequency and the response can go to infinity in the absence of any damping.

3. **Phase:** A ±90° phase difference of the response with respect to the applied force at the natural frequency.

The experimental observations are in line with the theory discussed in Chapter 2.

7.4.4 Step 4: Mode Shape Extraction

The amplitude at 49.63 Hz in the FRF plot (Figure 7.12) is made up of a complex number (6.759 – 166.4*j*). The real part is close to zero, as expected, at the natural frequency, so it can be ignored, and the imaginary part of –166.4 can be used for the mode shape for the mass movement. This simply means that the mass movement is ±166.4 (±1 on a normalized scale) in one complete cycle with the time period $T = 1/49.63$ s. The value of 166.4 and its unit (ms^{-2}/N) is not important for the mode shape. This is because mode shape at a natural frequency just defines the shape of deformation (refer to Section 3.8). Example 7.2 in Section 7.5 explains this concept of the mode shape more clearly.

7.4.5 Step 5: Half-Power Point (HPP) Method for Estimation Modal Damping

Chapter 2 discussed the theory of using the free decay response for a SDOF system for the estimation of damping using logarithmic decrement, but it is generally useful for a SDOF system or when the system is vibrating at a single frequency. This is because in practice most of the systems are MDOF systems and have a number of natural frequencies, so the free decay motion may not be just associated with a single natural frequency. The free oscillation could have influence from many natural frequencies. Hence, a different approach is generally used for the estimation of damping. It is good to use all the FRF data from all measured points on a system at a natural frequency to estimate the system damping at that natural frequency, but this process requires a complex mathematical estimation method. This is included in most of the commercially available software on the experimental modal analysis. However, the half-power point (HPP) method is a simple and a good alternative for most practical purposes, which is discussed here.

The amplitude of the FRF of the displacement to the force in Equation (7.1) in terms of the nondimensional form is written as

$$Y_{ND}(f) = \frac{ky(f)}{F(f)} = \frac{1}{\sqrt{\left(1-\left(\frac{f}{f_n}\right)^2\right)^2 + \left(2\zeta\frac{f}{f_n}\right)^2}} \tag{7.8}$$

where $Y_{ND}(f)$ is the nondimensional amplitude of the FRF. At the natural frequency, $Y_{ND}(f_n) = \frac{1}{2\zeta}$, and in terms of power, $Y_{ND}^2(f_n) = \left(\frac{1}{2\zeta}\right)^2$, and so the half-power point will be $\frac{1}{2}\left(\frac{1}{2\zeta}\right)^2$ and the nondimensional amplitude becomes $\frac{1}{\sqrt{2}}\left(\frac{1}{2\zeta}\right) = 0.707\left(\frac{1}{2\zeta}\right)$. Substituting this half-power point value into Equation (7.8) gives

$$\frac{1}{2}\left(\frac{1}{2\zeta}\right)^2 = \frac{1}{\left(1-\left(\frac{f}{f_n}\right)^2\right)^2 + \left(2\zeta\frac{f}{f_n}\right)^2} \tag{7.9}$$

On simplification the solution of Equation (7.9) gives two roots for the frequency f: $f_1 = (1-\zeta)f_n$ and $f_2 = (1+\zeta)f_n$. They are depicted in the nondimensional power and amplitude plots of FRF in Figure 7.13. The damping ratio can then be estimated as

$$\zeta = \frac{f_2 - f_1}{2f_n} \tag{7.10}$$

Thus, the HPP method is used to estimate the system damping using the measured FRF amplitude plot, as shown in Figure 7.12. The frequencies

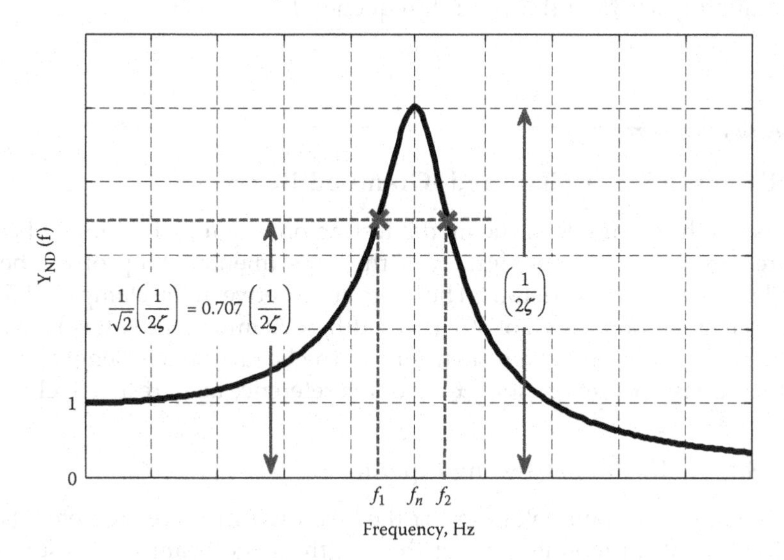

FIGURE 7.13
Nondimensional FRF plot showing peak and half-power amplitudes.

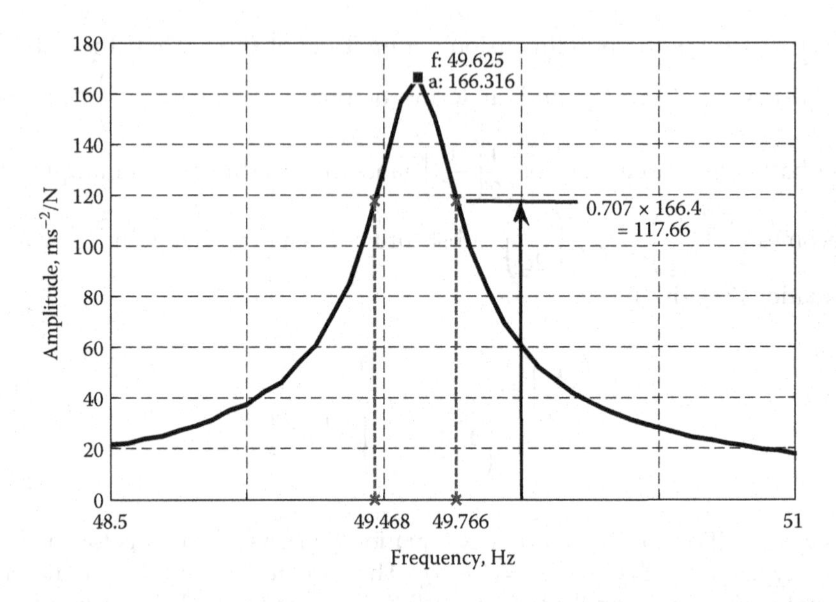

FIGURE 7.14
Zoomed view of the FRF plot in Figure 7.12.

related to the half-power amplitude at the natural frequency, 49.63 Hz, are found to be f_1 = 49.468 Hz and f_2 = 49.766 Hz. These frequencies are also marked in the zoomed view of the FRF plot in Figure 7.14. The damping for the system using Equation (7.10) is estimated as 0.003 (0.3%). Hence, the system damping is 0.3% at the natural frequency, f_n = 49.63 Hz.

7.5 Example 7.2: A Clamped–Clamped Beam

This is another example of the modal testing on a clamped–clamped beam. Figure 7.15 shows the photograph of the experimental setup of the beam together with clamping details. The beam span between the clamps is 2.54 m with a cross-sectional area of 30 mm (width) × 20 mm (depth). It is made up of aluminum. The physical dimensions of the beam, namely, length-width-depth, are assumed to be along *x-y-z* of the reference axes, respectively.

7.5.1 Step 1: The Modal Testing Experiment

Figure 7.16 shows the photograph of the beam with 13 accelerometers (sensitivity 100 mV/g) mounted along the length of the beam to measure the response in the *z*-direction (depth), i.e., lateral direction. The instrumented hammer with the force sensor has a sensitivity of 1.1 mV/N, as shown in

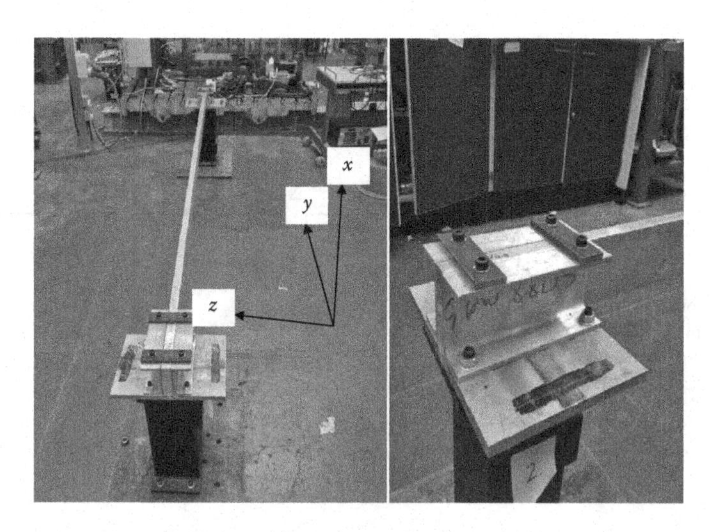

FIGURE 7.15
Example 7.2: Clamped–clamped beam.

FIGURE 7.16
Beam with accelerometers and instrumented hammer for modal testing.

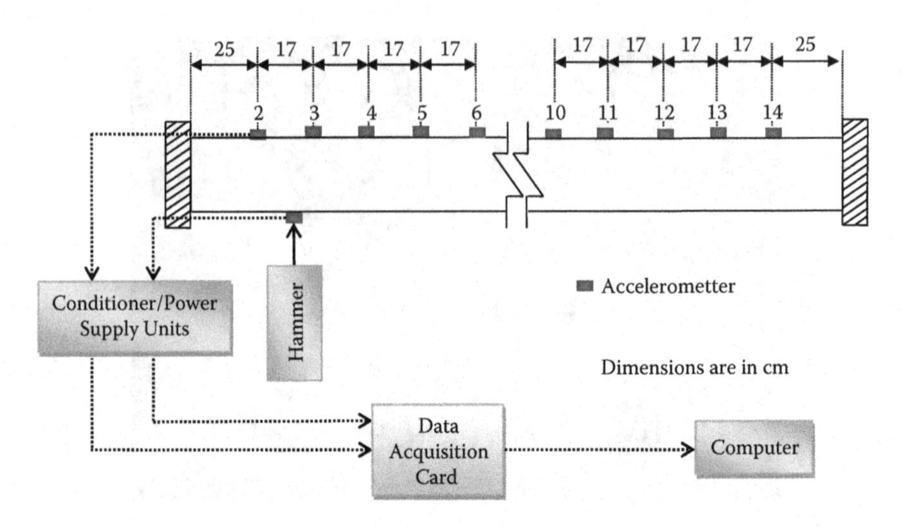

FIGURE 7.17
Schematic view of the experiment for the clamped–clamped beam.

Figure 7.16, and is used to excite the beam for the modal test. The schematic of the accelerometer locations on the beam together with associated instruments, data acquisition (DAQ) device, and the impulse hammer is shown in Figure 7.17. A number of impulses are given to excite the beam. The impulses together with the vibration responses from all 13 accelerometers are simultaneously recorded into the PC through a 16-input-channel 16-bit DAQ device using the sampling frequency of 5000 Hz. A typical measured impulsive excitation force and the measured response at location 3 are shown in Figure 7.18.

7.5.2 Step 2: Computation of the Amplitude Spectra, FRFs, and Coherences

Averaged acceleration spectra are calculated for both the force and the 13 response data initially, and then the FRFs and the coherences are calculated for each accelerometer datum with respect to the applied force. These calculations are carried out as per Equations (6.25) and (6.27). Since the applied force and the decay responses are short-duration phenomena, special care is needed for the signal processing to compute the averaged spectra, FRFs, and coherences. A trigger level of 1.0 N has been used as the starting point of acquisition of data for the force and the corresponding responses, as shown in Figure 7.19 to compute the FT using Equation (6.16). The time length seen in Figure 7.19 corresponds to the segment size $N = 8192$. This process is repeated for all the applied forces and the corresponding acceleration responses from all 13 accelerometers. The averaged spectra, FRF data, and coherences are then calculated using Equations (6.14), (6.25), and (6.27), respectively. Typical spectra for the applied force and the

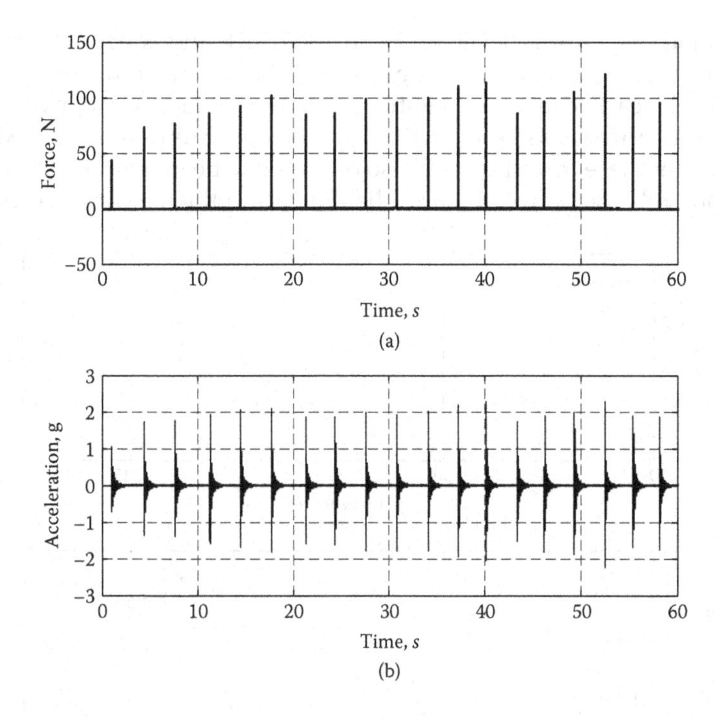

FIGURE 7.18
Typical measured data. (a) Applied force by instrumented hammer. (b) Beam acceleration response measured by accelerometer at location 3.

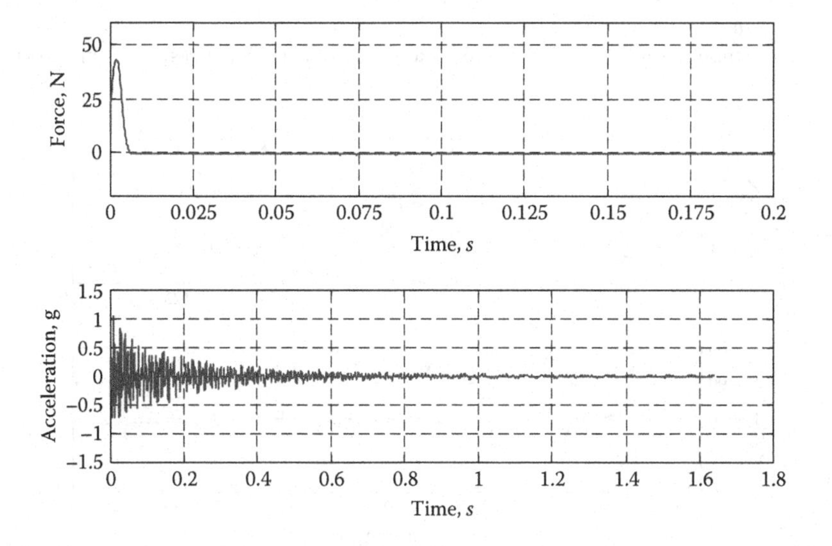

FIGURE 7.19
Trigger force and the corresponding response used for the FFT estimation for each applied force for averaging purposes.

acceleration response at location 3 and their FRF and coherence plots are shown in Figures 7.20 to 7.22, respectively. The spectrum of the excitation force shows nearly constant excitation up to the frequency range of 200 Hz, and there are three peaks seen in the response spectrum and in the FRF plot. Coherence at these three peaks is also observed to be more than 0.8, indicating a good relation between the applied force and the measured response.

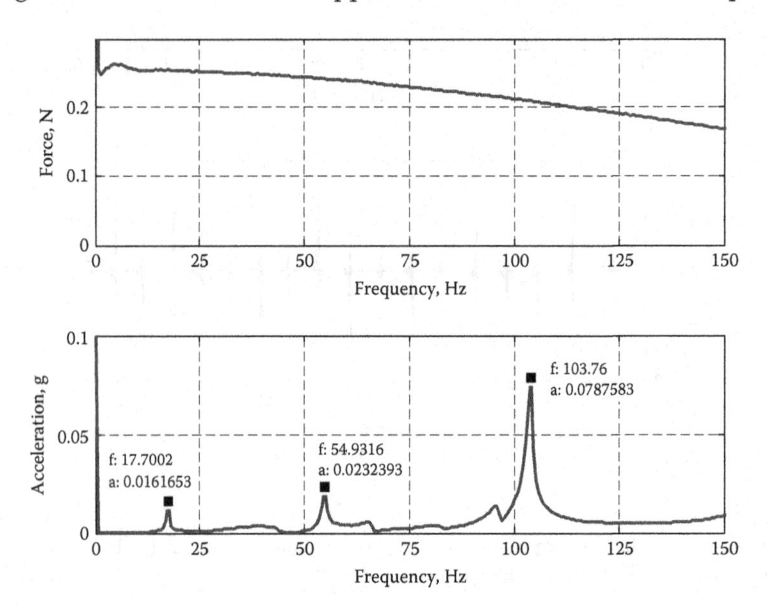

FIGURE 7.20
Averaged amplitude spectra of the applied force (a) and the beam response (b) shown in Figure 7.18.

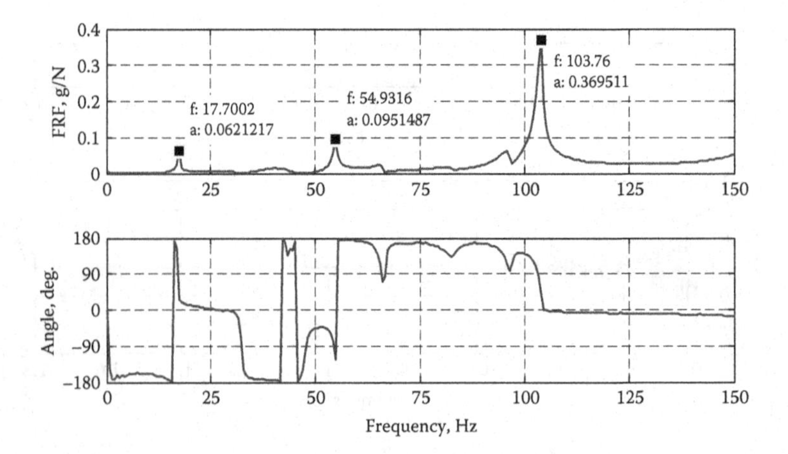

FIGURE 7.21
Measured FRF (inertance) at the location 3. (a) FRF amplitude. (b) FRF phase.

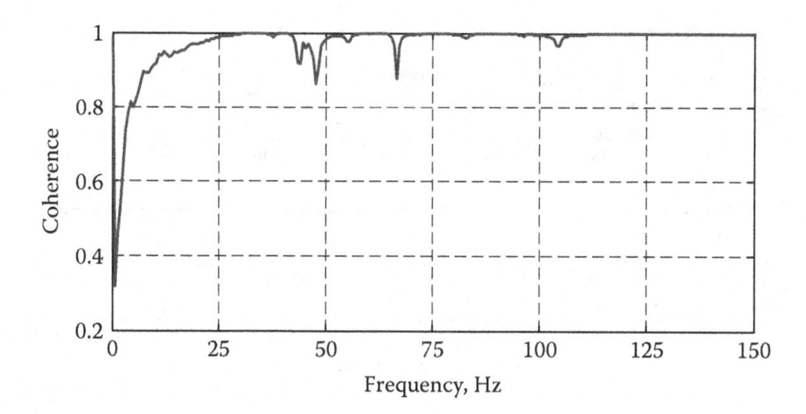

FIGURE 7.22
Coherence between the applied force and the response shown in Figure 7.18.

7.5.3 Step 3: Natural Frequency Identification from the FRF Plots

Now the three distinct peaks in the response spectrum and the FRF plot in Figures 7.20 and 7.21 are assumed to be the natural frequencies for the experimental beam to begin the identification process. The frequency values for these three peaks are $f_{p1} = 20$ Hz, $f_{p2} = 80$ Hz, and $f_{p3} = 200$ Hz. Now each peak is analyzed based on the SDOF system, and hence the following examinations need to be done (as discussed in Section 7.3 (step 3)) for each peak of all 13 measurement locations using the FRF plots to confirm whether it is a natural frequency or not.

1. Real part must pass through zero.
2. Imaginary part must be either a +*ve* or −*ve* peak.
3. Phase angle of the measured response with respect to the applied force at the frequency peak must pass through ±90°.

The majority of the measured locations must meet the above criteria except when the measurement location is close to a node (zero deflection) location for any natural frequency/frequencies. It is to be noted that this approach is a very simple yet elegant procedure for the extraction of modal parameters from the experimental data, and at the same time quick and easy in assessment and generally works well for most cases.

The FRF at each measured location is examined for the above criteria for each frequency peak, and it is observed that all three frequencies satisfy the criteria to be natural frequencies. A few typical FRF plots are also shown in Figures 7.23 to 7.25 for better clarity. Hence, the frequencies are now identified as the first three natural frequencies in the lateral direction of the experimental beam.

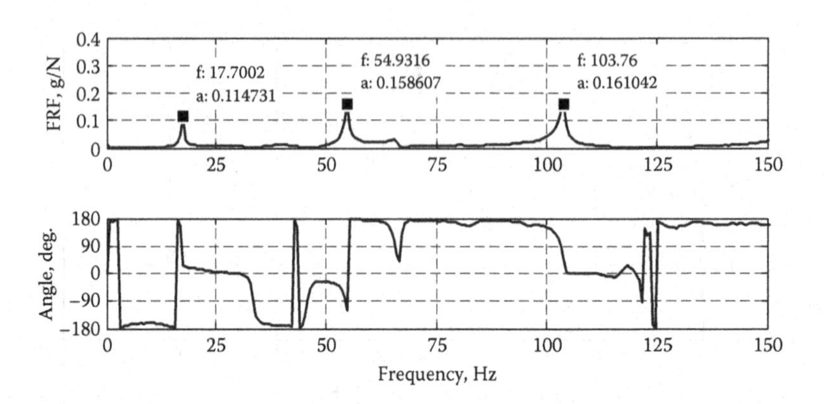

FIGURE 7.23
Measured FRF (inertance) at location 6.

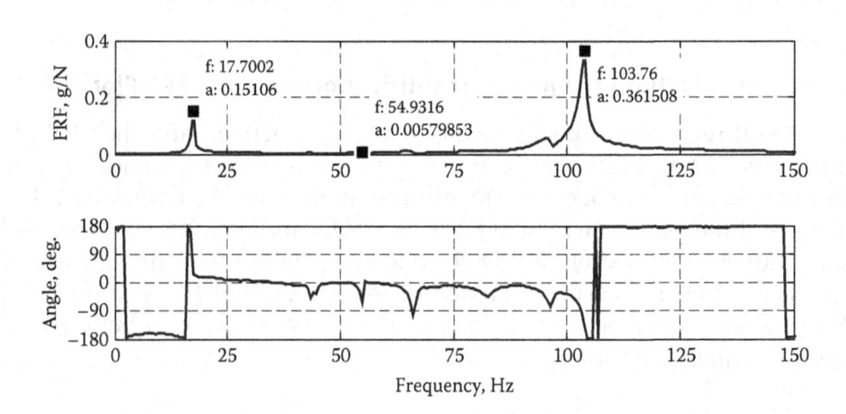

FIGURE 7.24
Measured FRF (inertance) at location 8 (beam center).

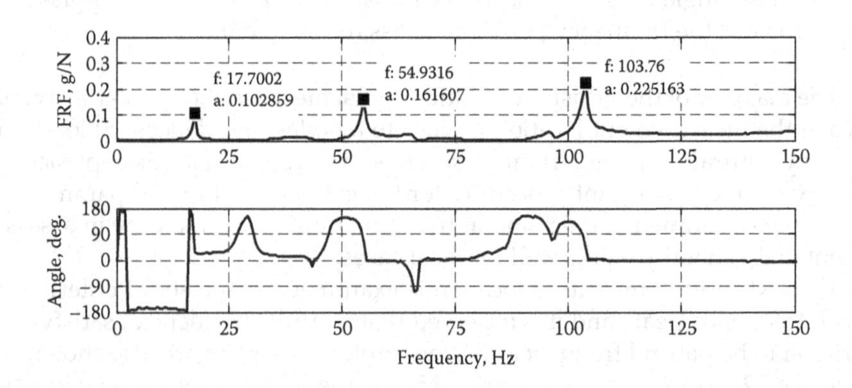

FIGURE 7.25
Measured FRF (inertance) at location 11.

7.5.4 Step 4: Extraction of Mode Shapes

As discussed earlier, the imaginary values of the FRF plot at each measured location for each natural frequency can be used for generating the mode shape. The mode shapes of the three identified natural frequencies directly measured from the imaginary part of the FRF plots are listed in Table 7.1. The normalized values of the mode shapes are also listed in Table 7.1.

It is to be noted that there are no measurements at the clamped locations (1 and 15) on either side of the beam. So, it is assumed that there are no deflections at the clamped locations. The mode shape data (Table 7.1) in terms of the imaginary values from the measured FRFs are denoted as vectors $\varphi_I = \begin{bmatrix} \varphi_{I1} & \varphi_{I2} & \varphi_{I3} \end{bmatrix}$, as the mode shapes for the first (mode 1), second (mode 2), and third (mode 3), respectively. Since the mode shape is the shape of deformation at each mode for the structure, the measured values are not important; hence, the mode shapes are normalized between +1 and –1 by dividing mode shape data at a natural frequency by their absolute maximum value. For the present case the normalized mode shape data for the three modes are estimated as $\varphi = \begin{bmatrix} \dfrac{\varphi_{I1}}{0.0659} & \dfrac{\varphi_{I2}}{0.1398} & \dfrac{\varphi_{I3}}{0.2685} \end{bmatrix} = \begin{bmatrix} \varphi_1 & \varphi_2 & \varphi_3 \end{bmatrix}$, which are also listed in Table 7.1. The measured mode shapes for mode 1, mode 2, and mode 3 are also shown graphically in Figure 7.26. The modal assurance criterion (MAC) in Equation (3.47) is also calculated between

TABLE 7.1

Measured Imaginary Values at the Natural Frequencies and the Normalized Mode Shapes

Measurement Location		Imaginary Values from Measured FRFs			Normalized Mode Shapes		
Position	Distance, cm	φ_{I1}	φ_{I2}	φ_{I3}	φ_1	φ_2	φ_3
1	0	0	0	0	0	0	0
2	25	+0.0156	–0.0211	+0.1534	+0.2367	–0.1509	0.5713
3	42	+0.0261	–0.0784	+0.2396	+0.3961	–0.5608	0.8924
4	59	+0.0385	–0.1252	+0.2313	+0.5842	–0.8956	0.8615
5	76	+0.0492	–0.1353	+0.1139	+0.7466	–0.9678	0.4242
6	93	+0.0573	–0.1195	–0.0550	+0.8695	–0.8548	–0.2048
7	110	+0.0633	–0.0718	–0.1978	+0.9605	–0.5136	–0.7367
8	127	+0.0659	–0.0031	–0.2492	+1.0000	–0.0222	–0.9281
9	144	+0.0637	+0.0664	–0.1783	+0.9666	0.4750	–0.6641
10	161	+0.0580	+0.1208	–0.0183	+0.8801	0.8641	–0.0682
11	178	+0.0461	+0.1398	+0.1534	+0.6995	1.0000	0.5713
12	195	+0.0368	+0.1387	+0.2685	+0.5584	0.9921	1.0000
13	212	+0.0247	+0.1015	+0.2600	+0.3748	0.7260	0.9683
14	229	+0.0133	+0.0491	+0.1459	+0.2018	0.3512	0.5434
15	254	0	0	0	0	0	0

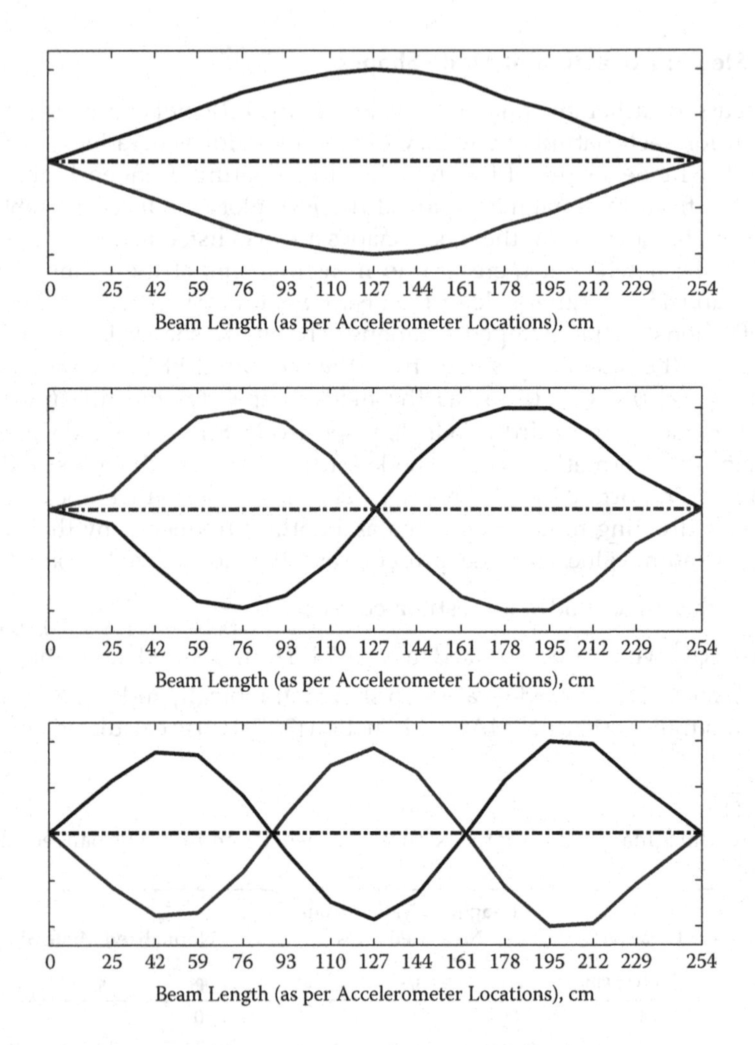

FIGURE 7.26
Mode shapes for the clamped–clamped beam of Example 7.2.

the experimental modes. It is shown in the 3D bar graph in Figure 7.27 and also tabulated in Table 7.2. It is obvious from Figure 7.27 and Table 7.2 that cross-coupling mode 1 with modes 2 and 3, and between modes 2 and 3 is nearly negligible; hence, the modes are orthogonal to each other.

7.5.5 Step 5: Estimation Modal Damping

Here again, the HPP method is used to compute the modal damping ratio at all 13 measurement locations for mode 1 to mode 3 using the measured FRF inertance plots. The estimated damping values are listed in Table 7.3.

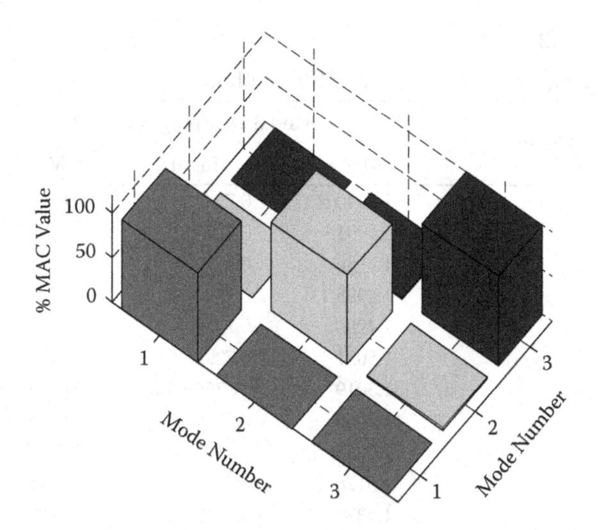

FIGURE 7.27
Bar graph for percent MAC values.

TABLE 7.2
MAC Values (%) between the Modes of Example 7.2

	Mode 1	Mode 2	Mode 3
Mode 1	100	0.0067	0.1187
Mode 2	0.0067	100	1.8183
Mode 3	0.1187	1.8183	100

The estimated damping ratios at each mode at all 13 measured locations may not be exactly the same due to different amplification and noise content at each measurement location. A few measurement locations for mode 2 and mode 3 either at the node locations or close to the nodal locations may show significantly high damping, as listed in Table 7.3. These values should be ignored. Therefore, it is always advisable to choose the lowest damping ratio at each mode as a conservative approach. Hence, the measured damping ratios at mode 1 to mode 3 are 1.494, 0.716, and 0.592%, respectively. These values are also shown in Table 7.3.

7.6 Industrial Examples

Three typical real industrial examples are discussed here briefly to further aid the understanding of in situ modal testing for the identification

TABLE 7.3

Damping Ratio Estimated by HPP Method

Measurement Location	Modal Damping Ratios (%)		
	Mode 1	**Model 2**	**Mode 3**
2	1.4940 (L)	0.7594	0.6161
3	1.4991	0.7223	0.6198
4	1.4986	0.7193	0.6258
5	1.4988	0.7190	0.6504
6	1.4986	0.7203	0.5916 (L)
7	1.5021	0.7294	0.6114
8	1.5019	21.7569 (NL)	0.6233
9	1.5062	0.7194	0.6468
10	1.5085	0.7162 (L)	3.7727 (NL)
11	1.5178	0.7177	0.5918
12	1.5279	0.7187	0.6102
13	1.5444	0.7234	0.6243
14	1.5773	0.7258	0.6344
Final (lowest) damping	1.4940	0.7162	0.5916

Note: L = lowest value, NL = a proximity to node location.

of the modal parameters. The following instruments are used for all three examples:

8-input-channel digital tape recorder
Charge accelerometer (sensitivity 100 pC/g)—7 nos.
Accelerometer signal conditioner—7 nos.
Impulse hammer with a force sensor (sensitivity 23 mV/N)—1 no.
Power supply unit for the force sensor—1 no.
Two-channel FFT analyzer

It is to be noted that the construction of the instrumented hammer was the same as shown in Figure 5.6. It was like a sledgehammer weighing around 3 kg with the hammer handle and was around 1 m long, often required for modal testing on large structures, machines, etc. The pendulum-type arrangement for the hammer was used to apply the impulsive loads for the discussed three examples, as is typically illustrated in Figure 7.28. Many different hammer heads were used for these examples depending on the requirement of the frequency band of excitation. The soft hammer head was used for Example 7.3 to excite the tube natural frequencies up to 150 to 200 Hz, and a still softer hammer tip was used for Examples 7.4 and 7.5 to excite frequencies up to a range of 30 to 35 Hz. It is because the structures were sufficiently large and expected to have a few natural frequencies at much lower frequency values.

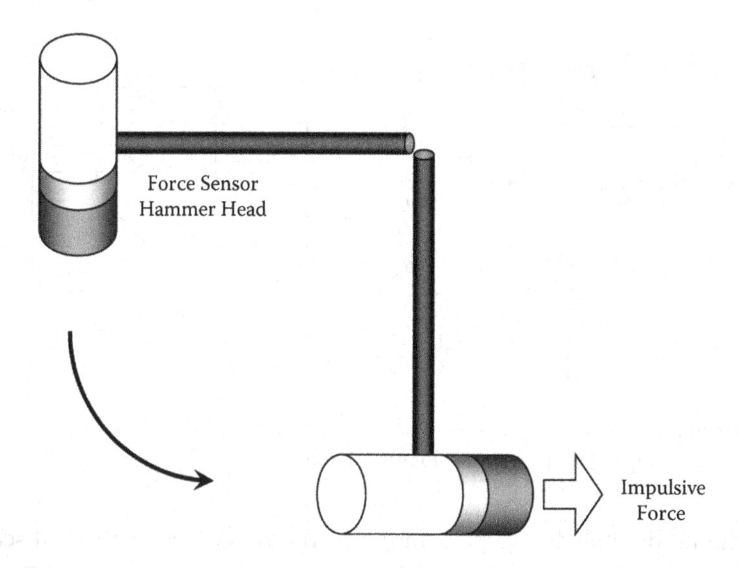

FIGURE 7.28
Pendulum arrangement used for the hammer to apply impulsive excitation to the large objects for modal testing.

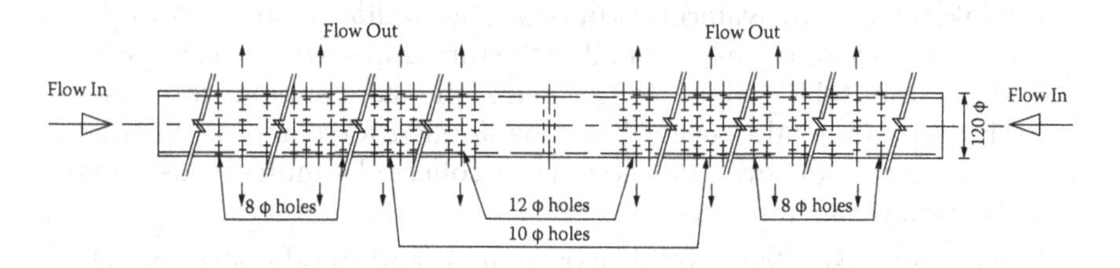

FIGURE 7.29
Typical configuration of the sparger tube.

7.6.1 Example 7.3: Sparger Tube

This is related to the earlier study (Sinha and Moorthy, 1999), which is discussed here. The moderator sparger tube is used to carry and distribute the moderator inside the reactor vessel. To achieve uniform distribution along the length, perforations of different sizes and pitches are made on the tube as shown in Figure 7.29. It is a tube of 120 mm diameter and 4980 mm long and is perforated with holes of different sizes (1350 holes of 12 mm diameter, 750 holes of 10 mm diameter, and 720 holes of 8 mm diameter) and at different pitches (see Figure 7.29).

1. **Experimental setup:** It was impossible to do modal testing on the tube inside the reactor vessel and also difficult to estimate the natural frequencies theoretically for the tube when submerged into

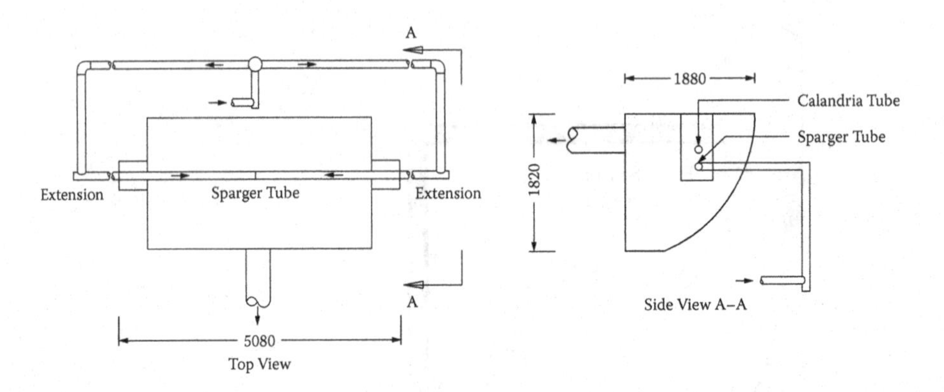

FIGURE 7.30
Experimental setup of the sparger tube for modal tests.

the moderator due to perforation in the tube. Hence, the full-scale experimental setup of the tube in Figure 7.30 was used for the modal testing. The outer tank simulates one-quarter of the reactor vessel.

2. **Instrumentation and modal experiments:** Modal tests were conducted on the tubes for the nonsubmerged (in air) and submerged (tank filled with water) conditions. The excitation location by the hammer was kept near one end of the tube and seven accelerometers along the tube length to measure the acceleration responses. The experimental data for both the applied force and acceleration responses from seven locations were collected simultaneously into the digital data tape recorder.
3. **Data analysis for modal parameters and mode shapes:** The recorded data in the tape recorder were played back and analyzed through a two-channel FFT analyzer. Measured responses at each location were analyzed with the applied force to estimate the spectrum, FRF, and coherence one by one for all seven measured locations. Each frequency peak in the FRF plots was then examined to identify the natural frequencies of the tube. Typical measured FRF plots for the tube with the nonsubmerged and submerged conditions are shown in Figures 7.31 and 7.32. The identified natural frequencies and modal damping for each identified mode are listed in Table 7.4, and corresponding mode shapes in Figure 7.33.

7.6.2 Example 7.4: Shutoff Rod Guide Tube

This is related to the earlier study (Sinha and Moorthy, 1999; Sinha et al., 2003) discussed here. It is a zircaloy tube of 91 mm outer diameter and 6570 m long and is perforated with 1392 holes of 6 mm diameter at different pitches (see Figure 7.34). It is used to guide movement of the shutoff rod inside the reactor

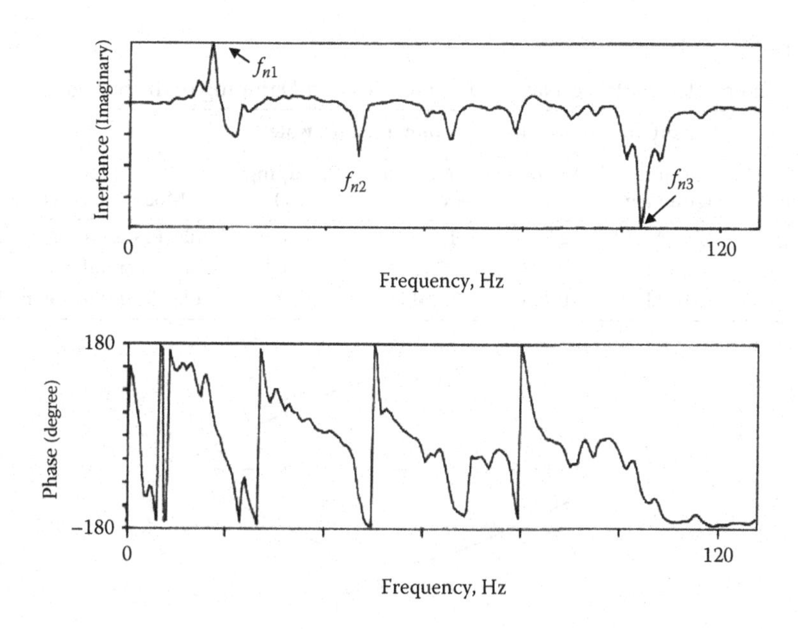

FIGURE 7.31
Experimental FRF (inertance) for the nonsubmerged sparger tube (in air) condition.

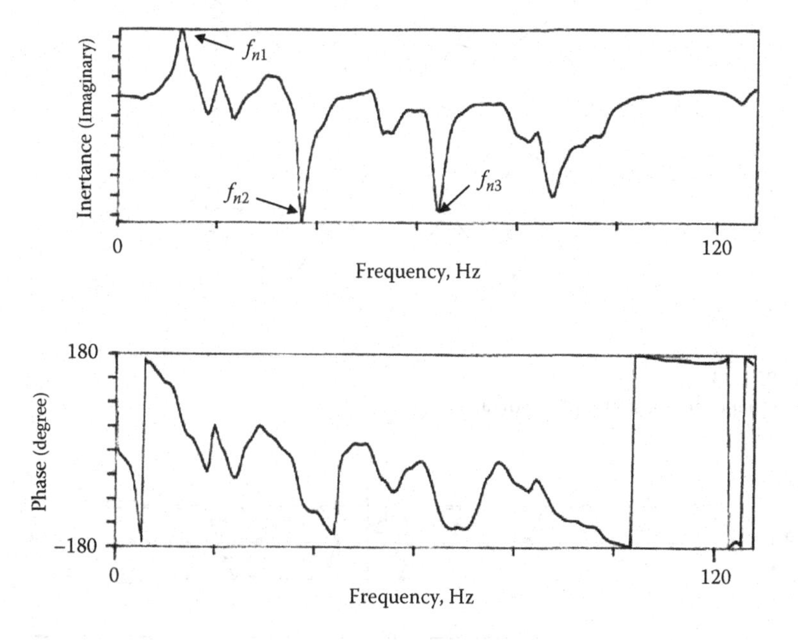

FIGURE 7.32
Experimental FRF (inertance) for the submerged sparger tube (in water) condition.

TABLE 7.4
Experimentally Identified Natural Frequencies and Damping of the Sparger Tube

Mode		Nonsubmerged (in air)		Submerged (in water)		
		Natural Frequency	Damping (%)	Natural Frequency	Damping (%)	Mode Interpretation
1	f_{n1}	17.24 Hz	2.730	12.65 Hz	3.660	Tube first flexural mode
2	f_{n2}	46.56 Hz	0.668	37.12 Hz	1.960	Tube second flexural mode
3	f_{n3}	103.9 Hz	0.496	64.04 Hz	1.200	Tube third flexural mode

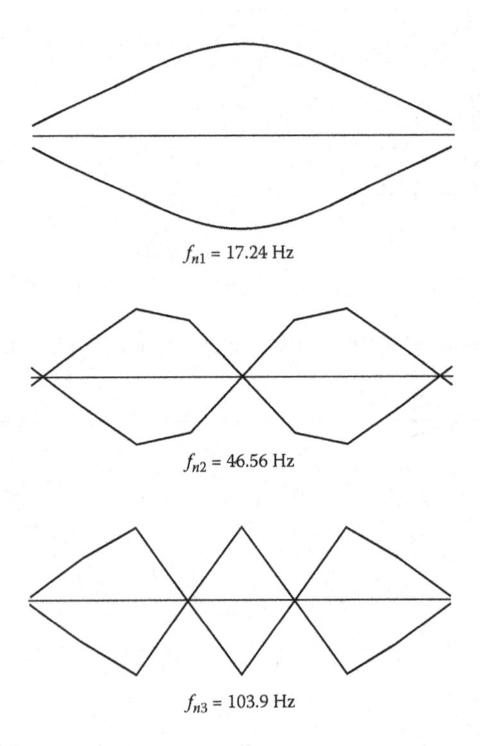

FIGURE 7.33
Mode shapes of the sparger tube (in air).

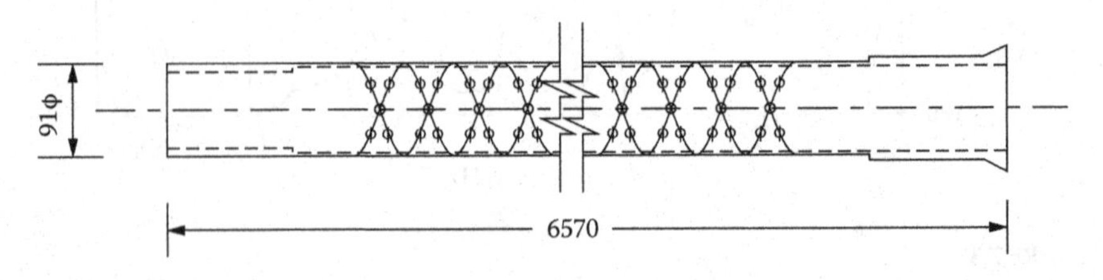

FIGURE 7.34
Typical configuration of the shutoff rod guide tube.

vessel, which controls the nuclear reaction. The schematic of the test setup is shown in Figure 7.35. The setup consists of the shutoff rod guide tube and the surrounding mild steel (M.S.) tank of 890 mm inner diameter and 5840 mm long, which is placed coaxially to the guide tube.

The modal experiments on the guide tube were conducted by impulse-response tests. The impulse was given at one location, and the response was obtained using accelerometers all along the length of the tube. The response was not measured on the tank. The test was conducted when the M.S. tank was empty as well as filled with water. Once again, the recorded modal data on the data recorder were analyzed on the two-channel FFT analyzer, and each frequency peak in the FRF plots was examined for the natural frequencies. A typical measured FRF (inertance) for the nonsubmerged (in air) condition is shown in Figure 7.36. The identified natural frequencies are listed in Table 7.5, and the corresponding mode shapes are shown in Figure 7.37.

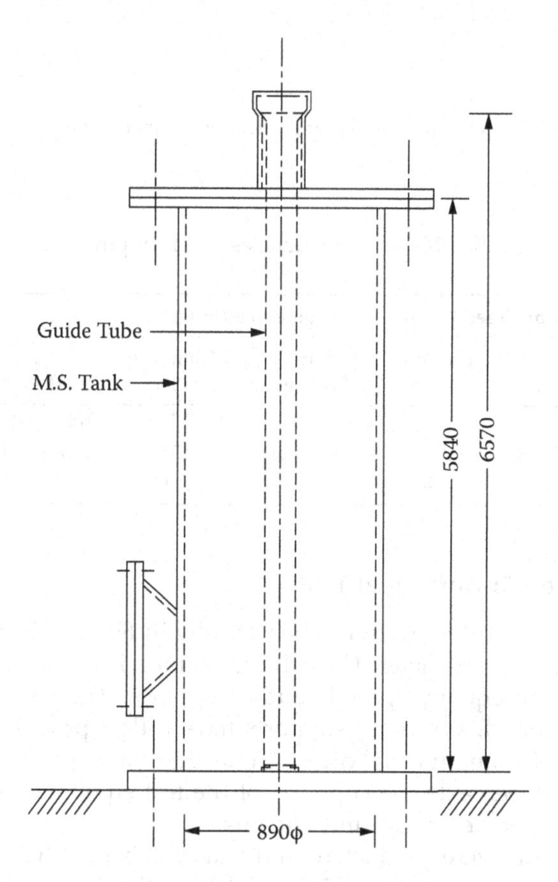

FIGURE 7.35
Experimental setup of the shutoff rod guide tube for modal tests.

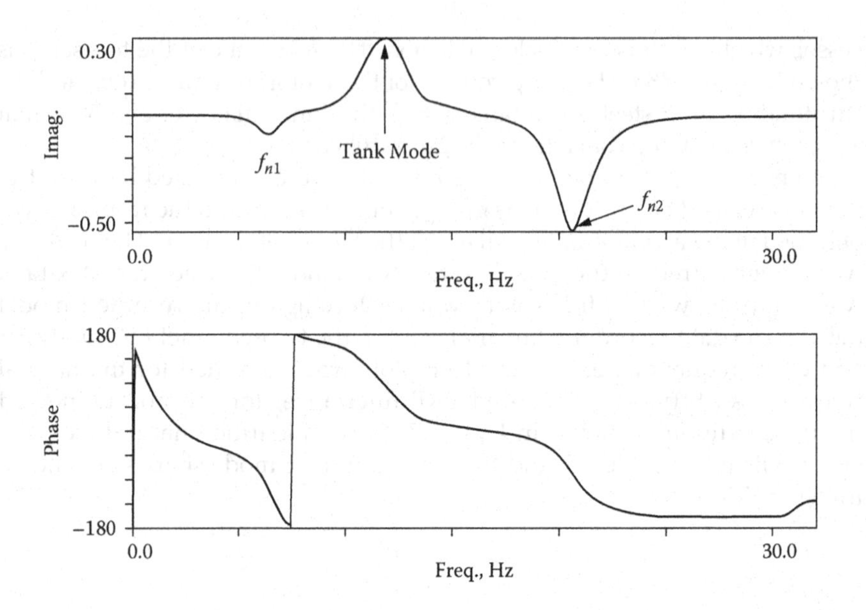

FIGURE 7.36
Experimental FRF (inertance) for the shutoff guide tube in air condition.

TABLE 7.5

Experimentally Identified Natural Frequencies and Damping of the Shutoff Rod Guide Tube

Mode		Nonsubmerged (in air)		Submerged (in water)		
		Natural Frequency	Damping (%)	Natural Frequency	Damping (%)	Mode Interpretation
1	f_{n1}	6.510 Hz	1.660	2.970 Hz	5.549	Tube first flexural mode
2	f_{n2}	20.997 Hz	2.311	9.493 Hz	2.282	Tube second flexural mode
3	f_{n3}	—	—	20.976 Hz	2.493	Tube third flexural mode

7.6.3 Example 7.5: Horizontal Tank

The large storage tank as shown schematically in Figure 7.38 is 25.5 m long and has a 4 m outer diameter. The cylindrical tank is mounted horizontally. It rests on seven equally spaced saddle supports. The central support is fixed to the ground, and other supports have rollers providing release for translation in the longitudinal direction for thermal expansion of the tank. The length between extreme supports of the test equipment is about 24 m and the filled tank weighs around 370 tons.

The modal tests were conducted on the tank to extract the modal parameters. More details are given in Sinha and Moorthy (2000). The experimental modal analysis was carried out using the impulse-response method.

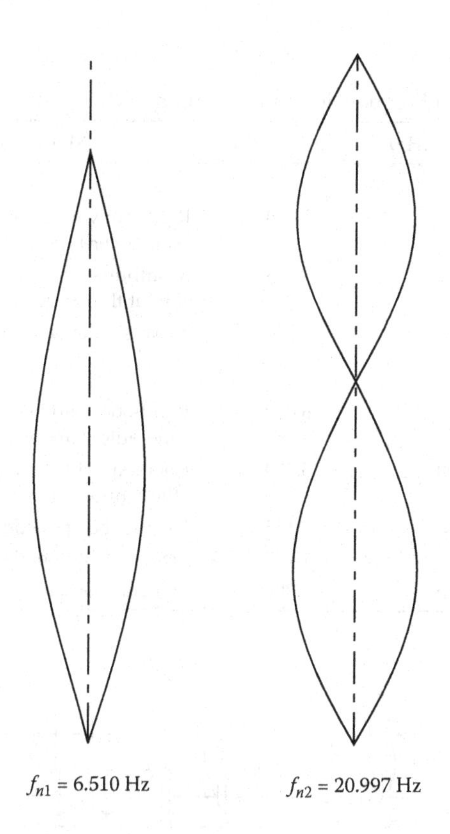

FIGURE 7.37
Mode shapes of the shutoff rod guide tube (in air).

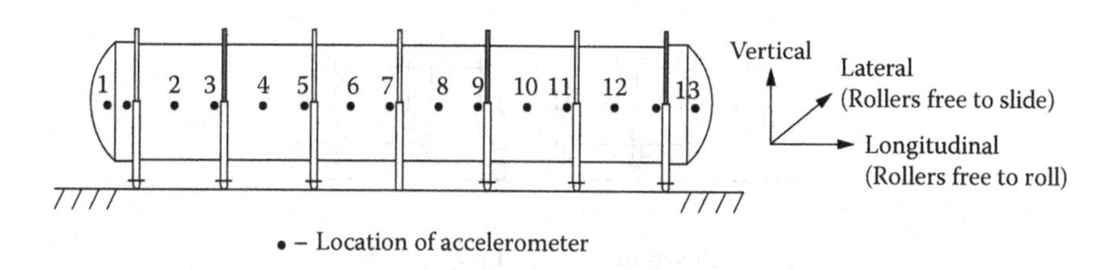

FIGURE 7.38
Schematic of the long horizontal tank.

The experimental data were processed through a two-channel FFT analyzer. The natural frequencies, mode shapes, and modal dampings were identified and estimated through the experimentally obtained frequency response functions (FRFs). The results are briefly summarized in Table 7.6. Few typical mode shapes along with their natural frequencies, as obtained experimentally in the longitudinal and lateral directions, are shown in Figures 7.39 and 7.40, respectively.

TABLE 7.6

Experimental Natural Frequencies and Damping of the Tank

Mode No.	Frequency (Hz)	Damping (%)	Mode Interpretation
Longitudinal Direction			
1	7.500	2.910	Rigid-body mode of the tank on its cantilever-type support
2	19.250	1.416	(Compression-expansion mode of tank about the central support)
3	22125	1.240	Expansion-expansion mode of the tank
Lateral Direction			
1	5.625	6.689	Rigid-body mode of the tank on its central and outer supports
2	19.000	5.220	Rotation of the tank due to twisting of the central support
3	25.250	1.155	Flexural beam mode 1 of the tank
4	26.875	1.660	Flexural beam mode 2 of the tank
5	32.875	2.202	Flexural beam mode 3 of the tank

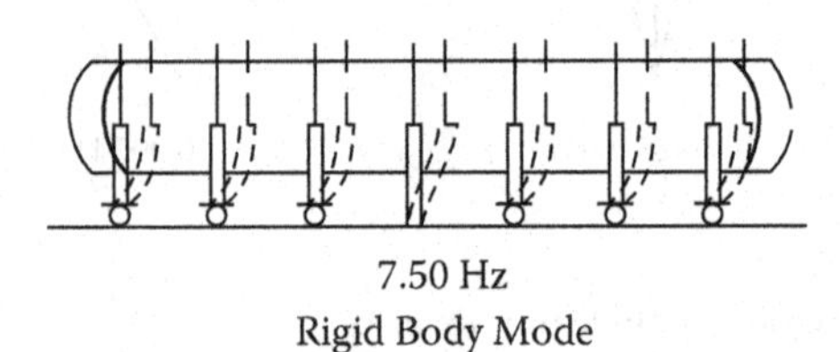

7.50 Hz
Rigid Body Mode

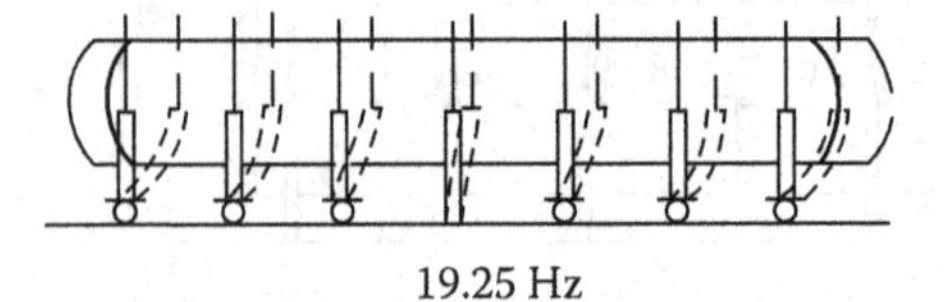

19.25 Hz
Unsymmetric Rigid Body Mode

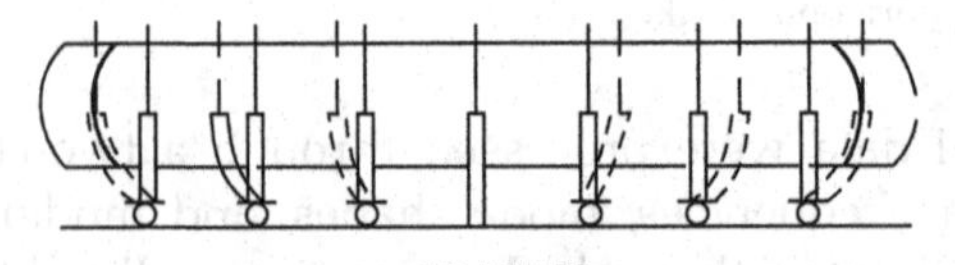

22.125 Hz
Expansion–Expansion Mode

FIGURE 7.39
Mode shapes of the tank in the longitudinal direction.

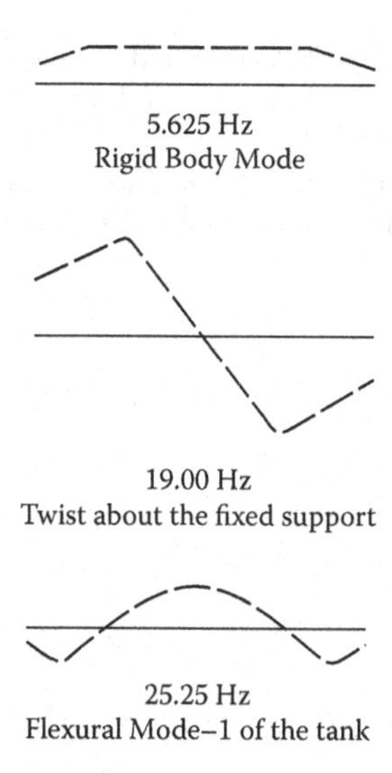

FIGURE 7.40
Mode shapes of the tank in the lateral direction.

7.7 Summary

A few important things are worth noting here. A very practical and fundamental approach for the extraction of the modal parameters (natural frequencies, modal dampings, and mode shapes) based on the concepts of the SDOF system has been discussed in this chapter. A total of five examples that include three actual field applications are discussed in this chapter to demonstrate the use of this simple yet elegant method. Many advanced methods already incorporated into a number of commercially available software on experimental modal analysis exist and can be used for better accuracy if required based on specific need of the application. But this is a quick, elegant, and very practical approach for many industrial applications, yielding fairly accurate results. It is also important to note that the simple impulse-response method using the instrumented hammer is fairly good enough for dynamic characterization of many big and small industrial structures, machines, and equipment.

References

Sinha, J.K., and Moorthy, R.I.K. 1999. Added Mass of Submerged Perforated Tubes. *Nuclear Engineering and Design* 193, 23–31.

Sinha, J.K., and Moorthy, R.I.K. 2000. Combined Experimental and Analytical Method for a Realistic Seismic Qualification of Equipment. *Nuclear Engineering and Design* 195, 331–338.

Sinha, J.K., Rao, A.R., Moorthy, R.I.K. 2003. Significance of Analytical Modelling for Interpretation of Experimental Modal Data: A Case Study. *Nuclear Engineering and Design* 220, 91–97.

8

Finite Element Model Updating

8.1 Introduction

The finite element (FE) method is the most appropriate tool for numerical modeling in structural engineering today. This method can handle complex structural geometry, large complex assemblies of structural components, and is also able to perform different types of analysis on them. Even with the great advances in the field of structural modeling, an initial FE model is often a poor reflection of the actual structure, particularly in the field of structural dynamics. This inaccuracy arises because of a number of simplifying assumptions and idealizations that have to be made while constructing the FE model that generally depends on engineering judgment. Such inaccuracy in the FE model a priori is well known in the scientific community and is generally highlighted when predictions from the model are compared to the modal test results. Although a range of experimental data such as frequency response functions or time domain data may be used, modal data consisting of natural frequencies, mode shapes, and damping ratios are the most common.

A simple example of the need for model updating is a cantilever beam, where often the beam is assumed to be rigidly fixed at the clamped end. However, during tests it is often found that the beam has either a small rotation ($\delta\theta$) or both a small rotation ($\delta\theta$) and a deflection (δy) at the clamped end. This is shown pictorially in Figure 8.1. If one has to construct the FE model without the knowledge of the experimental modal data, the natural assumption would be to include an ideal, fixed boundary condition, which may not be true. Even with such a simple structure the FE model is not reliable a priori, and based on intuition or engineering judgment, it is difficult to estimate the values of the boundary stiffnesses. Indeed, in large and structurally complex components there could be many more potential sources for the deviation between the FE model predictions and the measured data. Hence, a FE model, a priori, may be useful at the design stage only. However, after commissioning a structure, the initial FE model should be updated based on the experimental modal data obtained from in situ modal tests so that the FE model may be used with confidence for further analysis.

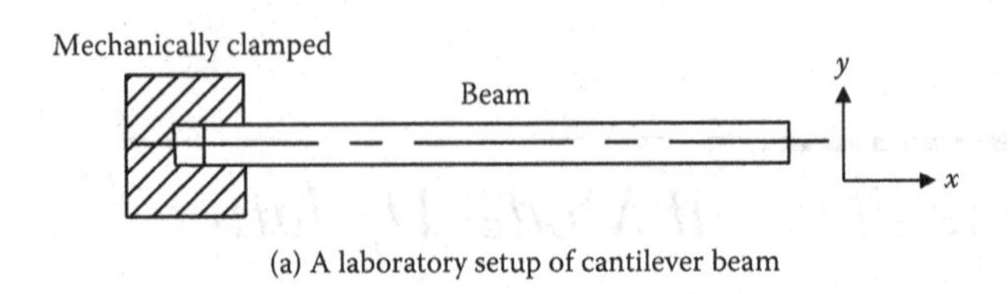

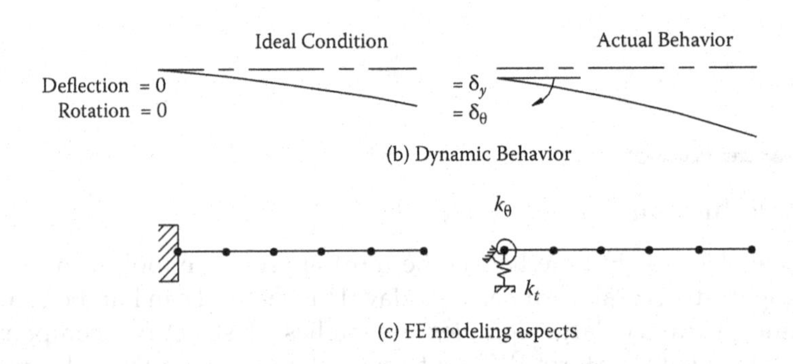

FIGURE 8.1
Cantilever beam and its behavior and FE modeling. (a) A laboratory setup of the cantilever beam. (b) Dynamic behavior. (c) FE modeling aspects.

8.2 Model Updating Methods

Model updating may be performed either by direct methods or by sensitivity methods. The direct methods produce exact results, so that the model predictions match the experimental modal data. However, the resulting updated FE model may not provide any physical meaning, and this is the main reason why these have not been generally used in practice and hence not discussed here. The sensitivity methods overcome the limitations of the direct methods but require an iterative solution. There are several sensitivity methods using either modal data or frequency response function data to update the FE model. Here, a very simple model updating method often used in practice is discussed for easy understanding; however, for details of other methods, a book on model updating by Friswell and Mottershead (1995) can be referred to.

8.3 Gradient-Based Sensitivity Method

In the gradient-based sensitivity methods of the model updating, the most important aspect is to define an error function between the computed and experimental modal data. The estimated parameters are obtained by minimizing the error function, which is usually a highly nonlinear function with respect to the updating parameters. In defining the error function, as well as in the construction of the sensitivity matrix (the derivative of the natural frequencies with respect to the updating parameters), the correct pairing of computed modal data with the experimental modal data is essential. This is important because the pairing of computed and experimental modal data based on the sequential order of mode numbers may not always be correct. The correct mode correlation and pairing between the experimental modes and the computed modes is generally established using the modal assurance criterion (MAC), and here Equation (3.46) is modified as

$$MAC_{jk} = \frac{\left|\varphi_{ej}^T \varphi_{ck}\right|^2}{\left(\varphi_{ck}^T \varphi_{ck}\right)\left(\varphi_{ej}^T \varphi_{ej}\right)} \tag{8.1}$$

where φ_{ej} and φ_{ck} are the experimental (measured) mode shape at the jth mode and the computed (analytical) mode shape at the kth mode, respectively. The value of the MAC is between 0 and 1. A value of 1 (100%) or close to 1 means the experimental and computed modes are well correlated. The experimental and computed mode shapes must contain the same number of elements. In general, the number of elements in the computed mode shapes is very large compared to the measured mode shapes. Hence, in practice, the MAC is calculated using only those elements of the computed mode shapes corresponding to the measurement locations, together with the measured mode shapes. This is called the coordinate MAC (CoMAC).

Another important task in model updating is the selection of the parameters to be updated (such as boundary stiffnesses, joint offsets, physical dimensions, material properties, etc.). The parameters should be chosen with the aim of correcting the recognized uncertainty in the model. Moreover, the computed natural frequencies, mode shapes, and response of the FE model should be sensitive to the updating parameters. Figure 8.2 shows a simplified flow diagram of this approach to model updating, where λ is a vector of eigenvalues, θ is the parameter vector, and $\mathbf{S}$ is the sensitivity matrix. The steps involved are discussed below.

8.3.1 Steps Involved

The objective function to be minimized in the following examples is the weighted sum of squares of the error in the eigenvalues, given by

$$J(\theta) = \sum_{i=1}^{m} W_{\varepsilon i}\left(\lambda_{ei} - \lambda_{ci}(\theta)\right)^2 = \left(\mathbf{z}_e - \mathbf{z}_c(\theta)\right)^T \mathbf{W}_\varepsilon \left(\mathbf{z}_e - \mathbf{z}_c(\theta)\right) \tag{8.2}$$

where λ_{ei} and λ_{ci} are the ith experimental and computed eigenvalues (natural frequency squared), $W_{\varepsilon i}$ are weighting factors that reflect the confidence level

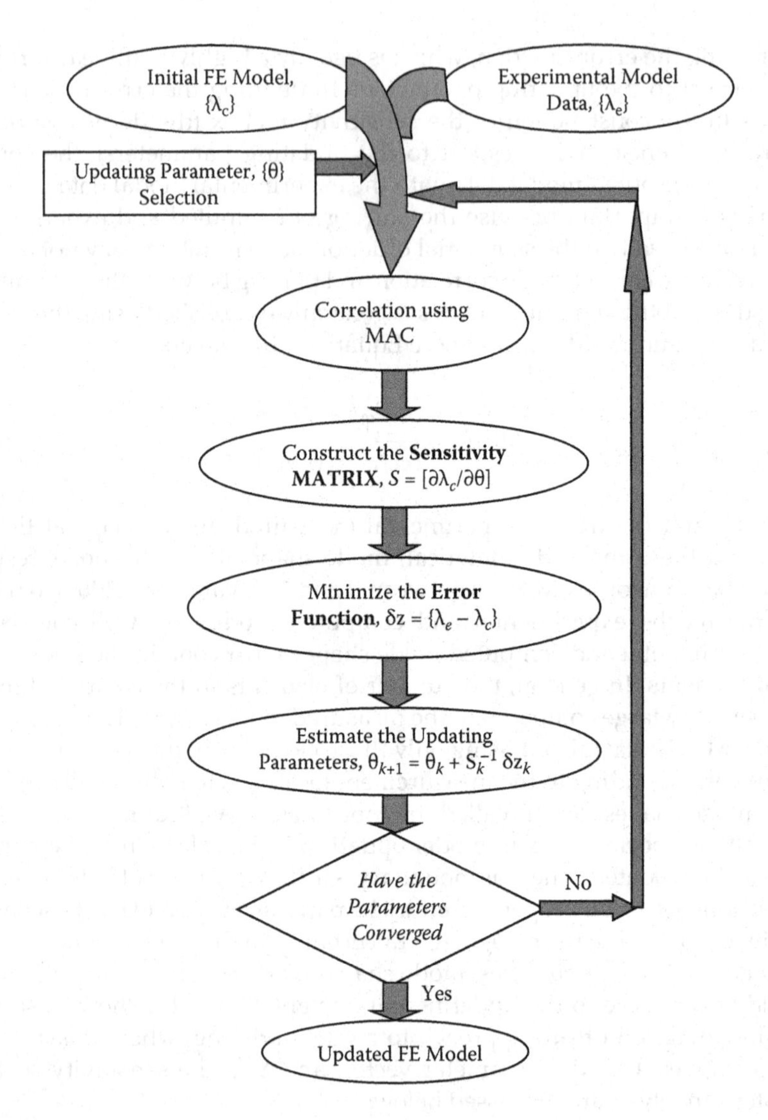

FIGURE 8.2
Simplified flow diagram for the model updating method.

in the measurements, and m is the number of measured natural frequencies used for updating. These eigenvalues and weighting factors are assembled into vectors,

$$\mathbf{z}_e = \left[\lambda_{e1}, \lambda_{e2}, \ldots, \lambda_{em}\right]^T, \mathbf{z}_c = \left[\lambda_{c1}, \lambda_{c2}, \ldots, \lambda_{cm}\right]^T, \mathbf{W}_\varepsilon = \operatorname{diag}\left(W_{\varepsilon 1}, W_{\varepsilon 2}, \ldots, W_{\varepsilon m}\right) \quad (8.3)$$

Often only the measured eigenvalues are used since the measured mode shapes usually contain more noise than the natural frequencies, and the mode shapes are not very sensitive to parameter changes. However, the mode shapes are vital to ensure that the correct modes are paired. $\theta = \left[\{\theta\}_1, \{\theta\}_2, ..., \{\theta\}_p\right]^T$ is the vector of the p uncertain parameters to be estimated. The weighting matrix $\mathbf{W}_\varepsilon$ must be positive definite and is often diagonal with the reciprocals of variance of the corresponding measurements along the diagonal.

The optimization given by Equation (8.2) is nonlinear and requires an iterative solution, usually by a gradient search technique. This requires the formulation and computation of the sensitivity matrix of the error function with respect to the updating parameters. Usually a local linearization is carried out using a Taylor series expansion of the computed eigenvalues as a function of the parameters, by retaining the first-order terms. A pictorial representation of this updating scheme is shown in Figure 8.3.

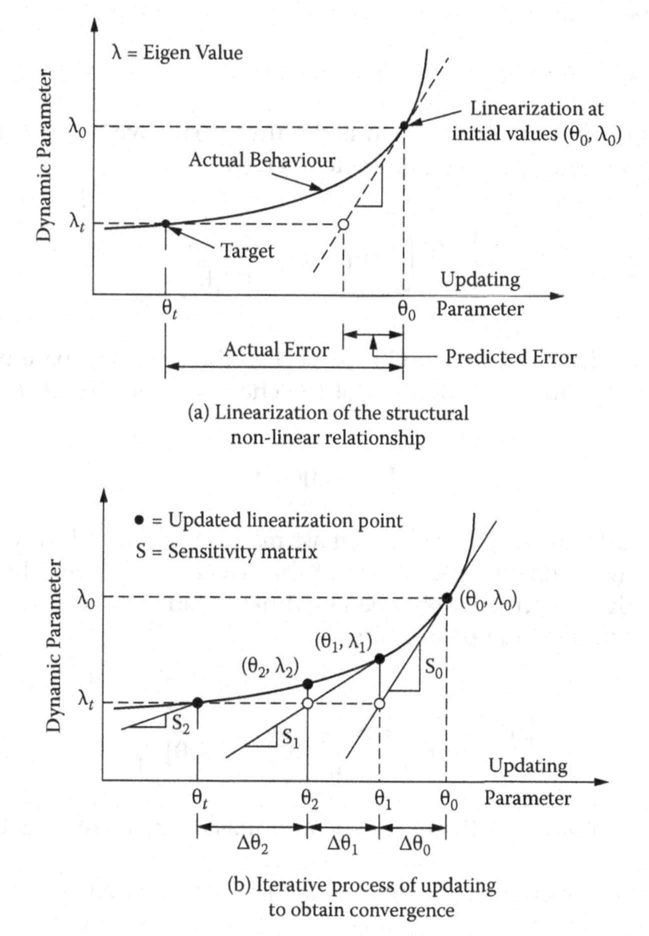

FIGURE 8.3
Graphic representation of the iterative scheme.

At the kth iteration,

$$\delta \mathbf{z}_k = \mathbf{S}_k \delta \theta_k \tag{8.4}$$

where $\delta \mathbf{z}_k$ is the output error,

$$\delta \mathbf{z}_k = \mathbf{z}_e - \mathbf{z}_{c,k} \tag{8.5}$$

$\delta \theta_k$ is the perturbation in the updating parameters,

$$\delta \theta_k = \theta_{k+1} - \theta_k \tag{8.6}$$

and θ_k is the current estimate of the parameters (that is, at the kth iteration). It is to be noted that the output vector in Equation (8.5) is at the current parameter estimate,

$$\mathbf{z}_{c,k} = \mathbf{z}_c(\theta_k) \tag{8.7}$$

$\mathbf{S}_k$ is the sensitivity matrix, which is the first derivative of the eigenvalues with respect to the updating parameters, given by

$$\mathbf{S}_k = \left[\frac{\partial \mathbf{z}_c}{\partial \theta} \right]_{\theta=\theta_k} \text{ or } [\mathbf{S}_k]_{ij} = \left[\frac{\partial \lambda_{ci}}{\partial \{\theta\}_j} \right]_{\theta=\theta_k} \tag{8.8}$$

The eigenvalue derivatives with respect to the updating parameters can be obtained by the differentiation of the characteristic structural dynamic equation

$$[\mathbf{K} - \lambda \mathbf{M}]\phi = \mathbf{0} \tag{8.9}$$

where $\mathbf{K}$ and $\mathbf{M}$ are the stiffness and mass matrices and λ and φ are the eigenvalues and normalized eigenvectors of the structural system. The elements of the sensitivity matrix are obtained by differentiation of Equation (8.9) with respect to each updating parameter:

$$[\mathbf{S}]_{ij} = \frac{\partial \lambda_i}{\partial \{\theta\}_j} = \phi_i^T \left[\frac{\partial \mathbf{K}}{\partial \{\theta\}_j} - \lambda_i \frac{\partial \mathbf{M}}{\partial \{\theta\}_j} \right] \phi_i \tag{8.10}$$

where λ_i and ϕ_i are the ith eigenvalue and mass normalized eigenvector of the structure.

The penalty function at the kth iteration is approximated as

$$J(\delta\theta_k) = (\delta \mathbf{z}_k - \mathbf{S}_k \delta\theta_k)^T \mathbf{W}_e (\delta \mathbf{z}_k - \mathbf{S}_k \delta\theta_k) \tag{8.11}$$

The solution of Equation (8.11) is obtained by minimizing J with respect to $\delta\theta$, which involves the differentiation of J with respect to each element of $\delta\theta$ and setting the result equal to zero. Finally, this leads to the following equation for the estimation of updating parameters after each iteration:

$$\boldsymbol{\theta}_{k+1} = \boldsymbol{\theta}_k + \left[\mathbf{S}_k^T \mathbf{W}_\varepsilon \mathbf{S}_k\right]^{-1} \mathbf{S}_k^T \mathbf{W}_\varepsilon (\mathbf{z}_e - \mathbf{z}_{c,k}) \tag{8.12}$$

8.3.2 Regularization

The updating of Equation (8.4) is often of the form

$$\mathbf{A}\boldsymbol{\theta} = \mathbf{b} \tag{8.13}$$

where $\mathbf{A} \in \Re^{m \times p}$, $\boldsymbol{\theta} \in \Re^{p \times 1}$, $\mathbf{b} \in \Re^{m \times 1}$, $m > p$, and the parameters $\boldsymbol{\theta}$ are required. The vector $\mathbf{b}$ consists of the measured data, which are generally contaminated by some random noise. Equation (8.13) is a least-squares problem and will have a solution given by

$$\boldsymbol{\theta} = \mathbf{A}^{\oplus}\mathbf{b} \tag{8.14}$$

where $\mathbf{A}^{\oplus}$ denotes the generalized inverse of $\mathbf{A}$. In general, the least-squares solution, θ, is unique and unbiased provided that rank ($\mathbf{A}$) = p. However, when $\mathbf{A}$ is close to being rank deficient and singular, Equation (8.13) will be ill-posed, and small noise in $\mathbf{b}$ may lead to a large deviation in the required vector, θ, from its exact value. In such cases, the common practice is to replace the ill-posed problem by a nearby well-posed approach. Such practice is known as *regularization*. Different regularization methods are available. However, the selection of either method in model updating is a subjective one and generally depends on the type of problems and the individual involved in the updating. Ahmadian et al. (1998) and Ziaei-Rad and Imregun (1999) gave an excellent review and a comparative study of different regularization methods used in model updating.

8.4 Example 8.1: A Simple Steel Bar

To aid in understanding, a simple steel bar of length 2 m and diameter 10 mm, shown in Figure 8.4, is considered here. The material properties, Young's modulus of elasticity (E) and density, are 200 GN/m^2 and 8000 kg/m^3, respectively. The bar is fixed at one end and undergoes only axial vibration. Steps required in model updating are illustrated below:

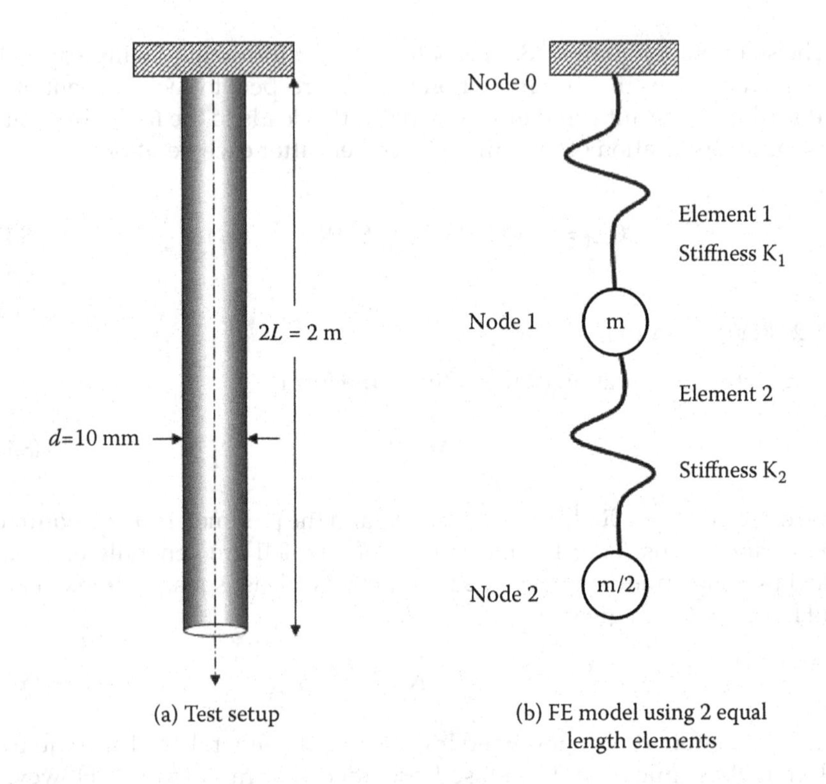

FIGURE 8.4
A simple test setup and its FE model. (a) Test setup. (b) FE model using two equal-length elements.

Step 1: Experimental modal analysis. Conduct modal testing and extract the modal properties—natural frequencies, modal dampings, and mode shapes. For the simulated experimental setup shown in Figure 8.4, the experimental modal parameters are

$$\text{Eigenvalues, } \lambda_e = \begin{Bmatrix} \lambda_{e1} \\ \lambda_{e2} \end{Bmatrix} = \begin{Bmatrix} \omega_{e1}^2 \\ \omega_{e2}^2 \end{Bmatrix} = \begin{Bmatrix} (2\pi f_{e1})^2 \\ (2\pi f_{e2})^2 \end{Bmatrix} = \begin{Bmatrix} 1.4645e07 \\ 8.5355e07 \end{Bmatrix}, \quad (8.15)$$

where natural frequencies $f_{e1} = 609.10$ Hz, and $f_{e2} = 1470.40$ Hz. Damping is found to be very small for both modes.

Mode shape (Eigenvectors),

$$\phi_e = \begin{bmatrix} \phi_{e1} & \phi_{e2} \end{bmatrix} = \begin{bmatrix} \phi_{e11} & \phi_{e12} \\ \phi_{e21} & \phi_{e22} \end{bmatrix} = \begin{bmatrix} -0.8921 & -0.8921 \\ -1.2616 & +1.2616 \end{bmatrix} \quad (8.16)$$

Step 2: FE Modeling. An initial FE model is constructed using accurate dimensional parameters and the material density, but error in the modulus of elasticity *by* assuming $E = 150$ GN/m^2. As shown in Figure 8.4(b), the bar is divided into two elements, and the stiffness of each element is

$$\mathbf{K}_1 = \mathbf{K}_2 = \frac{EA}{L}\begin{bmatrix} 1 & -1 \\ -1 & 1 \end{bmatrix} = 1.1781e07\begin{bmatrix} 1 & -1 \\ -1 & 1 \end{bmatrix} \text{N/m}, \qquad (8.17)$$

where $A = \frac{\pi}{4}d^2 = 78.5398\,\text{mm}^2$ is the area of cross section. Since node 0 is fixed, so the system dynamic equation can be written as

$$\begin{bmatrix} m & 0 \\ 0 & m \end{bmatrix}\begin{Bmatrix} \ddot{x}_1 \\ \ddot{x}_2 \end{Bmatrix} + \frac{EA}{L}\begin{bmatrix} 2 & -1 \\ -1 & 1 \end{bmatrix}\begin{Bmatrix} x_1 \\ x_2 \end{Bmatrix} = 0, \qquad (8.18)$$

where the mass per element, $m = \rho AL = 0.6283$ kg.

The Eigenvalue and Eigenvector analysis of Equation (8.18) gave the following results,

$$\text{Eigenvalues, } \lambda_c = \begin{Bmatrix} \lambda_{c1} \\ \lambda_{c2} \end{Bmatrix} = \begin{Bmatrix} \omega_{c1}^2 \\ \omega_{c2}^2 \end{Bmatrix} = \begin{Bmatrix} (2\pi f_{c1})^2 \\ (2\pi f_{c2})^2 \end{Bmatrix} = \begin{Bmatrix} 1.0983e07 \\ 6.4017e07 \end{Bmatrix}, \qquad (8.19)$$

where the computed natural frequencies are f_{c1} = 527.50 Hz, and f_{c2} = 1273.40 Hz, and the normalized Mode shapes (Eigenvectors),

$$\phi_c = \begin{bmatrix} \phi_{c1} & \phi_{c2} \end{bmatrix} = \begin{bmatrix} \phi_{c11} & \phi_{c12} \\ \phi_{c21} & \phi_{c22} \end{bmatrix} = \begin{bmatrix} -0.8921 & -0.8921 \\ -1.2616 & +1.2616 \end{bmatrix} \qquad (8.20)$$

Error in the FE model is present when compared to the results of the FE model with the experimental results (Equations (8.15) and (8.19)). Hence, the model requires updating.

Step 3: Updating parameter selection. In general, the person or group of persons involved in the experiments and modeling may be able to point out the source of errors in the modeling, and hence they can choose the updating parameters accordingly. However, for the present case, the obvious updating parameter is the modulus of elasticity (E), as the modeling has been done with error in this value for the purpose of demonstration. So the only updating parameter is $\theta = \{\theta_1\} = \{E\}$.

Step 4: Pairing of modes. Since there is only error in the modulus of elasticity (E), mode shapes are the same for both the experimental and analytical models, and their MAC using Equation (8.1) is

$$\text{MAC} = \begin{bmatrix} 1 & 0 \\ 0 & 1 \end{bmatrix} \tag{8.21}$$

Equation (8.21) can be graphically represented as shown in Figure 8.5. This indicates that first two experimental modes are 100% correlated with the first two computed modes.

Step 5: Error function. Now pairing of the experimental and computed modes is known, and hence the error function at the start of the iteration ($k = 0$) using Equations (8.15), (8.19), and (8.21) is defined as

$$\delta \mathbf{z}_{k=0} = \begin{Bmatrix} \lambda_{e1} \\ \lambda_{e2} \end{Bmatrix} - \begin{Bmatrix} \lambda_{c1} \\ \lambda_{c2} \end{Bmatrix}_{k=0} = \begin{Bmatrix} 1.4645e07 \\ 8.5355e07 \end{Bmatrix} - \begin{Bmatrix} 1.0983e07 \\ 6.4017e07 \end{Bmatrix}_{k=0} = \begin{Bmatrix} 0.3662e07 \\ 2.1338e07 \end{Bmatrix}_{k=0} \tag{8.22}$$

Here in Equation (8.22), the first experimental mode is well correlated with the first computed mode, and second experimental mode with the second computed mode. However, it may not always be the

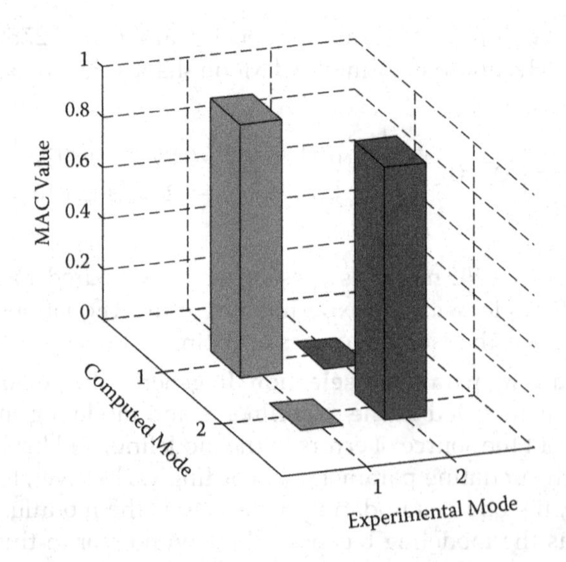

FIGURE 8.5
Bar graph representation of MAC values between the experimental and computed modes.

case for other problems, and it has to be paired strictly based on MAC value. For a complex experimental problem, the MAC value may not always be close to 1; it may be anything, say, 0.50 or higher. Hence, the person/group involved in model updating has to judiciously decide up to what lower (cutoff) value the mode pairs can be considered, as well as correlated.

Step 6: Construction of sensitivity matrix. There are two modes and one updating parameter, so the size of the sensitivity matrix **S** will be 2 × 1. The elements of the matrix **S** at the start of the iteration ($k = 0$) can be estimated using Equation (8.10) as

$$S_{11,k=0} = \left(\frac{\partial \lambda_1}{\partial \theta_1}\right)_{k=0} = \left(\frac{\partial \lambda_1}{\partial E}\right)_{k=0} = \phi_1^T \left[\frac{\partial \mathbf{K}}{\partial E} - \lambda_1 \frac{\partial \mathbf{M}}{\partial E}\right] \phi_1$$
$$= \phi_1^T \left[\frac{A}{L}\begin{bmatrix} 2 & -1 \\ -1 & 1 \end{bmatrix} - \lambda_1 \begin{bmatrix} 0 & 0 \\ 0 & 0 \end{bmatrix}\right] \phi_1 \tag{8.23}$$

$$S_{21,k=0} = \left(\frac{\partial \lambda_2}{\partial \theta_1}\right)_{k=0} = \left(\frac{\partial \lambda_2}{\partial E}\right)_{k=0} = \phi_2^T \left[\frac{\partial \mathbf{K}}{\partial E} - \lambda_2 \frac{\partial \mathbf{M}}{\partial E}\right] \phi_2$$
$$= \phi_2^T \left[\frac{A}{L}\begin{bmatrix} 2 & -1 \\ -1 & 1 \end{bmatrix} - \lambda_2 \begin{bmatrix} 0 & 0 \\ 0 & 0 \end{bmatrix}\right] \phi_2 \tag{8.24}$$

Substituting the computed mode shapes of first and second modes from Equation (8.20) into Equations (8.23) and (8.24),

$$S_{11,k=0} = \left[-0.8921 - 1.2616\right] \left[\frac{A}{L}\begin{bmatrix} 2 & -1 \\ -1 & 1 \end{bmatrix} - \lambda_1 \begin{bmatrix} 0 & 0 \\ 0 & 0 \end{bmatrix}\right] \begin{Bmatrix} -0.8921 \\ -1.2616 \end{Bmatrix}$$

$$= 0.0732e-3$$

$$S_{21,k=0} = \left[-0.8921 + 1.2616\right] \left[\frac{A}{L}\begin{bmatrix} 2 & -1 \\ -1 & 1 \end{bmatrix} - \lambda_1 \begin{bmatrix} 0 & 0 \\ 0 & 0 \end{bmatrix}\right] \begin{Bmatrix} -0.8921 \\ +1.2616 \end{Bmatrix}$$

$$= 0.4268e-3$$

At the start of the iteration, $k = 0$, the sensitivity,

$$\mathbf{S}_{k=0} = \begin{bmatrix} 0.0732e-3 \\ 0.4268e-3 \end{bmatrix}_{k=0} \tag{8.25}$$

Step 7: First iteration. The weighting factors defined in Equation (8.3) are assumed to be unity for both eigenvalues. Now every term of Equation (8.12) is known, and hence the first iteration has been carried out:

$$\theta_1 = \theta_0 + \left[\begin{bmatrix} 0.0732e-3 & 0.4268e-3 \end{bmatrix} \begin{bmatrix} 1 & 0 \\ 0 & 1 \end{bmatrix} \begin{Bmatrix} 0.0732e-3 \\ 0.4268e-3 \end{Bmatrix} \right]^{-1}$$
$$\begin{bmatrix} 0.0732e-3 & 0.4268e-3 \end{bmatrix} \begin{bmatrix} 1 & 0 \\ 0 & 1 \end{bmatrix} \begin{Bmatrix} 0.3662e07 \\ 2.1338e07 \end{Bmatrix} \tag{8.26}$$

Here the first iteration itself has yield that is the exact value of the modulus of elasticity (see Equation (8.26)). In many cases, it may require a few more iterations if the number of updating parameters is more than one. Further iterations were also done by repeating steps 1 to 6 using the updated FE model after each iteration. Convergence is shown in Figure 8.6.

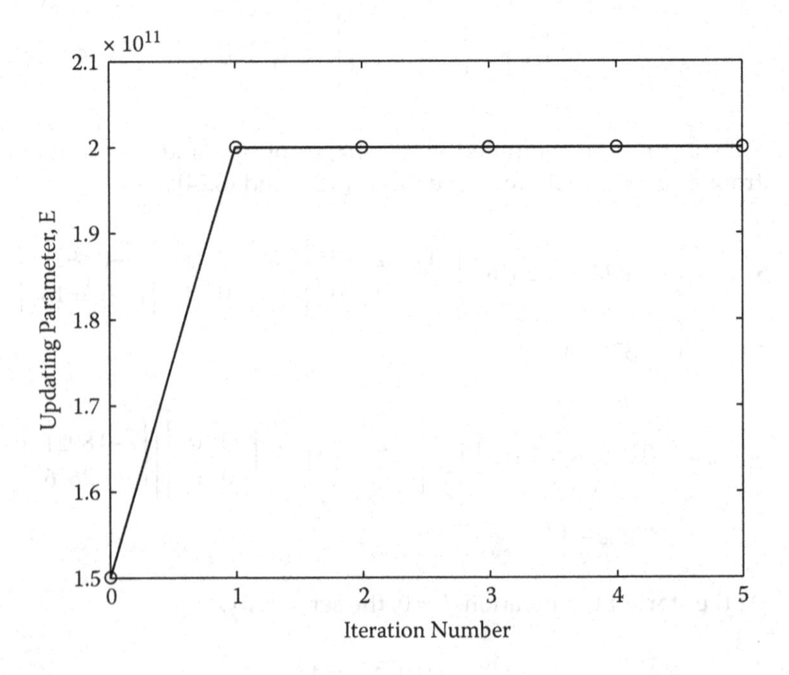

FIGURE 8.6
Convergence process.

8.5 Example 8.2: An Aluminum Cantilever Beam

This is an example of an aluminum cantilever beam with geometric parameters (996 × 50 × 25 mm) and material properties (E = 69.79 GN/m^2, density (ρ) = 2600 kg/m^3) (Sinha et al., 2002). Figure 8.7(a) shows a schematic of the test setup where the beam was mechanically clamped at one end to make it a cantilever beam. The modal parameters of the beam were obtained by the impulse-response method using a small instrumented hammer for excitation and an accelerometer of mass 3.5 g for the response measurements. The experimental frequency response functions were processed to obtain the modal parameters. Table 8.1 gives the identified experimental natural frequencies.

A FE model of the cantilever beam was constructed using two node Euler-Bernoulli beam elements, with each node having two degrees of freedom (DOFs) (one bending deflection, other bending rotation), as discussed in Chapter 3. The translational and rotational springs were also included in the model to simulate the boundary conditions at the clamped end of the beam, as the exact behavior of the clamped end was known. The FE model is shown schematically in Figure 8.7(b) and has 16 elements and 34 degrees of freedom. In the initial FE model, the boundary stiffnesses, k_t = 10 MN/m

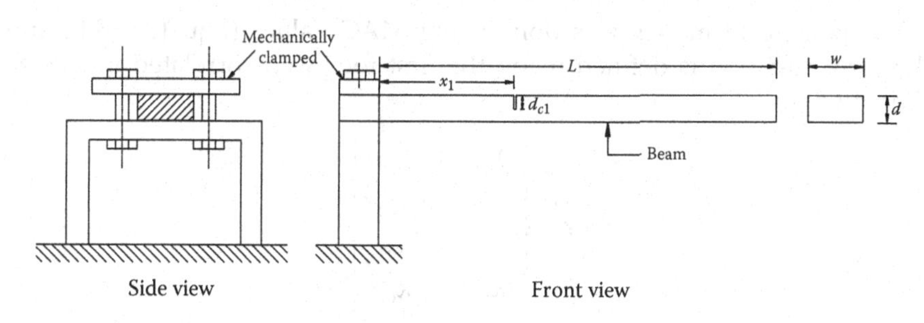

(a) Schematic of experimental setup

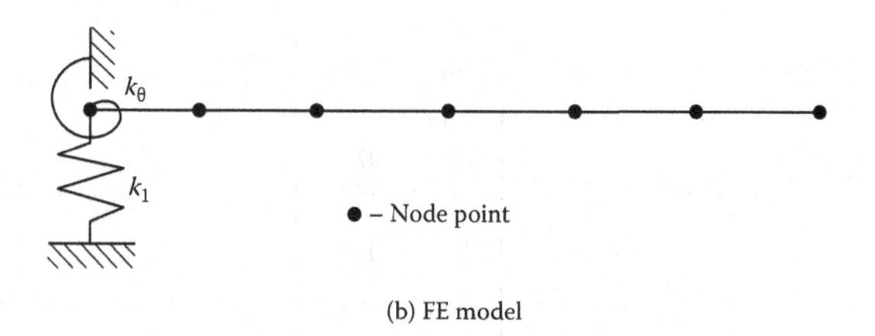

(b) FE model

FIGURE 8.7
Cantilever beam and its FE model. (a) Schematic of experimental setup. (b) FE model.

TABLE 8.1

Measured and Computed Natural Frequencies (Hz) of the Cantilever Beam

Flexural Modes	Experimental (a)	Initial FE Model (b)	% Error (a and b)	Updated Model (c)	% Error (a and c)
First mode, Hz	20.000	15.183	-24.083	19.900	-0.502
Second mode, Hz	124.500	106.524	-14.438	124.531	+0.025
Third mode, Hz	342.1875	308.006	-9.989	345.714	+01.031
Fourth mode, Hz	664.3750	599.800	-9.720	665.762	+0.209
k_t, MN/m	—	10.0	—	26.5	—
k_θ, kNm/rad	—	20.0	—	150.0	—

and $k_\theta = 20$ kNm/rad, were assumed. The computed natural frequencies are listed in Table 8.1. The error in the computed modes is up to 20% compared to the experimental results. Hence, the initial FE model needs updating. Since the geometrical and material properties are known accurately, the error in the boundary stiffnesses seems to be causing error in the model. Hence, these two parameters are chosen as the updating parameters.

$$\text{Updating parameters, } \theta = \begin{bmatrix} \theta_1 & \theta_2 \end{bmatrix}^T = \begin{bmatrix} k_t & k_\theta \end{bmatrix}^T$$

The pairing of modes was done using MAC values (Equation 8.1), and the error function is defined using the first four well-correlated modes as (Equation 8.5)

$$\delta\mathbf{z} = \begin{Bmatrix} \lambda_{e1} & - & \lambda_{c1} \\ \lambda_{e2} & - & \lambda_{c2} \\ \lambda_{e3} & - & \lambda_{c3} \\ \lambda_{e4} & - & \lambda_{c4} \end{Bmatrix}$$

The elements of the sensitivity matrix are estimated using Equation (8.10). The sensitivity matrix is

$$\mathbf{S} = \begin{bmatrix} \dfrac{\partial\lambda_{c1}}{\partial k_t} & \dfrac{\partial\lambda_{c1}}{\partial k_\theta} \\ \dfrac{\partial\lambda_{c2}}{\partial k_t} & \dfrac{\partial\lambda_{c2}}{\partial k_\theta} \\ \dfrac{\partial\lambda_{c3}}{\partial k_t} & \dfrac{\partial\lambda_{c3}}{\partial k_\theta} \\ \dfrac{\partial\lambda_{c4}}{\partial k_t} & \dfrac{\partial\lambda_{c4}}{\partial k_\theta} \end{bmatrix}$$

The weighting matrix was assumed to be a unit matrix. Having known elements of Equation (8.12), the iteration process has been carried out. The boundary stiffnesses, k_t = 26.5 MN/m and k_θ = 150 KNm/rad, are required to simulate the translation and rotation flexibility of the clamped support. Table 8.1 compares the measured and natural frequencies of the updated FE model of the experimental cantilever beam. Now the natural frequencies are within an error of 1%. Hence, the updated model truly represents the experimental setup.

8.6 Summary

An FE a priori for any system may not reflect the in situ or "as installed" dynamic behavior due to idealized conditions often assumed during modeling. Such an FE model needs to be updated using the in situ modal parameters of the system. The concept of the model updating method and the step-by-step procedures for the sensitivity-based gradient method are discussed here. Two simple examples on model updating are also illustrated in this chapter for better understanding. The use of the updated models is further discussed in Chapter 10.

References

Ahmadian, H., Mottershead, J.E., Friswell, M.I. 1998. Regularisation Methods for Finite Element Updating. *Mech. Syst. Signal Process.* 12, 47–64.

Friswell, M.I., Mottershead, J.E. 1995. Finite Element Model Updating in Structural Dynamics. Kluwer Academic Publishers, Dordrecht.

Sinha, Jyoti K, Friswell, M.I., Edwards S., 2002. *Simplified Models for the Location of Cracks in Beam Structures using Measured Vibration Data*. Journal of Sound and Vibration. Vol. 251. No. 1. pp. 13–38.

Ziaei-Rad, S., Imregun, M. 1999. On the Use of Regularisation Techniques for Finite Element Model Updating. *Inverse Problems Eng.* 7(5), 471–503.

9

A Simple Concept on Vibration-Based Condition Monitoring

9.1 Introduction

Machines, equipment, and structural components of any industrial plant need to be functional to meet plant objectives. Extended availability of all such objects is essential for productivity. However, it is a known fact that degradation in all objects takes place with time and may lead to catastrophic failures. Such failures are not healthy for plant safety, often result in huge financial losses, and sometimes claim the life of plant personnel as well. Considering these known facts, the plant is always equipped with a maintenance unit that looks into the health of machines and other components on a regular basis. The unit is mainly responsible for identifying health degradation or developing faults and the repair or replacement in a planned shutdown of the plant. The identification of such degradation/faults is perhaps the most difficult task. Take, for example, a turbogenerator (TG) set; it requires the knowledge of several parameters, including condenser pressure, bearing fluid pressure, and temperature during the machine's normal operation, and maybe material properties as well. Vibration-based condition monitoring (Sinha, 2002) is perhaps the most popular and well-recognized tool used in plants to meet the objective of predicting the faults in a cheaper manner. Vibration-based condition monitoring negates the need for monitoring of a large number of other parameters, because vibration response is more sensitive to any small structural or process parameter change. Hence, only vibration-based condition monitoring is discussed here. A few important issues need to be addressed before installation of a vibration-based condition monitoring system, such as:

1. Identification of critical machines and structural components related to safety and production
2. Types of transducers (sensors) that need to be used
3. Measurement locations, data processing, and storage

Apart from the above-mentioned items, the information gathered from the measured vibration data needs to be classified into different categories to address the needs of the different kinds of plant personnel. Different groups in a plant may not be familiar with the machines, equipment, and structural components (other than their functional use) and the vibration-based diagnostic techniques being used. However, their primary objective is to achieve optimum productivity without affecting the safety of the plant. So, the coordination or sharing of information among them is very important. To meet these requirements, vibration-based information is generally divided into the following three categories.

9.2 Operational Personnel

These are the people who are concerned with productivity, so for them the information required is whether or not to operate the plant or machines. For this requirement, several guidelines are available in the form of codes and standards, which are based on the experience gained over decades of plant operation. A few of them are discussed briefly.

9.2.1 Rotating Machines

The International Organization for Standardization (ISO) has recommended vibration severity limits for machines during normal machine operation. In particular, ISO 2372 (1974) code gives general guidelines for rotating machines with operating speeds ranging from 10 to 200 rev/s. Codes ISO 3945 (1980) and ISO 7919 (1986) provide vibration amplitude limits for large rotating machines. These codes recommend maximum overall vibration amplitudes measured at the recommended locations of machines. ISO 3945 (1980) gives vibration severity criteria for rotating machines like TG sets supported on rigid or flexible foundations, but it assumes a rigid rotor. It recommends the maximum value of the *root mean square* (RMS) *velocity* of vibration amplitude measured at the bearing pedestals in the frequency band of 10 to 1000 Hz. The vibration severity values as per ISO 3945 code are independent of the machine operating speed. The recent ISO 10816 code is the revised version, which suggests the use of different frequency band limits instead of the generic limit of 10 to 1000 Hz, irrespective of machine types and machine speed. The recommended frequency band limit should include possible faults of the machine. These codes are concerned with the vibration of non-rotating parts, i.e., bearing pedestals, of rotating machines. However, the ISO 7919 (1986) code is directly related to the vibration of the rotating shaft and is applicable to a flexible rotor supported on a flexible foundation. Part 1 of ISO 7919 gives the measurement guidelines that apply to shaft vibration measured close to the rotor bearings under normal operating conditions. It recommends both relative and absolute shaft displacements. ISO 7919 Part 2 (1986) gives the charts for the vibration severity limits to compare measured values and assess the machine's overall condition. Parkinson and McGuire (1995) gave an excellent review of the ISO codes related to rotating machines. So for the operational personnel, the trends of overall vibration values are sufficient.

9.2.2 Reciprocating Machines

The code VDI 2063 (1985) gives the measurement guidelines and vibration severity limits for the reciprocating machines with a power equal to or above 100 kW and a speed up to 3000 rpm. As per said code, the recommended frequency band is 2 to 300 Hz for the vibration measurement at bearing or fastening points. Vibration below 2 and above 300 Hz is disregarded. This is done because vibrations below 2 Hz are considered rigid-body movements of complete machines (no relative stress in the machine), while vibrations above 300 Hz are local vibrations of the machinery surface in the form of structure-borne noise. However, the recent ISO 10816-6 (BS7854-6), i.e., Part 6, is also related with reciprocating machines with power ratings above 100 kW. This has increased the frequency band to 2 Hz to 1 kHz.

9.2.3 Piping

The piping is also an essential part of any plant. It needs to be functional for the designed service loads and period. The piping is usually subjected to flow-induced vibration. ASME Code on Operation and Maintenance of Nuclear Power Plants (1990) recommends the overall vibration severity limit for piping vibration. The vibration limit can be evaluated for each configuration of piping depending on the following factors:

1. Concentrated weight, e.g., coupling valves and in the piping
2. Mass effect of fluid content and insulation
3. Boundary conditions
4. Bend (elbow, U-bend, etc.) in the piping

However, code gives a very conservative screening vibration velocity of 12.52 mm/s (0 to peak) using conservative values of all the factors for any configuration of piping.

9.2.4 Comparative Observations

There is the possibility that for some critical machines or structural components, related codes directly giving guidelines for vibration might not exist.

In such a case, either the manufacturer has to provide such details or the user may have to follow the trend of vibration amplitudes at different locations during their normal function. Perhaps comparison of day-to-day data may be useful for assessment of health. If two or more similar machines are available, then the best option is to do comparative vibration study for evaluation of the health of one machine with others.

9.3 Plant Maintenance Engineers

Plant maintenance engineers are responsible for overall plant safety. They may not be experts in vibration-based diagnosis, but they can be trained to understand and observe things such as the trend of the overall vibration, change in vibration spectra over a period of operation, etc. If these changes are persistent or increasing with time, then a shutdown may be planned and the machine experts called to identify and solve existing problems. It is also likely that the maintenance personnel will develop the required skills over a period of involvement in vibration monitoring that enables them to solve problems without consulting the machine experts.

9.4 Vibration Experts

Experts usually look at previous stored vibration data and may do some additional measurements and modal testing to identify the faults accurately so that remedial action can be taken effectively.

9.5 Condition Monitoring of Rotating Machines

To aid the philosophy of vibration-based condition monitoring, the well-established approach for the condition monitoring of rotating machines, especially turbogenerator (TG) sets, is discussed in detail. This can be extended to any other machine, equipment, or structural component.

In nuclear power plants and other power plants, several rotating machines are used. These machines mainly consist of three major parts—a rotor, journal bearings (fluid or antifriction bearings), and a foundation including supports, casing, piping, etc. TG sets are one such major rotating machine. A typical steam TG set, with schematic shown in Figure 9.1 (Sinha, 2002), generally

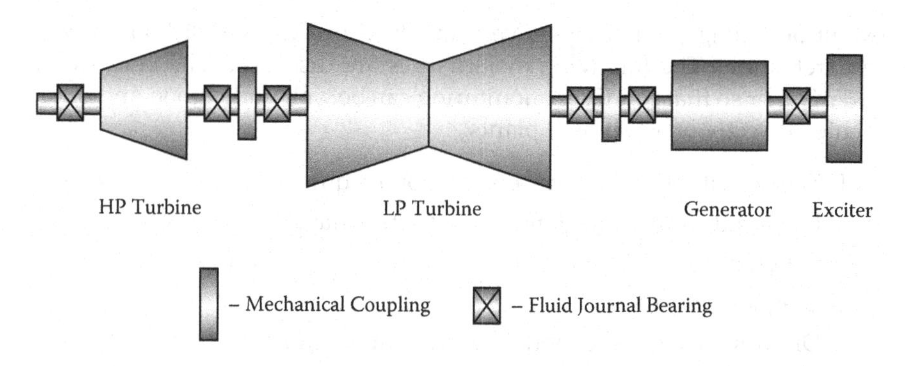

FIGURE 9.1
Schematic of a simple layout of a TG set.

consists of a high-pressure (HP) turbine, a low-pressure (LP) turbine, and an electric generator. In some high-capacity power plants, the TG sets also have a few intermediate-pressure (IP) turbines and more LP turbines that are not shown in Figure 9.1. The shafts of the individual systems are joined together by means of couplings, and the complete shaft is known as the rotor of the TG set. The shafts of each turbine are designed to have a number of rows of turbine blades along the shaft length. Each rotor of an individual system is normally supported by its own fluid journal bearings, which are supported on foundation structures that are often flexible.

Experience shows that faults develop in rotating machines during normal operation, for example, bends, cracks, or mass unbalance in the shaft (due to scale deposits or erosion in blades of the turbines). If a fault develops and remains undetected for some time, then, at best, the problem will not be too serious and can be remedied quickly and cheaply. However, at worst, it may result in expensive damage and downtime, injury, or even loss of life. Such a situation warrants the use of a reliable condition monitoring technique to reduce the downtime of the plant.

Condition monitoring requires the continuous observation of various parameters like vibration responses at different locations of the TG sets and many other process parameters, for example, condenser pressure, bearing fluid pressure, and temperature during the machine's normal operation. Vibration-based condition monitoring is possibly the most popular and well recognized in plants, perhaps because the machine vibration response is more sensitive to any small structural or process parameter change. Hence, only vibration-based condition monitoring (VCM) is discussed here. There are always two aspects of condition monitoring (CM): prediction of faults and their quantifications. In fact, the vibration-based prediction of faults for rotating machines is now well known and well understood for many known problems of rotating machines. Once a fault is suspected, the machine shutdown can be planned to identify the source of the fault alarm and its remedy. However, the quantification part (pinpointing the location of fault and its extent of damage) is perhaps a very involved process and still an ongoing research study. The following major issues are discussed in a very simplified manner so that condition monitoring can be well understood and implemented directly on the power plants:

1. Type of vibration transducers (sensors) required
2. Processing and management of vibration data
3. Different faults
4. Diagnosis of faults
5. Discussion about the model-based fault diagnosis (MFD)

9.5.1 Type of Vibration Transducers

Vibration is generally measured near the bearings in a rotating machine. To aid understanding, a small portion of shaft with a bearing pedestal instrumented with different types of vibration transducers is shown in Figure 9.2. Four types of transducers that have different purposes are shown in Figure 9.2. The purpose of these transducers is explained below.

The tachometer (tacho) signal is used to measure the rotor speed and also gives the relative phase reference with the signals from the other transducers. Russell (1997) highlights the importance of phase measurement, and its use in diagnosing machinery problems is discussed by Forland (1999). One tacho sensor is sufficient for the rotor of a machine.

The change in vibration amplitude due to a small change in structural or process parameters will be more significant in the horizontal and

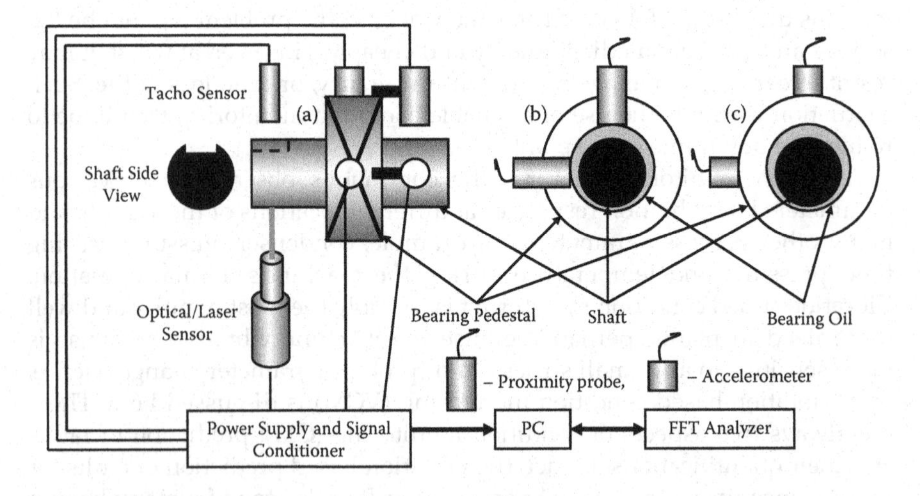

FIGURE 9.2
Mounting of different types of vibration transducers. (a) Front view of a bearing and shaft. (b) Side view: Accelerometers not shown for clarity. (c) Side view: Proximity probes not shown for clarity.

vertical directions. Hence, the vibration measurement in these two directions is important and is why the accelerometers and the proximity probes are mounted in both the horizontal (x) and vertical (y) directions. Similarly, an optical or laser sensor may be mounted in both directions; however, for clarity, only one is shown in Figure 9.2. The accelerometers measure the absolute acceleration of the bearing pedestals. The acceleration signal can be converted into a velocity and displacement signal. However, the proximity probe measures the relative displacement of the rotor with respect to the bearing pedestal, so obviously this will have more information about the shaft vibration. The absolute vibration of the shaft at the measurement location can be obtained by adding or subtracting the two signals depending upon the phase relation between them. Alternatively, the absolute shaft vibration can be directly measured using an optical or a laser sensor. All these signals are generally utilized to identify faults in machinery. The kind of data management and signal processing required for fault diagnosis is discussed in Section 9.5.2, 9.6, and 9.7.

9.5.2 Data Processing and Storage

The data obtained from all transducers shown in Figure 9.2 are analogue signals and continuously vary with time. A sample of these signals is shown in Figure 9.3, in which it is assumed that the response of the

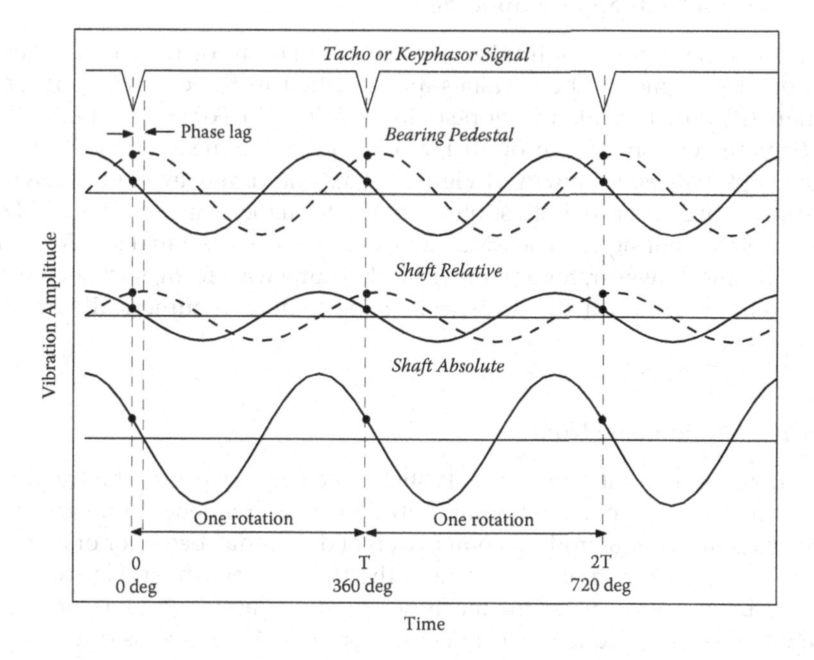

FIGURE 9.3
Time waveform measured from different transducers (_____, horizontal direction; _ _ _, vertical direction).

system is purely sinusoidal to aid understanding. These online measured analogue vibration data can be directly imported to a personal computer (PC) through suitable hardware (analogue-to-digital conversion [ADC]) and software by sampling the analogue data at a sufficiently high sampling rate. However, the storage of these data from all bearing locations during normal machine operation and transient operations, like run-up and run-down, would need a huge data storage capacity. Additionally, data handing and retrieval would be difficult. The normal practice is to process the data to get the information required for machine health monitoring and then store the processed data for future use and to identify whether or not a fault is becoming worse.

9.6 Normal Operation Condition

During normal machine operation, measurement of the following data at predetermined intervals is important.

9.6.1 Overall Vibration Amplitude

The overall vibration amplitude can be stored in terms of acceleration, velocity, or displacement. These values are specified in three ways: root mean square (RMS), 0 to peak (pk), or peak-to-peak (pk-pk). For simple illustration, a displacement sine wave of 50 Hz, indicating the meaning of RMS, pk, and pk-pk values, is shown in Figure 9.4(a). The time waveform shown in Figure 9.4(b) consists of both subharmonic and higher harmonics of 50 Hz in the displacement signal. The RMS value of a sine wave is equal to 0.707 times the pk value; however, for any other kind of time waveform, such as the one shown in Figure 9.4(b), the RMS needs to be computed numerically (refer to Chapter 6).

9.6.2 Vibration Spectrum

The time domain vibration signals should be converted into the frequency domain. The plots of vibration amplitude versus frequency (known as the spectrum) must be stored for comparison of day-to-day behavior of machine performance. The vibration spectra of the time waveforms in Figure 9.4 are shown in Figure 9.5. Since the machine speed was assumed to be 3000 rpm, the vibration amplitude at 50 Hz is referred to as 1X, 25 Hz as 0.5X, 100 Hz as 2X, and so on, as indicated in Figure 9.5. Change in the amplitude at these frequencies should be monitored, as these changes may be related to some kind of fault development.

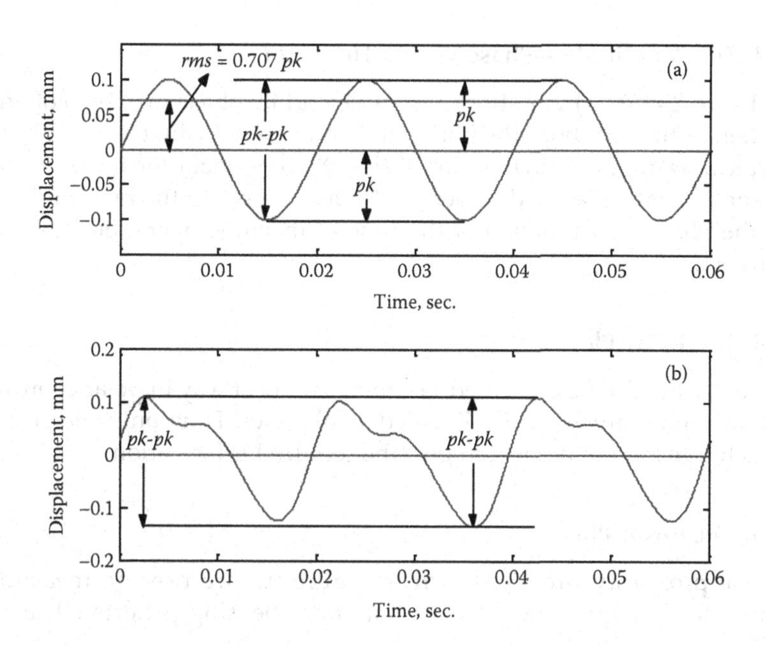

FIGURE 9.4
Typical displacement signals marked with peak, peak-to-peak, and RMS amplitude of vibration. (a) Sine wave. (b) Periodic wave.

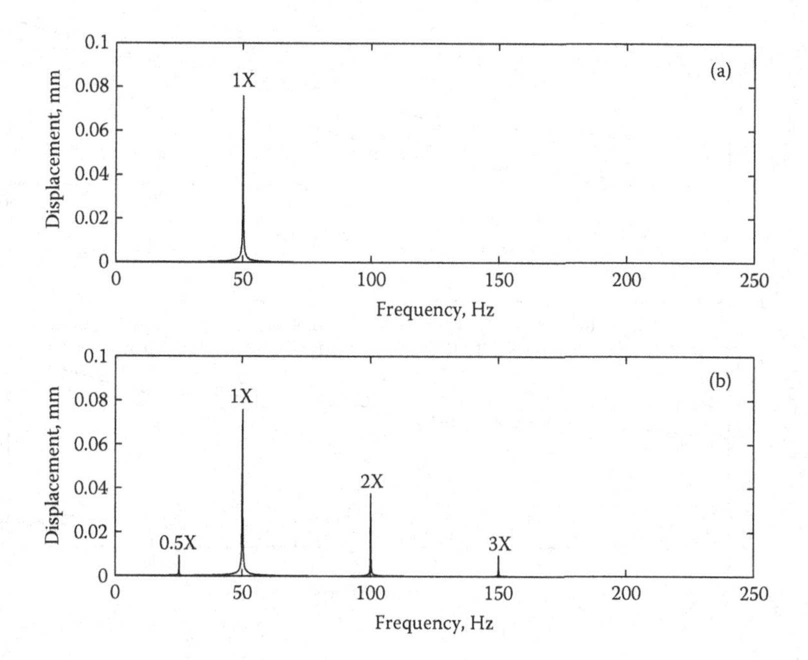

FIGURE 9.5
Spectra of the displacement signals in Figure 9.4.

9.6.3 The Amplitude—Phase versus Time Plot

The 1X (or 2X, 3X,...) vibration amplitude and its phase may be plotted as a function of time for both the horizontal and vertical vibrations at a bearing. A typical example of such a plot of the 1X component for the vertical and horizontal shaft relative displacement with respect to the bearing housing and the phase as a function of the time of machine operation is shown in Figure 9.6.

9.6.4 The Polar Plot

Figure 9.6 can also be presented in a more compact way in polar coordinates as shown in Figure 9.7, which is called a polar plot. Thus, only one polar plot for each bearing is sufficient to give the required information.

9.6.5 The Orbit Plot

The x-y proximity probes shown in Figure 9.2 are used to measure the relative shaft displacement to the stationary bearing pedestal. The signal

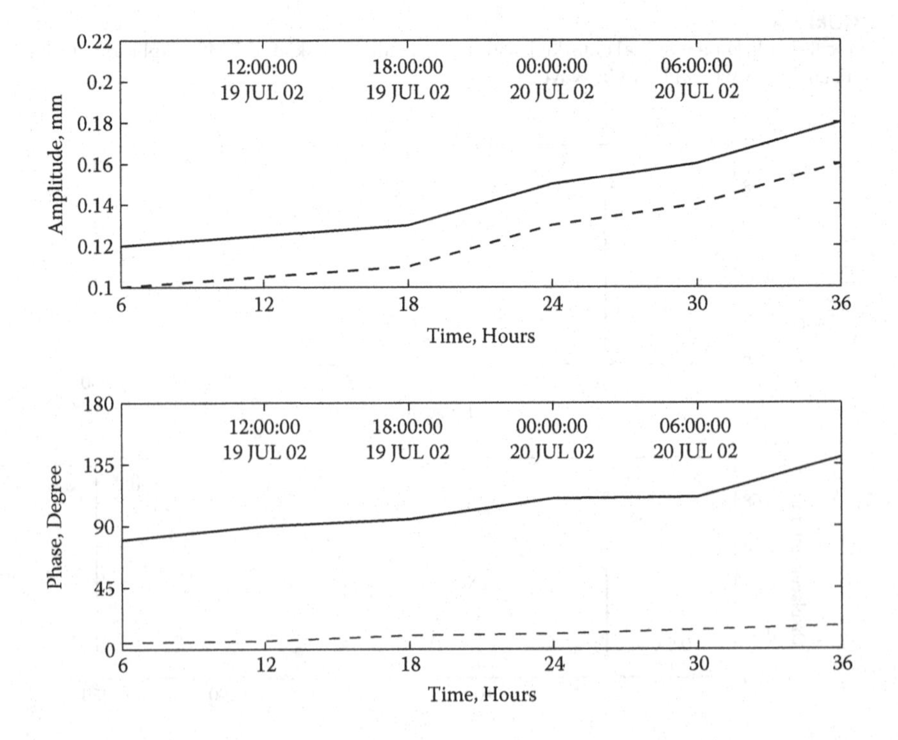

FIGURE 9.6
Variation of amplitude and phase of 1X component of shaft relative displacement with time (_____, horizontal; _ _ _, vertical).

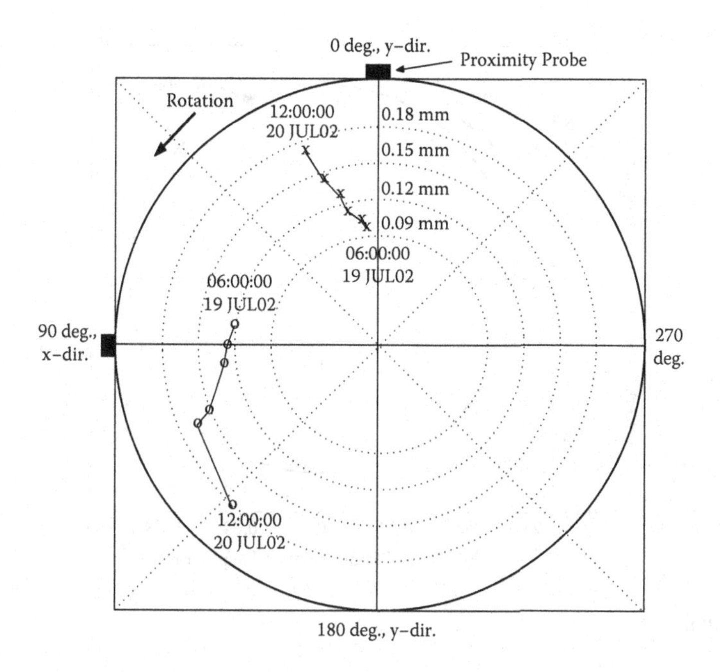

FIGURE 9.7
Polar plot representing the amplitude-phase versus time data in Figure 9.6 (x, horizontal; o, vertical).

from such a noncontacting proximity probe consists of the following two components:

1. A *DC signal* that is proportional to the average shaft position relative to the probe mounting. The *x-y* proximity probes give the shaft position in the bearing, and a change in shaft position during normal operation would indicate a load change or maybe bearing wear.
2. An *AC signal* corresponding to the shaft dynamic motion relative to the probe mounting. The *x-y* probe signals give the shaft centerline's path as the shaft vibrates. For illustration, Figure 9.8 shows an orbit plot obtained from the time waveform in Figure 9.3 for the shaft relative displacement measured by *x-y* probes. A perfect circular orbit plot indicates that the supporting foundation stiffness is the same in the vertical and horizontal directions, and the rotor response is at 1X only. This plot is the forward orbit plot as the path of the rotor oscillation is in the direction of rotor rotation with respect to a key phasor; otherwise, it may be a reverse orbit plot.

The orbit plot can be plotted either for the unfiltered time waveform or for the filtered component of 0.5X, 1X, 2X, etc., for each revolution of shaft. The shaft position and the orbit plots provide useful information about shaft malfunctions.

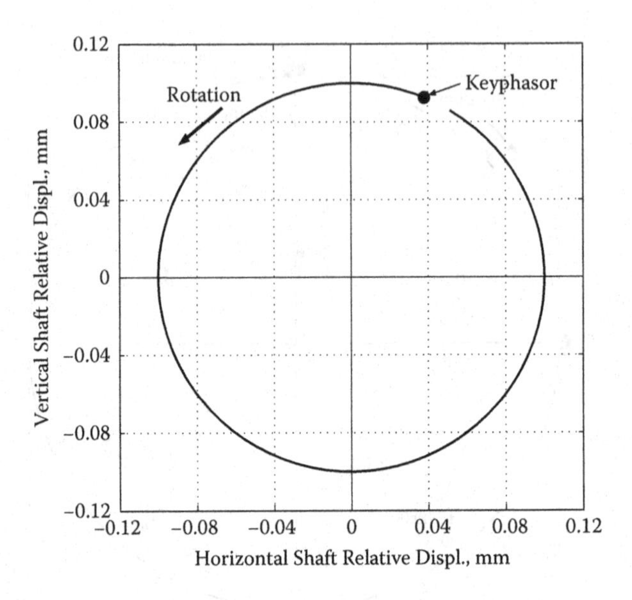

FIGURE 9.8
Orbit plot.

9.7 Transient Operation Conditions

The start-up and run-down of a machine are considered transient operation conditions. The measured data during this condition are very useful for confirming fault identification and its subsequent diagnosis. The time domain data for a complete transient operation must be stored, and the following data analysis should be performed.

9.7.1 The 3D Waterfall Plot of Spectra

A 3D waterfall plot is a plot of the spectra of the *x*-*y* proximity probes at a bearing with shaft rotating speeds on the third axis. A simple such plot for a machine start-up from 2000 to 3000 rpm for the vertical (*y*) proximity probe is shown in Figure 9.9. The appearance or disappearance of any frequency components (multiples of 1X) or even a small abnormal change in the vibration amplitude at any frequency component can be easily recognized by this 3D waterfall plot of the spectra.

9.7.2 The Shaft Centerline Plot

The plotting of DC signals of the *x*-*y* proximity probes at a bearing with machine rotating speed gives the shaft average centerline. This is useful for identifying changes in bearing load and bearing wear, as well as for calculating

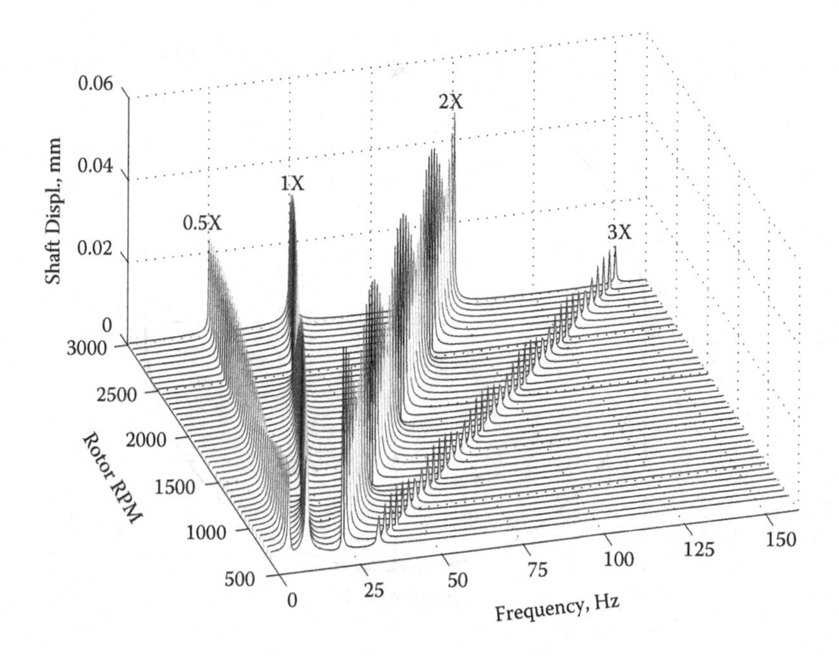

FIGURE 9.9
The waterfall plot of spectra during machine run-up.

the average eccentricity ratio and the rotor position angle. The measured eccentricity helps in the understanding of fluid-induced stability and is a parameter that can be directly used in the reliable mathematical modeling of fluid bearings. An assumed shaft centerline plot during machine start-up is shown in Figure 9.10. In Figure 9.10 the vertical and horizontal bearing clearance is 0.44 mm. At the start of the machine rotation the rotor was resting at bottom (180°), and then slowly moved up to 0.10 mm with an angle of 225°. This indicates that the rotor is rotating in the counterclockwise (ccw) direction, and the eccentricity (ε) of the journal center to the bearing center is 0.12 mm.

9.7.3 The Orbit Plot

The observation of orbit plots, both raw (unfiltered) and filtered 0.5X, 1X, 2X, etc., signals, with the change in machine speed during transient operation is important.

9.7.4 The Bode Plot

The measured data have to be order tracked to extract the 1X and higher harmonics with respect to the tacho signal. Order tracking gives the vibration amplitude-phase relationship of 1X, 2X, etc., with the change in the machine's rotating speed. The plot of vibration amplitude and its phase with

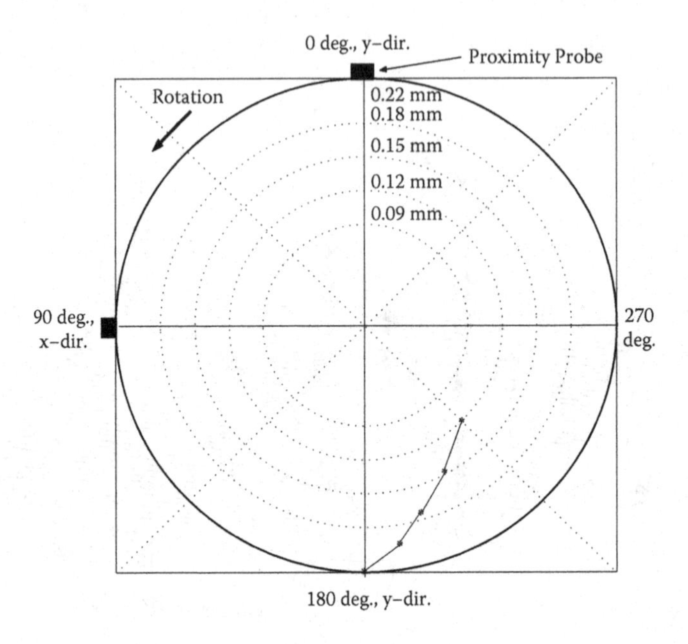

FIGURE 9.10
The shaft centerline plot.

speed is known as the Bode plot. A typical Bode plot of the 1X frequency component for an assumed machine, run-down from 2400 rpm to 360 rpm, is shown in Figure 9.11. As can be seen from Figure 9.11, for 1X, the plot is analogous to the frequency response function of standard modal tests. The change in the rotor speed during a run-down is in fact similar to swept-sine excitation to the system from an external shaker. There are a total of four resonant peaks—two in the vertical and two in the horizontal direction in the run-down frequency range. These resonant peaks are known as machine critical speeds. The change in phase at these critical speeds is generally dependent on the unbalance phase. But in case of a pure unbalance problem, if the phase is ± 90° at the critical speeds in direction (vertical), and in the one direction (vertical), (horizontal) then in the other direction it would be 0° or 180°. Changes in phase at the critical speeds are also indicative of the development of some faults.

9.8 Instrumenting TG Sets for Condition Monitoring

Considering the safety aspects of TG sets, the accepted practice in most modern power stations is to permanently install different vibration transducers at all possible locations in the TG set to monitor the vibration signals indicative of the

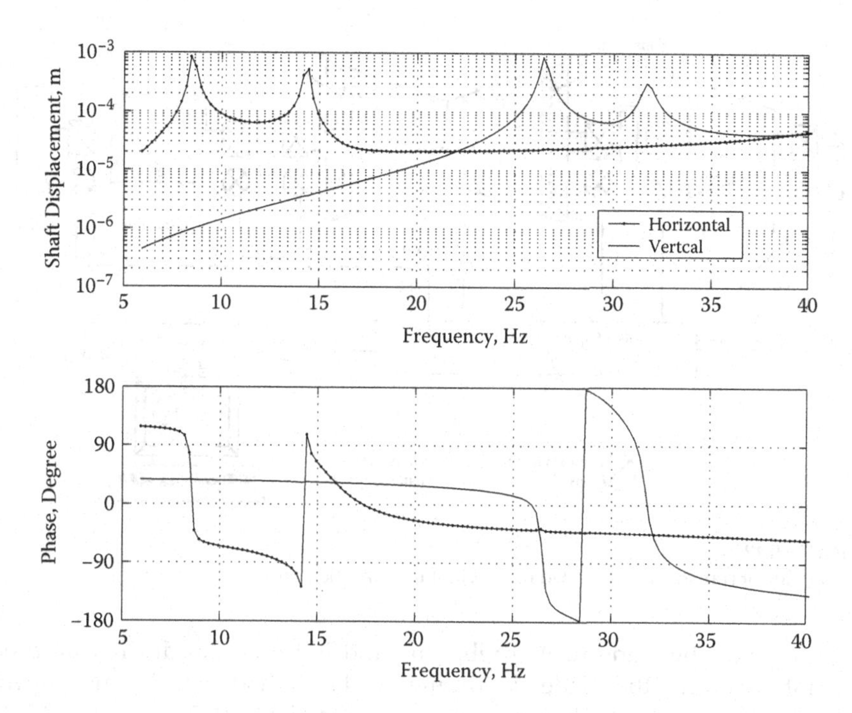

FIGURE 9.11
The Bode plot of 1X shaft displacement for a machine run-down.

machine performance during its normal and transient operations. A simplified diagram of a TG set instrumented with different types of transducers is shown in Figure 9.12. The installation of the accelerometers, proximity probes, or any other transducers at a bearing should be similar to that shown in Figure 9.2.

It is well known that the vibration measurements near bearings are generally indicative of rotor status. But experience shows that the vibration of turbine blades in different stages generally does not affect the vibration measurements near the bearings. In particular, the monitoring of the blades in the last stage of an LP turbine is important, because their fundamental frequencies are usually as low as two to three times the machine rotating speed and may be excited by small fluctuations in machine output power or operating condition. Indeed, the catastrophic breakdown of TG sets observed in many power plants in the past was due to the failure of the last stage blades. However, these blades could not be removed, as they contribute significantly to the total output of any LP turbine, which is a significant contribution that cannot be overlooked. A better design is needed, and in fact, the manufacturers have been in a constant search for better designs. Apart from the requisite design modification, it is also important to monitor the last stage blade vibration so that its failure leading to a catastrophic breakdown of the machine can be avoided. As of now, there is no reliable method available that

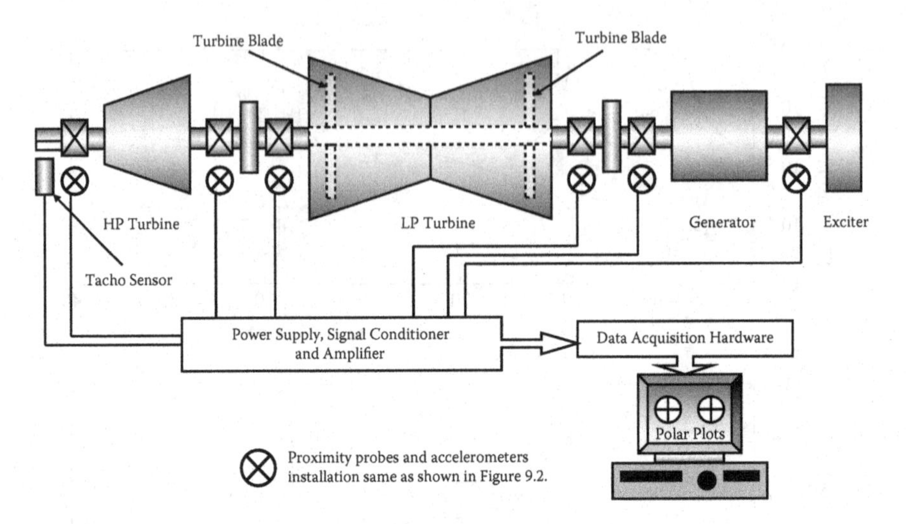

FIGURE 9.12
Schematic of transducers installed on a TG set for condition monitoring.

can monitor the blade-related vibration and its condition during machine operation except the blade tip timing (BTT) method. So, the instrument required for the blade vibration is not shown in Figure 9.12.

A variety of these transducers are available commercially, and the selection of the correct transducer is important, depending on the application. The selection of the transducers generally depends on the frequency range of the vibration measurement and the amplitude level. There are no absolute rules for the selection, which requires considerable experience in vibration measurement.

The measured signals should be processed as discussed in Section 9.2 and stored. The hardware and software required for data acquisition, management, processing, display, and storage can be either be developed in-house or a commercially available system. The advantage of a user-developed code is that the code can be modified to meet any proposed or required modification in the future. In fact, a smart system for data handling, processing the measured data, and pattern recognition through training the system for automatic online detection of faults may be considered.

9.9 Types of Faults

There can be many faults that develop or keep developing in a rotating machine during its operation. However, the failure of machines in the last few decades has always identified the probable cause for the failure in the post-failure study. Many failures have been related to the system design, and many others to human error—deviation in assembly and inaccuracy in manufacturing, as well as variation in the machine operation conditions. So, there have been consistent efforts, for many years, on design modifications and the development of fault identification tools that can solve problems in the lowest possible time if a fault is identified. The machine performance shows that there has been a significant improvement in design, as machines rarely fail due to design problems nowadays. However, other kinds of problems due to human error and deviation in operating conditions need to be avoided. Well-recognized faults resulting from other kinds of errors than the design are listed below:

1. Mass unbalance
2. Shaft bent or bow
3. Misalignment and preloads
4. Crack
5. Shaft rub
6. Fluid-induced instability
7. Turbine blade vibration

9.10 Identification of Faults

The nondestructive and nonintrusive identification of these faults is discussed in this section. Many studies, by both experiment and numerical simulation, have been carried out to characterize faults. In experiments, faults have been introduced one by one in a laboratory-scale rotating rig by researchers to study the effects on different parameters. Many researchers have numerically simulated these experimental observations or behaviors of several industrial problems of rotating machines to understand the mechanism of the system malfunction, known as MFD. These identification procedures are outlined here in a simple manner.

9.10.1 Mass Unbalance

In practice, rotors are never perfectly balanced because of manufacturing errors such as porosity in casting, nonuniform density of material, manufacturing tolerances, and loss or gain of material (e.g., scale deposit on blades of steam turbines) during operation. Obviously, these result in a high amplitude of rotor vibration due to the centrifugal force generated by the rotating rotor. So the control of machinery vibration is essential in today's industry to meet the requirement for rotating machinery vibration to be within specified levels of vibration during machine operation.

Generally, a predominant 1X component of the shaft relative or absolute displacement during normal operation may be an indication of rotor unbalance. The polar plots of the 1X component for both horizontal and vertical directions near all bearings may show increase in amplitude with and without any significant change in phase with time. The Bode plot of the 1X component for the machine run-down (transient operation) should look similar to Figure 9.11. The accepted practice for unbalance estimation is known as the *sensitivity method*, which is explained in Section 9.12.

9.10.2 Shaft Bow or Bend

Bent or bowed shafts may be caused in several ways, for example, due to creep, thermal distortion, or a previous large unbalance force. The forcing caused by a bend is similar, although slightly different, to that caused by mass unbalance. The shaft vibration spectra are generally dominated by the 1X component, but the 1X amplitude and its phase in polar plots (or amplitude-phase versus time plots) may vary significantly with time during normal machine operation like mass unbalance. The best way to detect the presence of the shaft bow is through the transient operation of machine. There may be a significant change in the 1X amplitudes and phases when machine passes through the critical speeds during machine transient operation. These changes generally depend on the shaft bow angle.

The bent shaft can be corrected and then rebalanced after bend correction. Subsequently, the machine can be returned to operation. Alternately, if the bend is small and the correction is costly, then estimated corrective masses may be added to respective planes of the rotor without bothering about the bend in the shaft. This will reduce the machine vibration at the operating speed, but may not balance the rotor at all frequencies in the machine run-down speed range. The work by Edwards et al. (2000) will be a great help for such cases.

9.10.3 Misalignment and Preloads

The preload and misalignment in a rotor system are interrelated. The most general source of preload is a unidirectional steady-state force acting on the fluid bearings in a rotor system due to the shaft self-weight. Other sources of preloads are due to misalignment in the shaft at coupling locations, or relatively small offsets in the bearing's position. In practice, the shaft misalignment may be of three types—parallel, angular, or coupled misalignment (Figure 9.13). It is generally accepted that a significant 2X vibration response of the shaft relative displacement is a major feature of misalignment, and the orbit plot consisting of 1X and 2X signals may be close to the shape of a figure eight if misalignment causes excessive loads.

Estimation of both the multiplane rotor unbalance and the shaft misalignment at the coupling for rotating machines has been developed recently and

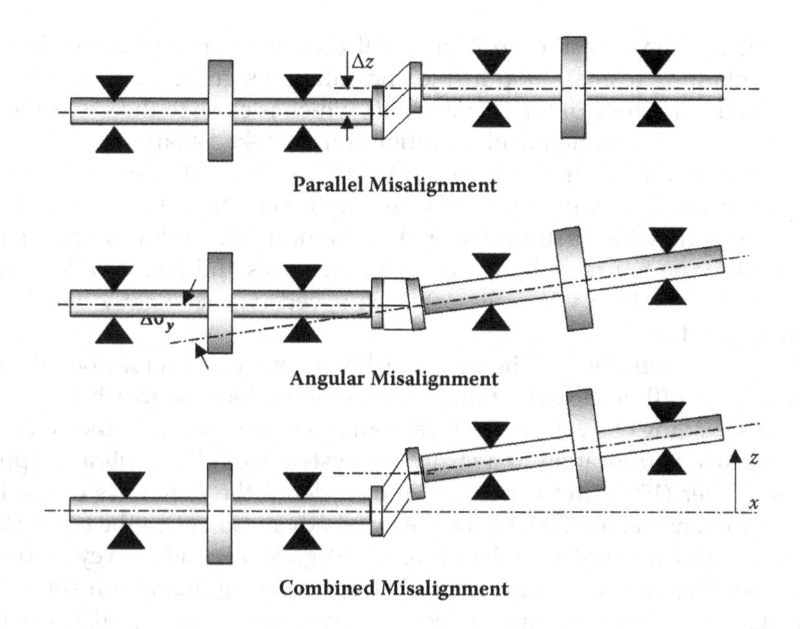

FIGURE 9.13
Possible types of rotor misalignment.

validated on a small experimental rig (Sinha, 2002; Sinha et al., 2003, 2004a; Lees et al., 2003). However, machines like TG sets, where misalignment is considered as the preloads at bearings. Hence, an alternate method was suggested to estimate the increase in bearing loads at bearings (Sinha et al., 2004b) that will give the amplitude of misalignment (Lees, 2003). If misalignment is detected in the rotor, then it needs to be aligned, and this may be a time-consuming process. Campbell (1993) discusses different alignment methods used in practice.

9.10.4 Crack

If a transverse crack develops in a shaft, the stiffness of the shaft will vary from high to low to high in a complete revolution of the shaft caused by breathing (opening and closing) of the crack due to the rotor self-weight. This behavior of the crack shaft also generates a 2X component similar to shaft misalignment. But unlike the misaligned shaft, both the amplitude and phase of the 1X and 2X components change with the time of machine operation due to the propagation of the transverse crack. This can be observed in either the polar plots or the amplitude-phase versus time plots. The orbit plot consisting of 1X and 2X components may change from a single loop to double loops like a figure eight. However, during transient operation of the machine, the shaft vibration may be very high when the machine passes through nearly half of the machine critical speed. At this particular moment the shape of the orbit plot will change from a figure eight to a loop containing a small loop inside, indicating a significant change in the phase and amplitude of the 2X vibration, which is close to the system natural frequency. Experimental observations on a cracked shaft supported on a rigid foundation during machine run-up are similar to the above-described shape and are shown in Figure 9.14. As can be seen from Figure 9.14, a single loop (with small distorted shape) in the orbit plot at lower speeds has changed to the figure eight as the speed increases and then to a loop containing another loop when the rotor rig speed reaches nearly half of the critical speed.

Online quantification of the exact crack size and location may be difficult; however, the offline identification of crack size and location may be relatively simple. Doebling et al. (1996, 1998) gave an extensive survey on the detection of crack size and location in a structural system from their vibration properties. Wauer (1990) and Gasch (1993) considered the dynamics of cracked rotors. Sabnavis et al. (2004) gave the latest advancement in the crack shaft detection and diagnosis. Sinha et al. (2002) gave a broad survey of previous modeling approaches for cracks before proposing their own simplified model for crack modeling and detection. Penny and Friswell (2002) have further advocated the usefulness of this simplified model for cracks in condition monitoring. Sinha et al. (2004c) gave one possible strategy for offline estimation of the location and size of cracks in a rotor. There could be many other possibilities.

9.10.5 Asymmetric Shaft

An asymmetric shaft is used in many industrial applications. In such cases, the shaft stiffness varies in a similar way to a cracked shaft during rotation. Hence, these shafts would also generate 2X vibration, but the amplitudes and the phases at 1X and 2X are expected to be constant with time for the rotor system having asymmetric shaft. So, it is important to know the difference in the 1X and 2X behavior of a cracked shaft, shaft misalignment, and an asymmetric shaft. The ratio of 1X and 2X components, both amplitudes and phases, will not change with time for an asymmetric shaft. However, change in both components will occur for a cracked shaft if the crack propagates with time, and the responses at 1X and 2X may be similar to the asymmetric shaft in the case of shaft misalignment. Furthermore, 2X components of shaft vibration would appear right from the start of machine operation in the case of an asymmetric shaft and misalignment. Thus, it is important to measure the vibration response when the shaft is aligned, especially for a machine with an asymmetric shaft, to distinguish if the 2X component is due to an asymmetric shaft or misalignment. Otherwise, Bode plots of 1X and 2X components during machine transient operation may distinguish these cases.

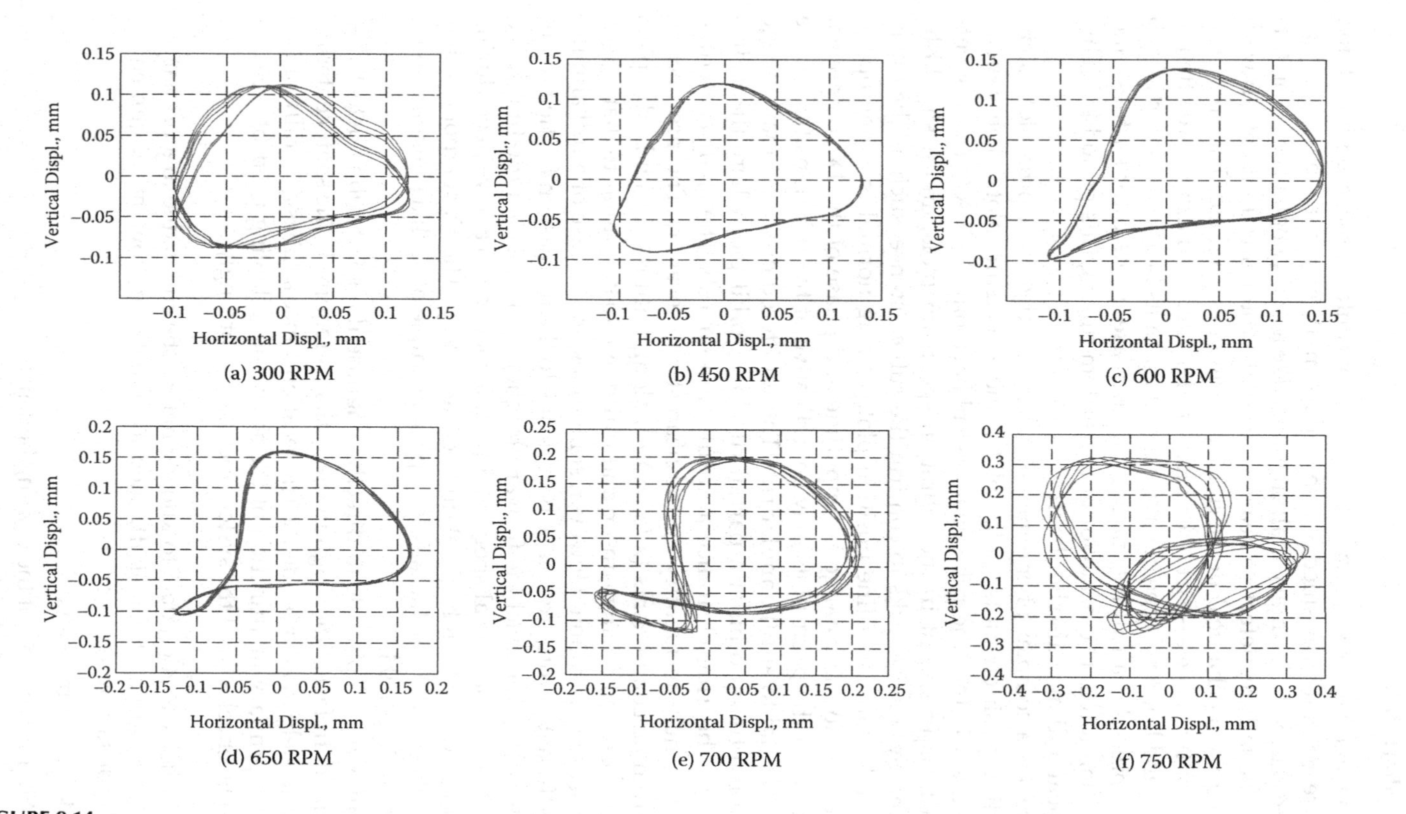

FIGURE 9.14
Orbit plots at different rotor speeds during the small rig run-up.

9.10.6 Shaft Rub

Rotating machine rubs occur when a rotor contacts the machine stator. Rubs are generally classified as a secondary malfunction, as they may typically be caused by primary malfunctions, like a poorly balanced rotor, turbine blade failure, defective bearings or seals, rotor misalignment, bowed shaft, either mechanical or thermal, deformed casing, etc. The shaft rub may be of two types—partial or full annular rub. Partial rubs are often of the "hit and bounce"-type behavior of a rotor in its bearings. A partial rub usually precedes a full rub causing an increase in the rotor stiffness and the friction force. The behavior of the system during this period is highly nonlinear and may also be chaotic.

Since it is a complex phenomenon, it is necessary to use orbit plots, shaft centerline position plots, polar plots, and spectrum plots. The general observation is that the rub generates subharmonic X/2, X/3 components of shaft displacement in the vibration spectrum plots, and it should be followed by other checks to confirm the rub existence, such as the polar, orbit, and shaft centerline position plots during normal machine operation. The abnormal change in amplitude and phase of the 1X component with time in the polar plots may be indicative of the rub transition phase from a partial to a full annular rub. The unfiltered orbit plot may change from a multiloop orbit at a partial rub to an orbit plot where the radius equals the bearing clearance at a full annular rub. The shaft centerline plots may help in identifying the nearest rub location. The presence of subharmonic components (X/2 or X/3) along with the 1X and orbit plots may be observed during run-up and run-down to be similar to that of normal machine operation. However, sometimes, in the case of a full annular rub, self-excited rotor response occurs due to the increase in friction forces (counteracting the machine forward torque) at the rub at a rotating speed just near the first critical speed and remains locked to a low speed during the run-down or a high speed during the run-up. The 1X component may not be observed in such cases. A typical example of such a phenomenon is shown in Figure 9.15.

There is no simple rule to directly diagnose if a shaft rub is detected or confirmed. In fact, a thorough investigation is required. The shaft centerline plots may help in identifying the rub location. Since a rub is generally caused by some primary malfunction of the shaft, the following probable reasons should be checked and the system rectified accordingly:

1. Earlier vibration data to know whether the rub is due to a primary malfunction of the shaft, such as unbalance, bow, misalignment, blade failure, etc., or not
2. Bearing lubrication
3. Alignment of all of the bearing housings
4. Bearing seals

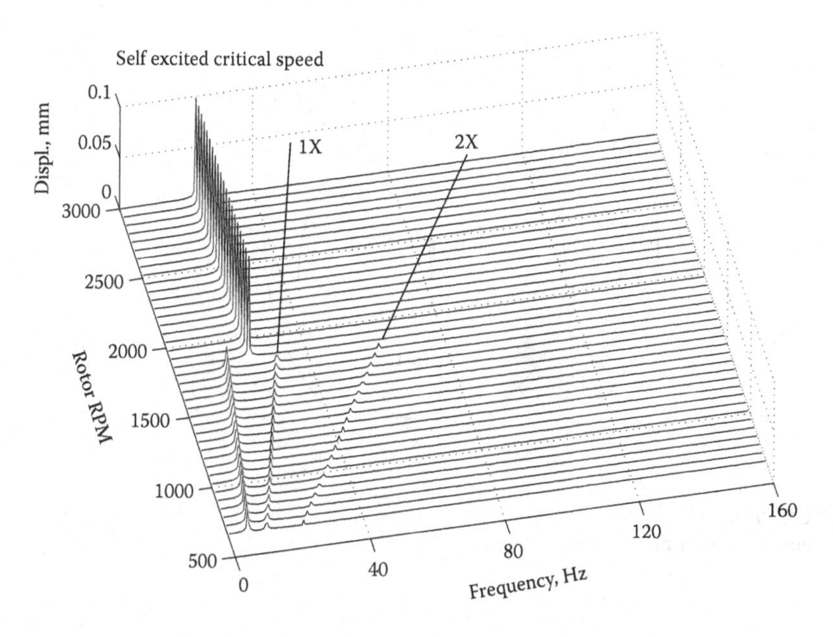

FIGURE 9.15
A typical case of a full annular rub during a run-up.

9.10.7 Fluid-Induced Instability

A fluid-induced instability is commonly referred to as *oil whip*, which results from *oil whirl*. Both typically occur in fluid bearings. This is actually a fluid-induced self-excited phenomenon leading to lateral shaft vibration. Oil whip occurs when the shaft rotates in the bearing and the fluid surrounding the journal (see Figure 9.16) also rotates at some circumferential speed ($\lambda\Omega$) relative to the shaft speed (Ω), where λ is known as the fluid average velocity ratio. The ratio λ is generally a nonlinear function of the fluid bearing radial stiffness, which depends on the shaft eccentricity ratio. It has been observed experimentally and analytically that the fluid bearing radial stiffness increases with the journal radial deflection and the velocity ratio λ decreases from maximum to zero; i.e., journal is in direct contact with the wall of bearing. The fluid resonance frequency of the fluid in the bearing during shaft rotation is normally slightly less than half of the shaft rotating speed, say 0.48Ω. If during machine operation or start-up the fluid circumferential speed comes close to the fluid resonance frequency, an instability of the shaft occurs that is known as oil whirl. This is generally identified by the presence of a λX (usually 0.45 to 0.48X or less) component in the spectrum and the forward circular orbit plot for the filtered λX component irrespective of the asymmetric stiffness in the vertical and the horizontal direction. If the rotor system natural frequency is equal to the fluid resonance, then the system would be completely unstable. The orbit plot may then be of the order

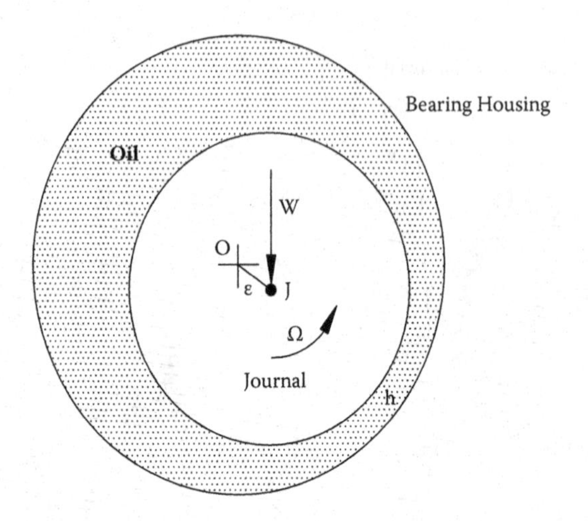

FIGURE 9.16
Schematic of a simple journal bearing.

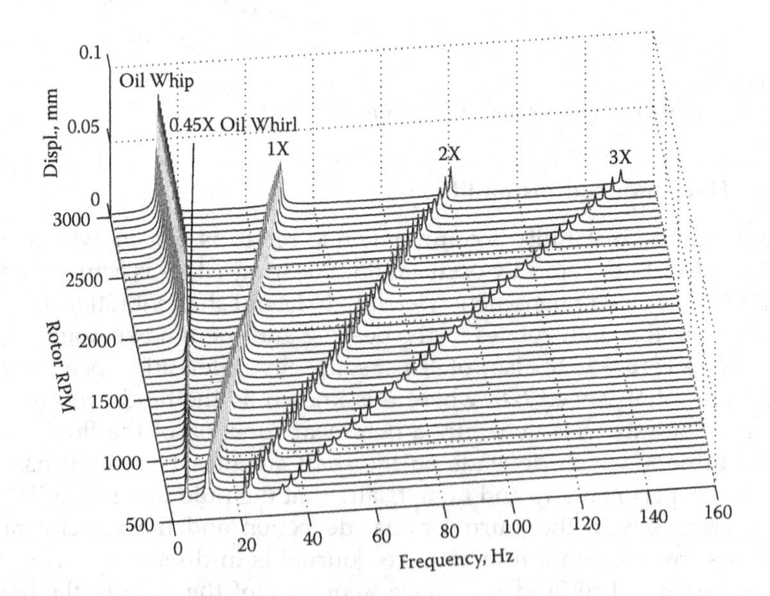

FIGURE 9.17
A typical case of oil whirl–oil whip during a machine run-up.

of the bearing clearance. This phenomenon is known as oil whip. Oil whip can occur at the first lateral mode as well as at higher modes of the rotor. Oil whip generally occurs during machine start-up. It initiates with oil whirl and then gets locked to the oil whip when the shaft speed passes through a critical speed. A typical example during a machine start-up is shown in Figure 9.17.

Rotor systems in TG sets have many fluid bearings, and if a fluid-induced instability is observed during machine operation, it is important to identify the bearing that is causing such an instability. The phase relation at the frequency of instability (λX) of the shaft relative displacement between the measurements at all the bearings can be used to identify the bearing. For the bearing that is suspected as the source of instability, its shaft relative displacement at the λX frequency component will have a phase lag with respect to the shaft relative displacement at other bearings.

In practice, this effect is usually considered while designing the systems to avoid the occurrence of this phenomenon. The important factors are short and stiff shafts and better designs of the bearings, instead of plain cylindrical bearings. Perhaps the scope of modifying the shaft design may be limited; however, improvement in the fluid bearing may be possible. Significant improvements have been achieved in bearing design. These improvements include the use of elliptical or lemon-shaped bearings with grooves and multilobes, rough journal surfaces, and bearings with mobile parts (such as tilting pad bearings), and the option of fluid antiswirl injections in bearings instead of just plain cylindrical bearings to avoid increase in the fluid circumferential velocity (Muszynska and Bently, 1996). These types of fluid bearings are generally used in machines. However, there is always a possibility of some system failure causing a drop in the fluid pressure, leakage, or human error during bearing assembly that may lead to fluid-induced instability.

9.10.8 Mechanical Looseness

Bolts, joints, and the bearing assembly may loosen over a period of normal machine operation. This looseness generates subharmonics (X/2 or X/3) and higher harmonics of the shaft displacement during normal operation of machines.

9.10.9 Blade Vibration

There is little literature on the measurement of blade vibration in condition monitoring, and absolutely no rule on vibration severity limits exists. A few measurement schemes that have been attempted on TG sets are discussed here.

1. **Blade passing frequency:** This is monitoring the amplitude of the blade passing frequency by measuring the responses at the casing near different stages of turbine blades or at bearing. But it may not be a justified way for condition monitoring for blades, as their amplitudes are generally found to be related to the 1X responses, which are related to the rotor unbalance or the preloads in the bearings due to a misaligned shaft. This condition may result in additional

loading to the blades. Hence, it may or may not give any information directly related to the blades.

2. **Casing vibration:** The casing vibration near the last stage of the LP turbine was also observed to show a trace of blade natural frequency only during fluctuations in condenser pressure due to changes in load. However, this measurement scheme may not be reliable enough as it contains very limited information.
3. **Shaft torsional vibration:** In recent years, the measurement of shaft torsional vibration was found to be a reliable way of detecting the blade natural frequencies during machine normal and transient operations. It was experimentally verified on small rigs, and the mistuning effect on the blade natural frequency was also observed. The potential of the shaft torsional vibration could be used for condition monitoring for turbine blades.
4. **Blade tip timing (BTT) method:** This is an in-direct approach to measuring and monitoring the deflection of each blade in the bladed discs during machine operation. It is an intrusive method that uses a number of noncontact sensors circumferentially mounted on the casing just above the blades' tip to measure the blade arrival time. The measured arrival times acquired by the different sensors are then converted to the blade vibration displacement (Russhard, 2010).

Perhaps the blade vibration can now be measured by monitoring the shaft torsional vibration using required instrumentation and analysis. The amplitude of the blade vibration can also be estimated, and its fatigue failure life may then be estimated. However, the recent concern is about the detection of crack in the blades, which is still a subjective matter among researchers and practicing engineers. This is because, in general, the natural frequency of any stage is not a unique number. It often depends on the root fixity of each blade, and the frequency of any stage is usually in a frequency band, say, 10% of mean frequency. So, in general, for actual machines, say, TG sets, the shaft torsional vibration itself will show a banded natural frequency of blades. Hence, the change in natural frequency due to crack cannot be distinguished. Hence, the present effort is on pattern recognition techniques, but no results have been published so far.

9.10.10 General Comments

The identification procedures for known faults in rotating machines are explained above. However, it is often observed that more than one fault appears simultaneously, so the identification may not be straightforward, and it may require considerable experience in condition monitoring.

9.11 Condition Monitoring for Other Rotating Machines

The discussion so far has been oriented toward the condition monitoring of turbogenerators. In principle, these methods are applicable to other types of rotating machines—pumps, compressors, electric motors, gas turbines, etc. ISO 2372 and 3945 are also applicable to overall vibration severity limits for all types of horizontal machines. The Hydraulic Institute (1983) gave the vibration severity limits for all kinds of rotating machines. The revised ISO10816-1 (1995) code includes both horizontal and vertical rotating machines. Many of these rotating machines have rotors that are supported through antifriction ball or roller bearings. Faults in such bearings also develop, and detection of these faults is also important. The most common techniques for detection of faults in antifriction bearings are crest factor and kurtosis measurements. The envelope or, more precisely, the amplitude demodulation at the carrier frequency (usually the rotating speed of a machine) may locate the exact nature of faults by identifying the bearing characteristic frequencies directly related to the defect in each part of a bearing.

9.11.1 Detection of Fault(s) in Antifriction Bearings

Any defect in antifriction bearing causes an impulsive loading consisting of high frequencies up to a few kHz during each revolution of the shaft. It may perturb the bearing pedestal acceleration responses, which can be quantified by the following nondimensional parameters:

1. **Crest factor (CF):** This is the ratio of the peak to the RMS acceleration (Section 6.5) of the bearing pedestal response. If a machine's response is purely sinusoidal at its rpm, the value of CF will be 1.414. The value increases to a high value due to the presence of an impulsive signal in the response due to the bearing defect. However, it has been observed that the CF is partially insensitive to changes in bearing load and speed.
2. **Kurtosis (Ku):** Statistical approach that has proven the most successful method of quantifying the characteristic changes due to the bearing defect, unlike the CF. The kurtosis is one of the most successful methods for this requirement. It is defined as the fourth-order statistical moment (Section 6.5) of the bearing pedestal acceleration response. The kurtosis value for the sine wave is 1.5, whereas the value is 3.0 for the waveform with Gaussian amplitude distribution. Hence, the measured value of 3.0 of an undamaged bearing indicates an acceleration waveform with a Gaussian amplitude distribution. A small increase from 3.0 often indicates the bearing may need some lubrication or there may be a minor defect; however, if the value is more than 5–6 and increasing with time, this indicates the propagation of a fault in the bearing.

9.11.2 Characteristic Frequencies of a Ball Bearing

A ball bearing consists of an inner race, outer race, and number of balls. The inner race is directly mounted on the shaft, and the outer race rests on the bearing housing or machine pedestal. A simple schematic of a ball bearing is shown in Figure 9.18. The following are the characteristic frequencies of each part.

Let d_b, d_i, and d_o be the diameter of the balls, the inner race (OD), and the outer race (ID), respectively; n is the number of balls. Then:

$$\text{Frequency related to the ball defect, } f_b = \frac{d_p}{d_b} f_r \left(1 - \left(\frac{d_b}{d_p}\right)^2 \cos\beta^2\right) \text{Hz} \tag{9.1}$$

$$\text{Frequency related to defect in the inner race, } f_i = \frac{n}{2} f_r \left(1 + \frac{d_b}{d_p} \cos\beta\right) \text{Hz} \tag{9.2}$$

$$\text{Frequency related to defect in the outer race, } f_o = \frac{n}{2} f_r \left(1 - \frac{d_b}{d_p} \cos\beta\right) \text{Hz} \tag{9.3}$$

$$\text{Frequency related to defect in the cage, } f_o = \frac{f_o}{n} \text{Hz} \tag{9.4}$$

where d_p is the pitch circle diameter, β is the contact angle for the balls, and f_r is the relative speed (Hz) between the inner race (ID) and the outer race (OD), which is generally equal to the rpm of the shaft. The same formula is also valid for roller-type bearings.

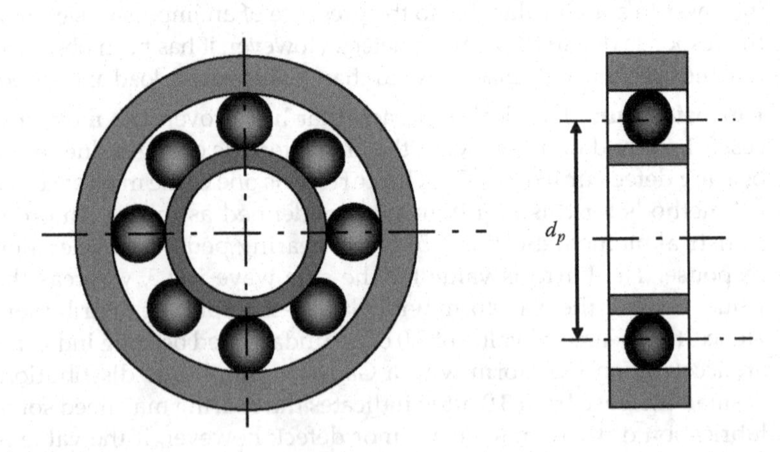

FIGURE 9.18
A simple schematic of a ball bearing.

9.11.3 Concept of Envelope Analysis

Let us assume that a horizontal centrifugal pump is supported on two bearing pedestals through ball bearings. The pump rpm is assumed to be *N*. Let us assume that the characteristic bearing frequencies are f_b, f_i, and f_o for the ball bearings. Now it is also assumed that the bearing has an inner race defect. The PSD of the bearing pedestal response should show the inner race–related frequency, f_i, but it is often observed that such frequency does not appear in the PSD when the defect is minor. Any defect in the antifriction bearing generally gives impact loading during each rotation of the shaft, which often excites the bearing housing's natural frequency, which may be of the order of 1 to 5 kHz. Usually this high-energy and high-frequency component modulates the low-energy frequency or frequencies. The modulated signal is typically shown in Figure 9.12(a). If the high-frequency component is the bearing housing's natural frequency, f_v Hz, then the PSD may appear as shown in Figure 9.20(a). The power spectral density (PSD) of such a modulated signal is typically shown in Figure 9.12(a), which has the frequency peak at f_v with side bands related to the defect inner race frequency, f_i. The high frequency (f_v) is known as the carrier frequency. Removal of such high frequency from the signal is important to know the modulated low-energy frequencies in the signal. The boundary of the modulated signal is shown in Figure 9.19(a) and also shown separately in Figure 9.19(b). This is called the envelope (demodulated) signal, which removes the effect

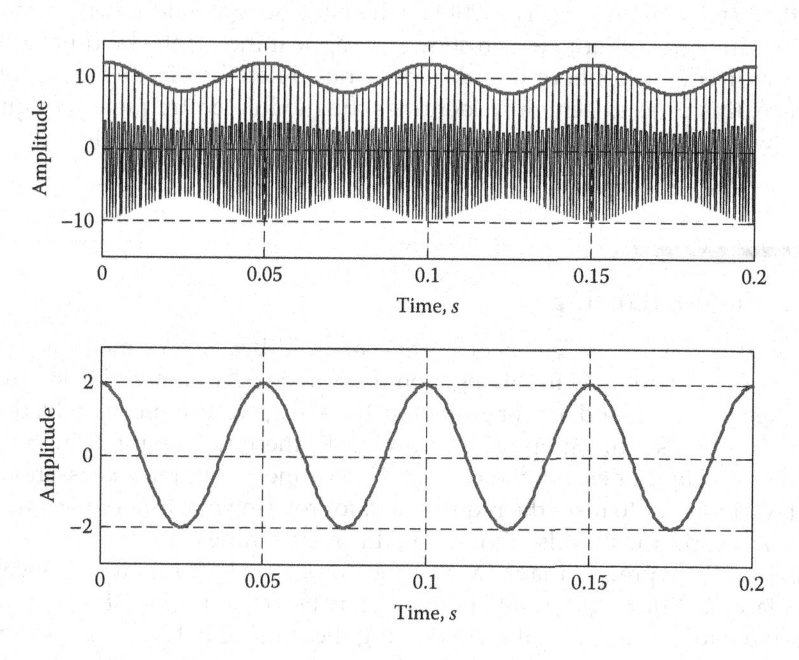

FIGURE 9.19
(a) Raw modulated signal. (b) Envelope signal.

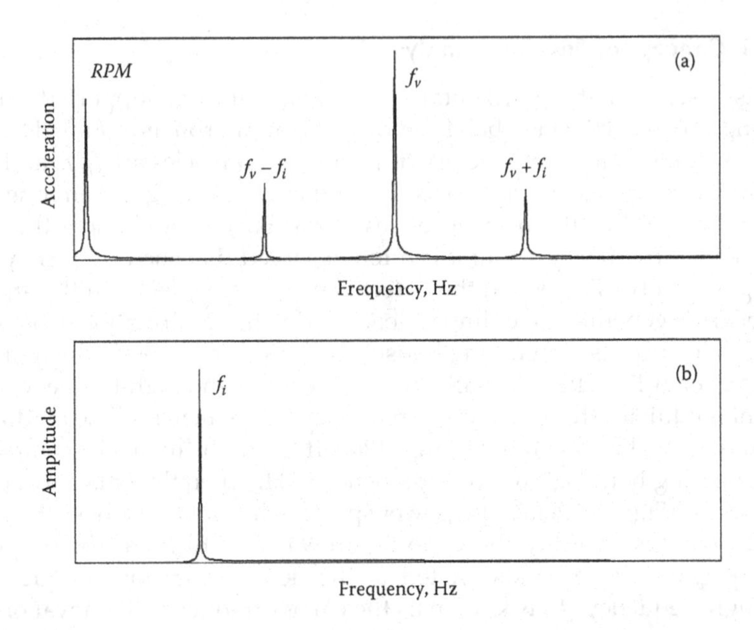

FIGURE 9.20
PSD of the bearing pedestal acceleration response. (a) Raw modulated signal. (b) Envelope signal at the carrier frequency, f_v.

of the carrier frequency. The PSD of the envelope (demodulated) signal is shown in Figure 9.20(b), which shows the appearance of the modulated frequency/frequencies; here it is the inner race related to frequency, f_i, only. Hence, this process generally identifies the modulated frequency/frequencies, which helps in the detection of the bearing defect accurately.

9.12 Field Balancing

In practice, rotors are never perfectly balanced because of manufacturing errors such as porosity in casting, nonuniform density of material, manufacturing tolerances, and loss or gain of material (e.g., scale deposit on blades of steam turbines) during operation. Obviously, these will result in high rotor vibration amplitudes. So, the control of machinery vibration is essential in today's industry to meet the requirement for rotating machinery vibration to be within specified levels of vibration during machine operation.

Generally, a predominant 1X component of the shaft relative or absolute displacement during normal operation may be an indication of rotor unbalance due to the centrifugal force causing the rotor unbalance. The accepted practice for unbalance estimation is known as the *sensitivity method*, which is explained here.

9.12.1 Single Plane Balancing—Graphical Approach

To aid understanding, a simple example of a short-length rotor with bladed disc, typically found in pumps, fans, compressors, etc., is considered here. A typical simplified form of a representative rotor with a balance disc and a bearing is shown in Figure 9.21.

9.12.1.1 Example 9.1

Let us consider vibration is just measured at a bearing of a small rotor. Now it is assumed that the shaft is rotating anticlockwise at 3000 rpm. The amplitude of vibration is 30 µm at 50 Hz (3000 rpm) at an angle of 20° from the reference signal of the tacho signal. A stroboscope can also be used to measure this angle. The time waveform of the measured vibration signal may look like the one shown in Figure 9.22(a). Figure 9.22(b) shows the *x*-axis in the shaft rotation angle for two complete rotations, which clearly indicates a phase angle of 20°. A tacho sensor measures the reference location marked on the shaft during each rotation. The unbalance response (**Ov**) of 30 µm @ 20° at an operating speed of 50 Hz (3000 rpm) is observed due to the unbalance (centrifugal) force, $m_{ub}r\omega^2$, where m_{ub} is the rotor unbalance mass assumed at the balance plane, r is the radius from the shaft center for the unbalance mass, and $\omega = 2\pi f$ is the shaft angular speed. Now it is important to know the behavior of the machine to calculate the unbalance mass (m_{ub}) and its location (both radial, r, and angular, θ_{ub}, locations) to reduce the rotor unbalance. To know the change in the machine dynamics behavior let us assume that a small trial mass m_t = 5 g added to the balance plane at radius r. The vibration (**Ov** + **Tv**) now becomes equal to 15 µm @ 80°. Both **Ov** and **Ov** + **Tv** are marked on the polar plot in Figure 9.23, where the radius of the circle indicates the vibration amplitude together with the phase angle. The arrow **Tv** (= 25.98 µm @ 30° with **Ov**)

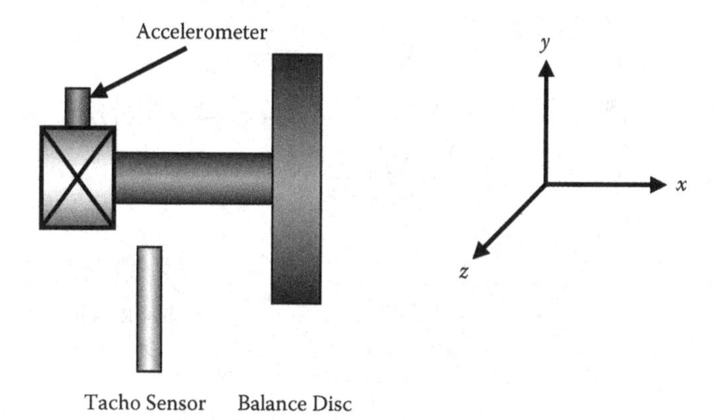

FIGURE 9.21
Schematic of a rotor with a balance disc together with measuring instruments.

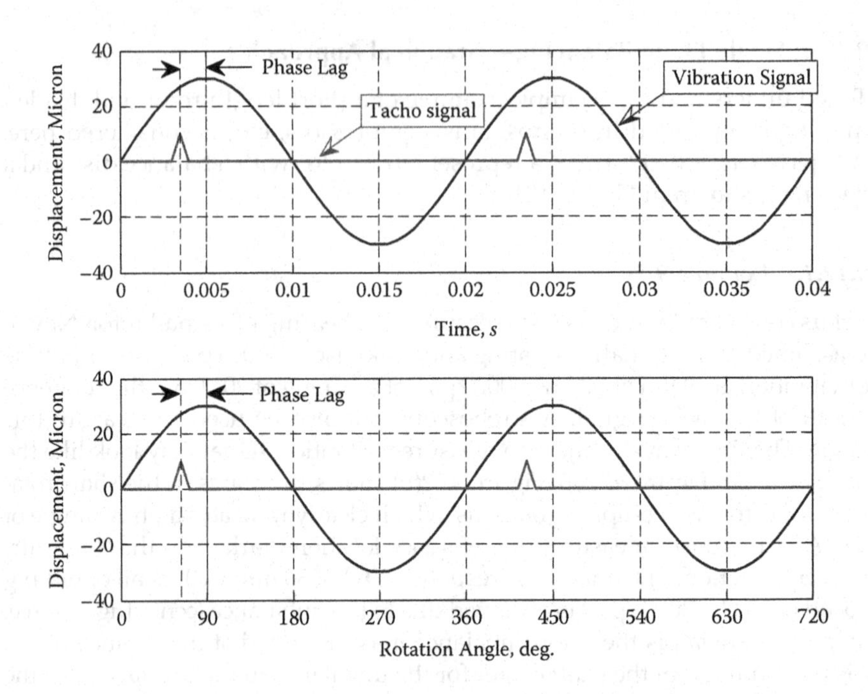

FIGURE 9.22
Measured vibration displacement signal together with tacho signal.

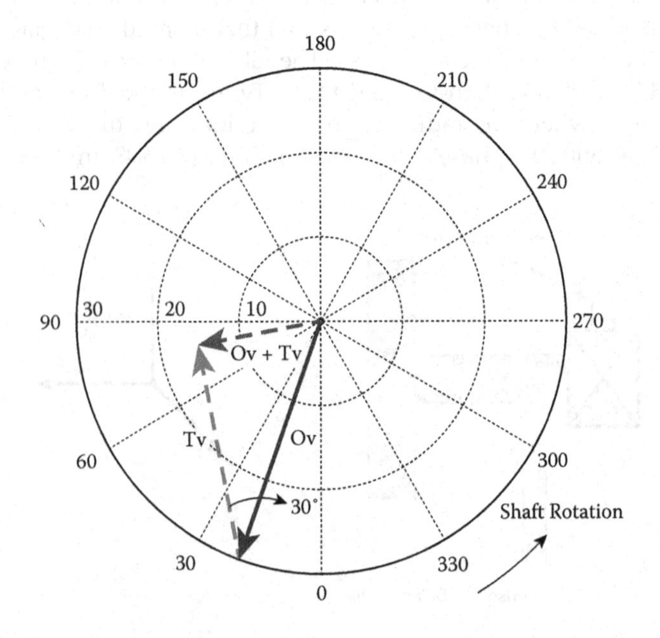

FIGURE 9.23
Polar plot: A graphical estimation method for the rotor unbalance for Example 9.1.

indicates the change in machine dynamics due to the trial mass; hence, the sensitivity (S) for the machine is defined as

$$S = \frac{|\mathbf{Tv}|}{m_t} = \frac{|\mathbf{Ov}|}{m_{ub}} \tag{9.5}$$

Thus,

$$m_{ub} = \frac{|\mathbf{Ov}|}{|\mathbf{Tv}|} m_t = \frac{30}{25.98} 5 = 5.77 \text{ g}$$

The clockwise angle of the arrow **Tv** with the arrow **Ov** is nearly 30°. Hence, the calculated unbalance mass of 5.77 g should be added to the balance disc plane at the radius r at an angle of 30° away from the trial mass location in the opposite direction to the shaft rotation (clockwise) to balance the rotor. The trial mass should then be removed. This will complete the balancing process for a small shaft and the vibration would be expected to reduce drastically.

9.12.1.2 Example 9.2

This is another example of a small rotor with a balancing disc of radius of 50 mm. The rotor is rotating at 3000 rpm in the anticlock direction. The measured vibration (**Ov**) on the bearing in the vertical direction is found to be 52 μm @ 30°, which therefore requires balancing. The vibration amplitude (**Ov** + **Tv**) changed to 90.21 μm @ 60° when a 1 g trial mass was added to the balancing disc at a radius of 50 mm @ 90° from the tacho reference. The unbalance mass and its location for the rotor balancing are now estimated as follows using the polar plot shown in Figure 9.24. The amplitude of arrow **Tv** is measured to be 52 μm from the polar plot, and then the unbalance mass is calculated as $m_{ub} = \frac{|\mathbf{Ov}|}{|\mathbf{Tv}|} m_t = \frac{52}{52} 1 = 1$ g. The arrow **Tv** makes a 120° angle with the arrow **Ov** in the clockwise (opposite to rotor rotation) direction in the polar plot (see Figure 9.24). Hence, the estimated mass of 1 g has to be added on the disc at the radius of 50 mm at an angle 120° from the trial mass location in the opposite direction of shaft rotation to balance the rotor. The trial mass must be removed during the balancing process. The disc with the trial mass and the unbalance mass is shown in Figure 9.25.

9.12.2 Single Plane Balancing—Mathematical Approach

Now a rotor with a balancing disc supported through two bearings, as shown in Figure 9.26, is considered here.

Step 1: Definition of unbalance planes and amplitudes. Ideally, the unbalance in the rotor will be distributed all along the length, but in practice, it is assumed to be distributed at a number of balancing planes.

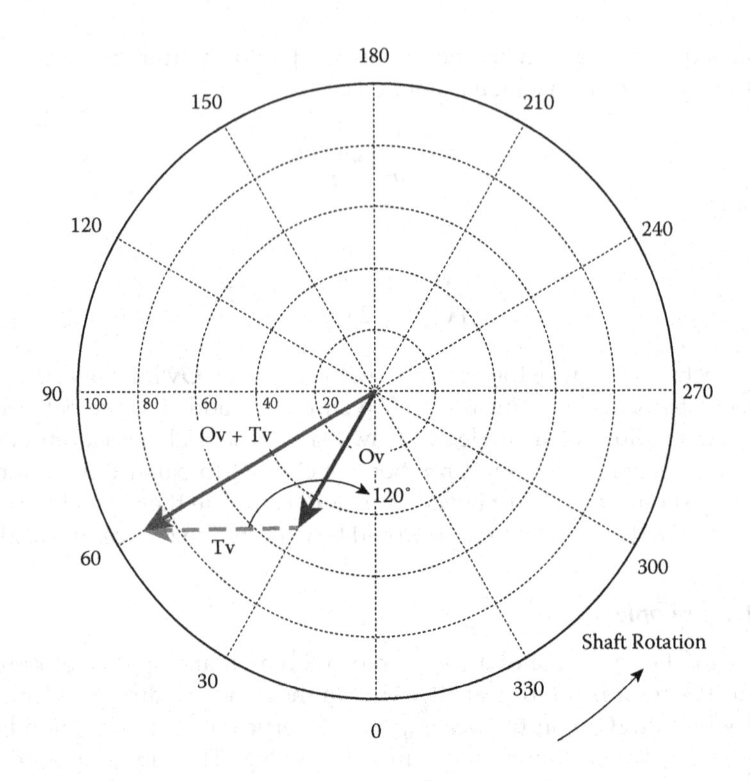

FIGURE 9.24
Polar plot for the rotor unbalance estimation for Example 9.2.

Here, there is only one balancing disc, and hence the unbalance is assumed to be in one plane with the unbalance amplitude m_1r_1 (unbalance mass multiplied with the distance between the mass and the geometric center), at an angle θ_1 with respect to the tacho reference. This is called the residual unbalance of the rotor, which is responsible for the unbalance response. The unbalance force is defined as $m_1r_1\omega^2e^{j\theta_1}$, where ω is the angular rpm of the rotor. If this residual unbalance is known, the mass (m_1) can then be added at the disc radius (r_1), at an angle $(180 + \theta_1)$, to balance the rotor. Hence, the unbalance is defined as $m_1r_1e^{j\theta_1}$, which can be expressed as the complex quantity $e_1 = e_{r,1} + je_{i,1}$. This is the quantity that needs to be estimated.

Step 2: Response measurement. There are three principal directions used in engineering to define the object. Likewise in dynamics, the response measurements in three directions are important to define the dynamic behavior of a system. But in rotor dynamics, particularly for unbalance responses, it is believed that an unbalance influences the vibration in the vertical and longitudinal directions only. Hence, the measurements in these two directions (shown in Figure 9.26) are

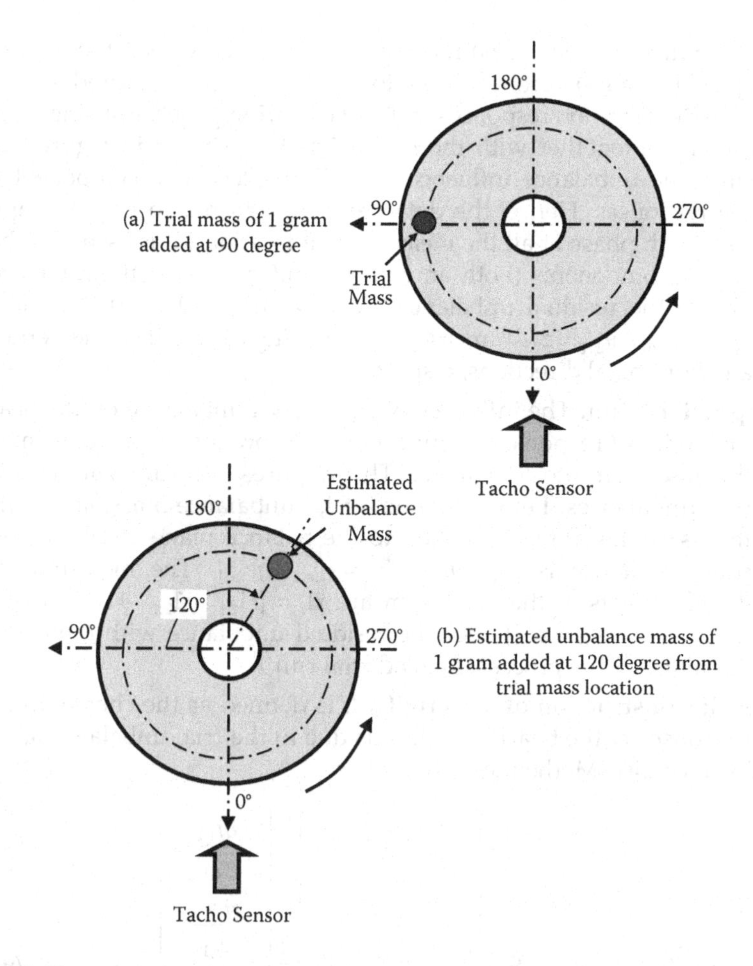

FIGURE 9.25
Disc showing location of trial mass and estimated unbalance mass. (a) Trial mass of 1 g added at 90°. (b) Estimated unbalance mass of 1 g added at 120° from trial mass location.

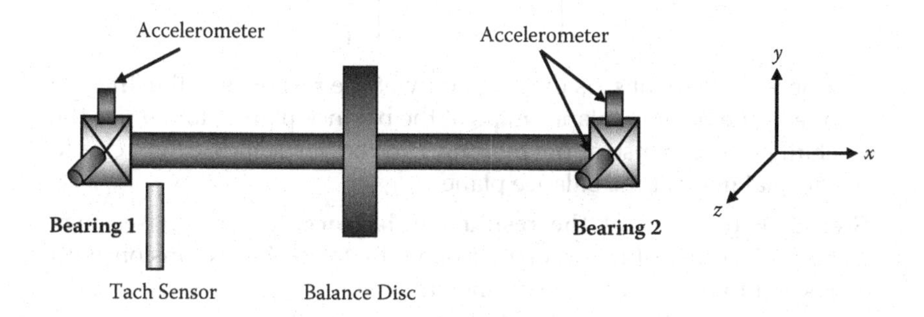

FIGURE 9.26
Schematic of a rotor with a balance disc together with measuring instruments.

generally essential. Assuming the rotor rpm is N, then the angular speed (in rad/s) becomes equal to $\omega = 2\pi N/60$. It is required to measure the vibration responses at the two bearing pedestals using accelerometers together with the tacho signal, as shown in Figure 9.26, since the unbalance influences the 1X displacement component of the responses. Hence, the data are order tracked to get 1X amplitude with phase at all the measurement locations. It is assumed that the 1X components (both amplitudes and phases) at the rotor rpm due to the residual unbalance are $\mathbf{r0} = [r0_{1,0} \;\; r0_{2,0} \;\; r0_{3,0} \;\; r0_{4,0}]^T = [y_{1,r0} \;\; z_{1,r0} \;\; y_{2,r0} \;\; z_{2,r0}]^T$ measured at bearings 1 and 2 in the vertical and horizontal directions, respectively.

Step 3: Trial run. The influence of the known unbalance on the bearing pedestal responses is important to know if one has to estimate the unknown rotor unbalance. This requires trial runs with known trial unbalances. Let us assume a trial unbalance mass, $m_{t,1}$, at the radius r_1 and angle $\theta_{t,1}$ added to the balance plane 1. The known trial unbalance is $e_{t,1} = m_{t,1} r_1 e^{j\theta_{t,1}} = e_{t,r,1} + je_{t,i,1}$. The measured 1X displacements at the rotor rpm are $\mathbf{r1} = [r1_{1,1} \;\; r1_{2,1} \;\; r1_{3,1} \;\; r1_{4,1}]^T = [y_{1,r1} \;\; z_{1,r1} \;\; y_{2,r1} \;\; z_{2,r1}]^T$ for the residual unbalance with the added unbalance. This process is called trial run 1.

Step 4: Construction of sensitivity. It is defined as the change in the responses at the bearing pedestals due to the trial unbalance at the balance disc. Mathematically,

$$\text{Sensitivity matrix, } \mathbf{S} = \begin{bmatrix} \dfrac{y_{1,r1} - y_{1,r0}}{e_{t,1}} \\ \dfrac{z_{1,r1} - z_{1,r0}}{e_{t,1}} \\ \dfrac{y_{2,r1} - y_{2,r0}}{e_{t,1}} \\ \dfrac{z_{2,r1} - z_{2,r0}}{e_{t,1}} \end{bmatrix} = \begin{bmatrix} \dfrac{\partial r_{1,1}}{e_{t,1}} \\ \dfrac{\partial r_{2,1}}{e_{t,1}} \\ \dfrac{\partial r_{3,1}}{e_{t,1}} \\ \dfrac{\partial r_{4,1}}{e_{t,1}} \end{bmatrix} = \left[s_{i,j} \right]_{4\times 1} \tag{9.6}$$

where an element $s_{i,j}$ is the sensitivity of the response at the ith DOF due to the trial unbalance mass at the balance plane j, and $\partial_{r1,1}$ is the change in response at the bearing 1 in the vertical direction due to the trial mass at the balance plane 1.

Step 5: Estimation of the residual unbalance. The responses may be written as a first-order truncated Taylor series expansion with respect to the unbalance parameters,

$$\frac{\partial \mathbf{r}}{\partial \mathbf{e}} \delta \mathbf{e} = \delta \mathbf{r} \tag{9.7}$$

Equation (9.7) for the present example can be written as

$$\frac{\partial \mathbf{r}}{e_{t,1}} e_1 = \mathbf{r}0 \tag{9.8}$$

$$\Rightarrow \mathbf{S}e_1 = \mathbf{r}0 \tag{9.9}$$

In Equation (9.9), the sensitivity matrix **S** and the residual responses **r**0 are known; hence, the residual unbalance can be estimated. However, the solution may not be straightforward, because all the parameters of Equation (9.9) are complex quantities. Equation (9.9) can be solved as discussed below:

$$\text{Residual unbalance, } e_1 = e_{r,1} + je_{i,1} = \begin{bmatrix} 1 & j \end{bmatrix} \begin{Bmatrix} e_{r,1} \\ e_{i,1} \end{Bmatrix} = \mathbf{Te} \tag{9.10}$$

Equation (9.4) becomes

$$\mathbf{S}_{4x1}\mathbf{T}_{1x2}\mathbf{e}_{2x1} = \mathbf{r}0_{4x1}$$

$$\Rightarrow (\mathbf{S}_T)_{4x2}\mathbf{e}_{2x1} = \mathbf{r}0_{4x1} \tag{9.11}$$

where $\mathbf{S}_T$ = **ST**. Separating the real and imaginary parts in Equation (9.11) as

$$\begin{bmatrix} real(\mathbf{S}_T)_{4x2} \\ imag(\mathbf{S}_T)_{4x2} \end{bmatrix}_{8x2} \mathbf{e}_{2x1} = \begin{bmatrix} real(\mathbf{r}0)_{4x1} \\ imag(\mathbf{r}0)_{4x1} \end{bmatrix}_{8x1} \tag{9.12}$$

$$\Rightarrow \mathbf{S}_s\mathbf{e} = \mathbf{r}0_s \tag{9.13}$$

now Equation (9.8) contains only real quantities, and it is directly solved as

$$\mathbf{e} = (\mathbf{S}_s^T\mathbf{S}_s)^{-1}\mathbf{S}_s^T\mathbf{r}0_s \tag{9.14}$$

Example 9.2 is considered again. Here **r**0 = $y_{1,r0}$ = **Ov** = 52 μm @ 30° = $4.5108 \times 10^{-5} + 2.6037 \times 10^{-5}j$ m, $e_{t,1} = m_{t,1}r_1e^{j\theta_{t,1}} = e_{t,r,1} + je_{t,i,1}$ = 1 g × 50 mm @ 90° = $3.0616 \times 10^{-5} + 5.0 \times 10^{-5}j$ kg-m, and **r**1 = $y_{1,r1}$ x = **Ov** + **Tv** = 90.21 μm @ 60 degree = $4.5113 \times 10^{-5} + 7.8121 \times 10^{-5}j$ m. Hence, the sensitivity matrix (Equation (9.6)) is computed as

$$\mathbf{S} = \frac{y_{1,r1} - y_{1,r0}}{e_{t,1}} = 1.0417 - 0.0001j$$

$$\Rightarrow \mathbf{S}_T = \mathbf{S}\,\mathbf{T} = [1.0417 - 0.0001j][\,1 \quad j\,]$$

$$= [\; 1.0417 - 0.0001j \quad 0.0001 + 1.0417j \;]$$

$$\Rightarrow \mathbf{S}_S = \begin{bmatrix} 1.0417 & 0.0001 \\ -0.0001 & 1.0417 \end{bmatrix}$$

Hence, the unbalance is computed as (Equation (9.14)) $\mathbf{e} = (\mathbf{S}_s^T\mathbf{S}_s)^{-1}\mathbf{S}_s^T\mathbf{r}0_s$ = $4.330 \times 10^{-5} + 2.5 \times 10^{-5}j$ kg-m. Therefore, the unbalance mass m_1 = 1 g @ θ_1 = 30°. Hence, to balance the rotor this unbalance mass is added at an angle of θ_1 + 180° = 210°, which is exactly the same as the graphical method. However, the advantage of this mathematical approach is that it can include many numbers of measurements along the rotor in comparison to the graphical approach. This can further be extended to multiplane balancing.

9.12.3 Multiplane Balancing

In the earlier example, the field balancing of a rotor with a balance disc at a rpm has been discussed in a simplified manner. However, in practice, the rotating machinery like TG sets generally consist of multirotors supported through b number of bearings, which often require multiplane rotor balancing. The steps involved are again the same as in the earlier case, which is discussed below.

Step 1. Let the unbalance at the balance planes 1 to p be $\mathbf{e} = [e_1 \;\; e_2 \;\; \cdots \;\; e_p]^T = [m_1r_1e^{j\theta 1} \;\; m_2r_2e^{j\theta 2} \;\; \cdots \;\; m_pr_pe^{j\theta N}]^T$, where the masses, $[m_1 \;\; m_2 \;\; \cdots \;\; m_p]$, are to be added to the 1 to p balance discs at the radius of $[r_1 \;\; r_2 \;\; \cdots \;\; r_p]$ at the phase angle of $[\theta_1 \;\; \theta_2 \;\; \cdots \;\; \theta_p]$ with respect to the tacho location, respectively. The unbalance can be written in the complex form as

$$\mathbf{e}_T = \begin{bmatrix} 1 & 0 & \cdots & 0 & j & 0 & \cdots & 0 \\ 0 & 1 & \ddots & 0 & 0 & j & \ddots & 0 \\ \vdots & \vdots & \cdots & 0 & \vdots & \vdots & \cdots & 0 \\ 0 & 0 & \cdots & 1 & 0 & 0 & \cdots & j \end{bmatrix} \begin{Bmatrix} e_{1,r} \\ e_{2,r} \\ \vdots \\ e_{p.r} \\ e_{1,i} \\ e_{2,i} \\ \vdots \\ e_{p,i} \end{Bmatrix} = \mathbf{Te} \quad (9.15)$$

Step 2: Response measurement. Assuming the machine runs at speed N RPM, the order-tracked 1X component of the measured responses at b bearings due to the residual unbalances can be written as $\mathbf{r0} = [r0_{1,0} \;\; r0_{2,0} \;\; \cdots \;\; r0_{2b-1,0} \;\; r0_{2b,0}]^T = [y_{1,r0} \;\; z_{1,r0} \;\; y_{2,r0} \;\; z_{2,r0} \;\; \cdots \;\; y_{b,r0} \;\; z_{b,r0}]^T$, where $2b$ is the total DOFs.

Step 3: Trial runs. Add the trial mass, $m_{t,1}$, at the radius r_1 and the angle $\theta_{t,1}$ of disc 1 and then measure 1X displacements at the rotor rpm as $\mathbf{r1} = [r1_{1,1} \;\; r1_{2,1} \;\; \cdots \;\; r1_{2b-1,1} \;\; r1_{2b,1}]^T = [y_{1,r1} \;\; z_{1,r1} \;\; y_{2,r1} \;\; z_{2,r1} \;\; \cdots \;\; y_{b,r1} \;\; z_{b,r1}]^T$. Repeat this exercise for all balancing discs one after another. A total of p number of trial runs are required to get the vibration influence on each bearing to do additional mass on the balance discs one by one. Similarly, let us assume that the measured 1X displacement vectors are $\mathbf{r2}, \mathbf{r3}, \ldots, \mathbf{rp}$ for the added trail masses, $e_{t,2} = m_{t,2} r_2 e^{j\theta_{t,2}}$, $e_{t,3} = m_{t,3} r_3 e^{j\theta_{t,3}}, \ldots, e_{t,p} = m_{t,p} r_p e^{j\theta_{t,p}}$, on the balance planes (discs) 2, 3, …, p, respectively. It is important to note that when a trial mass is added to a balance plane, the trial mass on other balance planes must be removed.

Step 4: Construction of sensitivity. It is constructed as

$$\text{Sensitivity matrix, } \mathbf{S} = \begin{bmatrix} s_{1,1} & s_{1,2} & \cdots & s_{1,p} \\ s_{2,1} & s_{2,2} & \cdots & s_{2,p} \\ \vdots & \vdots & \ddots & \vdots \\ s_{2b,1} & s_{2b,2} & \cdots & s_{2b,p} \end{bmatrix} = \left[s_{i,j} \right]_{2bxp} \tag{9.16}$$

where an element $s_{i,j} = \dfrac{rj_{i,j} - r0_{i,j}}{e_{t,j}}$ is the sensitivity of the measured response for the ith DOF due to the trial unbalance at the balance plane j. The DOFs $i = 1, 3, 5, \ldots, (2b-1)$ are related to the vertical DOFs at the bearings 1 to b, respectively. Similarly, even DOFs $i = 2, 4, 6, \ldots, 2b$ are the horizontal (lateral) DOFs at the bearings 1 to b, respectively. The residual unbalance is then estimated using Equation (9.14).

9.13 Comments about Model-Based Fault Diagnosis (MFD)

Most numerical simulations by researchers have assumed very simple rotating machine models to understand the machine malfunctions. The foundation models for such numerical simulations have been, in general, assumed to be either rigid or a simple model. Hence, such an approach is good for understanding the rotor system phenomenon only. However, to really understand the machine faults and their solution, it is necessary to have or develop a good mathematical model of that machine, e.g., models of the rotor, the fluid bearings, and the foundation (often flexible) for TG sets. Inclusion of the foundation model is very important, as it is observed that the dynamics of the flexible foundation also contribute significantly to the dynamics of the complete machine.

Once a good model of the complete machine is known, the malfunction in the machine can be simulated accurately, and that will also provide sufficient information for quantifying the faults and then solving the problem. Lees et al. (2009) gave an excellent review of MFD methods.

References

ASME Code on Operation and Maintenance of Nuclear Power Plants. 1990. *Requirement for Pre-Operational and Initial Start-Up Vibration Testing of Nuclear Power Plant Piping System*, Part 3. ASME/ANSI-O&M.

Campbell, A.J. 1993. Static and Dynamic Alignment of Turbomachinery. *Orbit*, 24–29.

Doebling, S.W., Farrar, C.R., Prime, M.B. 1998. A Summary Review of Vibration-Based Damage Identification Methods. *Shock and Vibration Digest* 30(2), 91–105.

Doebling, S.W., Farrar, C.R., Prime, M.B., Shevitz, D.W. 1996. *Identification and Health Monitoring of Structural Systems from Changes in Their Vibration Characteristics.* Los Alamos National Laboratory Report, No. LA-13070-MS.

Edwards, S., Lees, A.W., Friswell, M.I. 2000. Experimental Identification of Excitation and Support Parameters of a Flexible Rotor-Bearing-Foundation System from a Single Run-Down. *Journal of Sound and Vibration* 232(5), 963–992.

Forland, C. 1999. Why Phase Information Is Important for Diagnosing Machinery Problems. *Orbit*, 29–31.

Gasch, R. 1993. A Survey of the Dynamics Behaviour of a Simple Rotating Shaft with a Transverse Crack. *Journal of Sound and Vibration* 160(2), 313–332.

Hydraulic Institute. 1983. *Hydraulic Institute Standard for Centrifugal, Rotary and Reciprocating Pumps*, 14th ed.

ISO 2372. 1974. *Mechanical Vibration of Machines with Operating Speeds from 10 to 200rps—Basis for Specifying Evaluation Standards.*

ISO 3945. 1980. *The Measurements and Evaluation of Vibration Severity of Large Rotating Machines, In Situ, Operating at Speeds 10 to 200rps.*

ISO 7919/1. 1986. *Mechanical Vibration of Non-Reciprocating Machines—Measurements on Rotating Shafts and Evaluation, Part 1: General Guidelines.*

ISO 7919/2. 1986. *Mechanical Vibration of Non-Reciprocating Machines—Measurements on Rotating Shafts and Evaluation, Part 2: Measurement and Evaluation of Shaft Vibration of Large Turbine-Generator Sets.*

ISO 10816-1. 1995. *Mechanical Vibration—Evaluation of Machine Vibration by Measurements on Non-Rotating Parts—Part 1: General Guidelines.*

ISO 10816-6. 1995. *Mechanical Vibration—Evaluation of Machine Vibration by Measurements on Non-Rotating Parts—Part 6: Reciprocating Machines with Power Ratings above 100 kW.*

Lees, A.W. 2003. Where Next for Condition Monitoring? *Key Engineering Materials* 245–246, 203–214.

Lees, A.W., Sinha, J.K., Friswell, M.I. 2003. Estimating of Rotor Unbalance and Misalignment from a Single Run-Down. In *5th International Conference on Modern Practice in Stress and Vibration Analysis*, University of Glasgow, Scotland, UK, September 9–11, 229–236.

Lees, A.W., Sinha, J.K., Friswell, M.I. 2009. Model Based Identification of Rotating Machines. *Mechanical Systems and Signal Processing* 23(6), 1884–1893.

Muszynska, A., Bently, D.E. 1996. Fluid-Induced Instabilities of Rotors: Whirl and Whip—Summary of Results. *Orbit*, March, 7–15.

Parkinson, A.G., McGuire, P.M. 1995. Rotordynamics Standards: New Developments and the Need for Involvement. *Proceedings of the Institute of Mechanical Engineers Part C—Journal of Mechanical Engineering Science* 209(C5), 315–322.

Penny, J.E.T., Friswell, M.I. 2002. Crack Modelling for Structural Health Monitoring. In *Proceedings of 3rd International Conference—Identification in Engineering Systems*, Swansea, UK, April 15–17, 221–231.

Russhard, P. 2010. Development of Blade Tip Timing Based Engine Health Monitoring System. EngD thesis, University of Manchester, UK.

Russell, P. 1997. Why Install a Keyphasor Transducer? *Orbit*, June, 34–35.

Sabnavis, G., Kirk, R.G., Kasarda, M., Quinn, D. 2004. Cracked Shaft Detection and Diagnostics: A Literature Review. *The Shock and Vibration Digest* 36, 287–296.

Sinha, J.K. 2002. Health Monitoring Techniques for Rotating Machinery. Ph.D. thesis, University of Wales Swansea, Swansea, UK.

Sinha, J.K., Friswell, M.I., Edwards, S. 2002. Simplified Models for the Location of Cracks in Beam Structures Using Measured Vibration Data. *Journal of Sound and Vibration* 251(1), 13–38.

Sinha, J.K., Lees, A.W., Friswell, M.I. 2003. Multi-Plane Balancing of a Rotating Machine Using Run-Down Data. In *Proceedings of 21st International Modal Analysis Conference*, Kissimmee, FL, February, paper 230.

Sinha, J.K., Lees, A.W., Friswell, M.I. 2004a. Estimating Unbalance and Misalignment of a Flexible Rotating Machine from a Single Run-Down. *Journal of Sound and Vibration* 272(2–3), 967–989.

Sinha, J.K., Lees, A.W., Friswell, M.I. 2004b. Estimating the Static Load on the Fluid Bearings of a Flexible Machine from Run-down Data. *Mechanical Systems and Signal Processing* 18, 1349–1368.

Sinha, J.K., Rao, A.R., Sinha, R.K. 2004c. Simple Modelling of a Rotor with Cracks. *Advances in Vibration Engineering* 3(2), 144–151.

VDI 2063. 1985. *Measurement and Evaluation of Mechanical Vibration of Reciprocating Piston Engines and Piston Compressors*.

Wauer, J. 1990. On the Dynamics of Cracked Rotors: A Literature Survey. *Applied Mechanics Reviews* 43(1), 13–17.

10

Case Studies

10.1 Introduction

Over the decades, vibration measurements and vibration-based diagnosis techniques have been widely used in practice, which has made these techniques to be considered universally accepted tools for enhancing performance and safety of plants. Here, selected case studies that were summarized in an article by Sinha (2008) are presented with additional information to aid in a better understanding of the relevance and applications of these techniques. Although most of these case studies are related to nuclear power plants, the vibration-based diagnostic approaches used in them are applicable to most plants and industries. The presentations and discussions have been kept relatively simple, so as to enhance their practicability for engineers and researchers working in this area.

10.2 Roles and Philosophy of Vibration Diagnostic Techniques (VDTs)

The vibration diagnostic techniques (VDTs) used in different roles are classified into the following categories:

1. Condition monitoring of turbogenerator (TG) sets and other rotating machines
2. Design optimization
3. Dynamic qualification due to in-service load condition
4. Seismic qualification
5. Machine installation and commissioning
6. Aging management for machines and structural components
7. Vibration isolation

To carry out any one of the objectives/roles listed above, it is always important to link the following five aspects for appropriate diagnosis so that suitable remediation or modifications can be made.

1. Understand the object(s) (e.g., structures, machines, etc.) and the objectives.
2. What kind of vibration instrument is required for the experiments and measurements? This will greatly depend on previous failure/malfunction histories (in the case of machinery).
3. Experiments: Is the requirement to measure responses alone due to the internal sources of excitation (e.g., excitation due to machine operation itself, wind excitation when measuring on structures, flow-induced excitation for pipe conveying fluids, etc.), or due to any of the well-known external excitation instruments in a controlled manner (shakers or instrumented hammer), or measurements of both responses and excitation force?
4. What kind of data processing is required for the measured vibration signals?
5. An easy remedial suggestion (solution). Based on experience, it has been observed that a "magic solution" is often anticipated. Magic solution means that the problem can be rectified with minimal changes in either the structural configuration or operation parameters, without affecting the global setup and intended performance. However, this is only possible if the measurements, data processing, and their correlation to the theory are fully understood.

Chapter 9 already details vibration-based monitoring and diagnosis, mainly for rotating machines, in great depth, where the link between the four aspects listed was clearly highlighted. Other case studies linking these four aspects and the objectives/roles of VDTs are also discussed in the following sections.

10.3 Dynamic Qualification due to In-Service Load Condition

Critical structural components in plants have to be qualified due to their service load conditions, e.g., pipe conveying fluids, equipment, electrical panels. For piping and electrical panels, the qualification procedures are perhaps straightforward and well defined. The ASME O&M code (1990) gives guidelines and vibration severity limits for different pipe configurations. The electrical panels are also qualified by severe testing on the shake-table.

Two typical such examples, shutoff rod guide tube (Sinha and Moorthy, 1999, Sinha et al., 2003) and moderator sparger tube (Moorthy and Sinha, 1998), shown in Figures 10.1 and 10.2, are typical reactor in-core components that are qualified for expected in-service loads (i.e., flow-induced vibration) using the following steps (Moorthy and Sinha, 1998):

1. Full-scale experimental model (see Chapter 7).
2. FE model of the experimental setup.
3. Modal experiments on the setup and signal processing to extract the modal properties (natural frequencies, mode shapes, and modal damping) (see Chapter 7).
4. Tuning of the FE model to match its dynamic behavior to the experimental model. This tuning involves the quantification of the added mass, which accounts for the dynamics of the submerged perforated tubes—the shutoff rode guide tube and the moderator sparger tube (Sinha and Moorthy, 1999).
5. The tuned FE model was then used for the dynamic qualification together with measured maximum deflection of the tube in the setup, as well as the expected deflection in the reactor conditions. The calculated maximum stress for the measured deflection of the tube from the updated/tuned FE model is then compared with the endurance limit of the tube material for the design qualification (Moorthy and Sinha, 1998).

These components were observed to be performing their design requirements very safely.

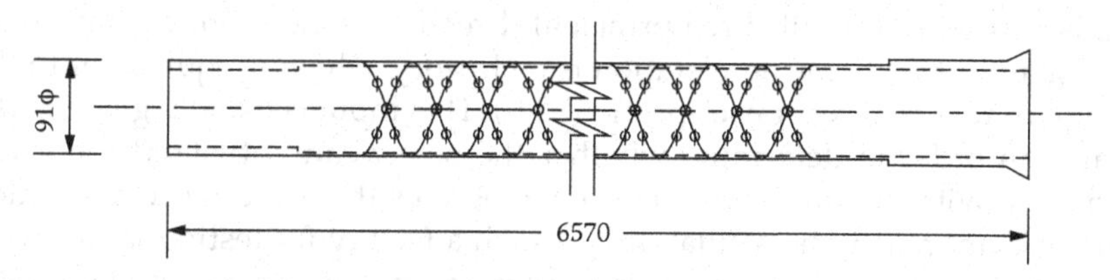

FIGURE 10.1
Shutoff Rod guide tube.

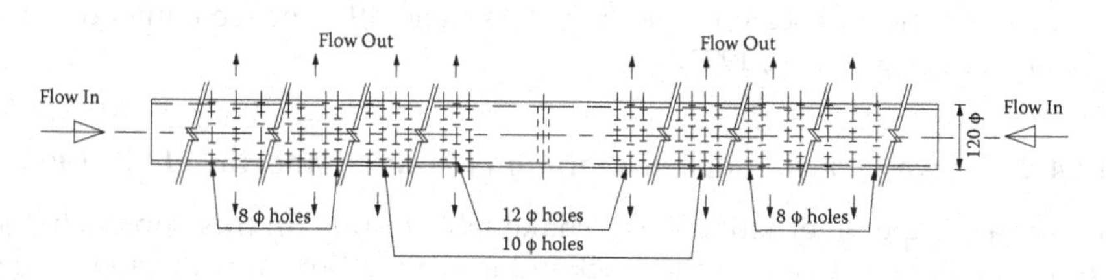

FIGURE 10.2
Sparger tube.

10.4 Seismic Qualification

In nuclear power plants (NPPs), it is mandatory to qualify the critical structures and equipment to their design seismic loads. These qualification procedures are well known (IAEA, 1992; ASME, 1981). The qualification exercise is usually conducted through analysis or by test on a shake-table or by comparison with previous experiences. Among several qualification approaches, the qualification by analysis using the finite element (FE) method is the most commonly adopted in practice. The reliability of the method totally depends on the FE model of the structures to be qualified. An FE model a priori is generally not found to be a true representation of the dynamic behavior of "as installed" structures even for structurally simple components. This is due to several simplified assumptions made during FE modeling, such as for joints and boundary conditions, decoupling of the secondary components, types of elements and the number of degrees of freedom (DOFs) used in FE mesh, etc. Hence, the following alternate methods that involve vibration experiments for seismic qualification certainly possess some advantages over the purely analytical approach, and this is briefly discussed here.

10.4.1 Shake-Table Method

This method has a distinct advantage in qualification, owing to the fact that testing can be done on the actual structures and machines, which eliminates the possible mathematical errors that may be associated with purely analytical methods. During this testing, the object (component, machine, or equipment) to be qualified is mounted on the shake-table, and then testing is carried out in all three principal directions. The input excitation for this qualification is the acceleration-time history, which is equivalent to the design or required response spectra (RRS). This input is usually given to the shaker in order to determine whether the object can withstand the given excitation without any failure. However, besides the exorbitance associated with this approach, the availability of such a facility for testing large structures is often rare. Owing to this limitation, the testing of a scaled-down model on an available shake-table is another alternative, but such a scaled-down model is known to show significant deviations from the dynamic behavior of the full-scale model, which is generally not recommended for dynamic testing (IAEA, 1992).

10.4.2 Railway Track–Induced Vibration Method (Moorthy et al., 1996)

This is a cheaper alternative to the shake-table testing. In this approach, the source of the excitation is the track-induced vibration transmitting to the wagon on which the object is mounted. Since the excitation source is not a controlled excitation (as in the case of the shake-table testing), the response

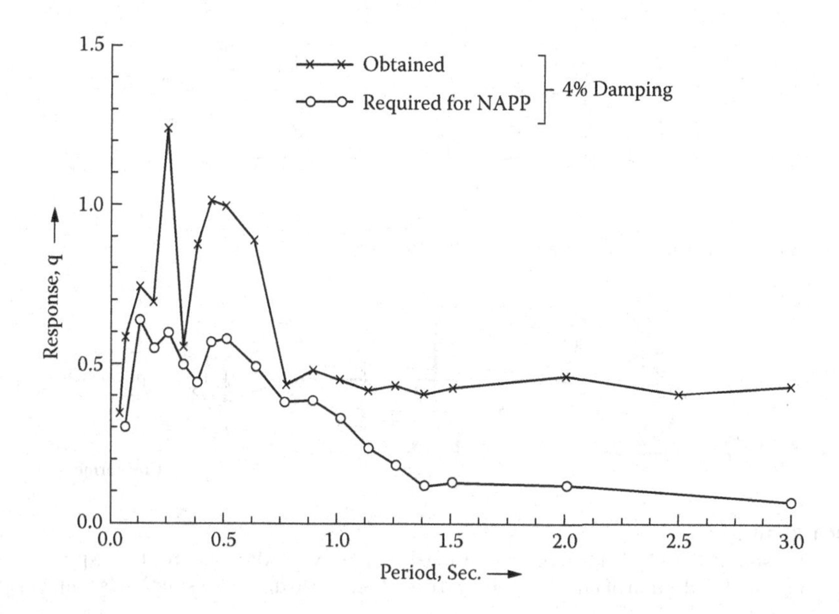

FIGURE 10.3
A typical computed response spectrum due to the railway track–induced vibration enveloping the required response spectrum.

spectra need to be computed for all three principal directions from the measured acceleration, and then compared with the RSS to qualify the object. Moorthy et al. (1996) have demonstrated the qualification of the Diesel Generator (DG) set using this approach. A typical computed response spectrum from the measured acceleration at the base of the DG set enveloping the RSS is shown in Figure 10.3, which confirms that the DG set can withstand much higher loads than the Safe Shutdown Earthquake (SSE) level of seismic loads without failure. Further details with regards to the procedure, measurements, and signal processing associated with this method are provided in Moorthy et al. (1996). It was also suggested that most of the structures and machines can be comfortably qualified by this method while transporting them from the manufacturers to the installation sites.

10.4.3 Direct Use of In Situ Modal Data (Sinha and Moorthy, 2000)

In this approach, the in situ modal test data have been directly used for the seismic response estimation. The direct use of the modal test data obtained from modal tests conducted on the as-installed structure (i.e., the extracted mathematical modal model) has overcome the limitations associated with the use of FE models alone for seismic analysis. Sinha and Moorthy (2000) have applied this approach to an approximately 24-m-long and 4 mm tank, supported on seven equally spaced saddle supports, as shown in Figure 10.4.

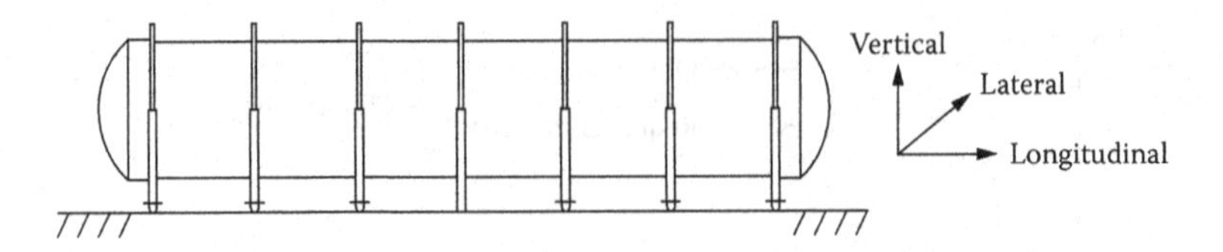

FIGURE 10.4
Tank on seven saddle (six roller and one fixed) supports.

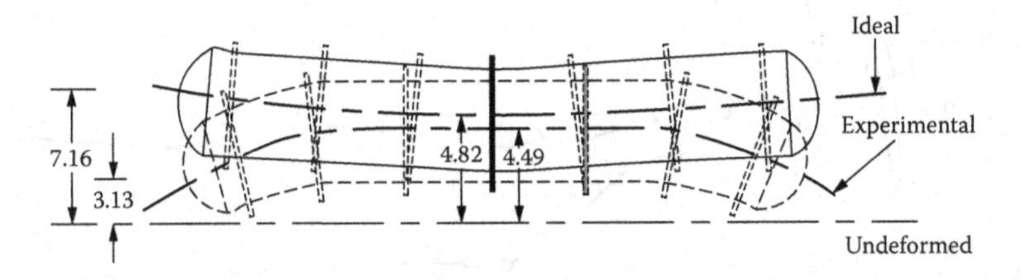

FIGURE 10.5
Comparison of the ideal (FE model with ideal boundary conditions) and the experimental (*in situ*) case of deflection of tank in the lateral direction due to design seismic loads (Dimension in Millimeter (mm)).

Authors have demonstrated the advantage of this approach by comparing with the deformation of the tank predicted by the ideal FE model. Initially, an FE model of the tank was constructed based on these ideal assumptions: (1) central support is completely fixed, and (2) rollers are free to roll in the longitudinal direction and also slide in the lateral direction. The responses are then calculated for the design seismic loads in the time domain using the direct integration (DI) method (Chapter 4). The maximum possible deflection pattern of the tank in the lateral direction based on the ideal FE analysis is shown in Figure 10.5.

Another method was also applied here, so as to understand the behavior. In situ modal tests conducted on the tank and the modal test results were already discussed in Chapter 7. The measured five modes are below 33 Hz, and hence only these modes are included in the analysis. The lumped mass matrix, **M**, is constructed for the tank using the measured locations as the measured DOFs. The measured mode shapes were then normalized using the mass matrix so that the decoupled equation of motion in the modal domain (Equation 4.24) can be written as

$$\ddot{p}_r(t) + 2\zeta_r\omega_{n_r}\dot{p}_r(t) + \omega_{n_r}^2 p_r(t) = \varphi_r^T \mathbf{F}(t) \tag{10.1}$$

where $p_r(t)$ is modal response, and ζ_r and ω_{n_r} $(= 2\pi f_{n_r})$ are the damping ratio and the natural frequency (rad/s) of the rth mode, respectively. The force vectors in the case of the seismic analysis, Equation (4.31), are given by

$$\mathbf{F}(t) = -\varphi_r^T \mathbf{M} a_g(t) \tag{10.2}$$

where $a_g(t)$ is the acceleration time history of the seismic load. It is also referred to as the ground motion. Equation (10.1) is then solved by the mode superposition method (Chapter 4) to estimate the responses at the measured DOFs directly from the measured modal properties. Figure 10.5 gives the typical comparison of the deflection pattern of the tank in the lateral direction estimated from the ideal FE model and from the measured in situ modal data. This clearly indicates that the prediction from the ideal analytical case is totally different from the actual behavior as installed of the tank. Hence, the confidence level obtained from qualifying by this method is definitely more than that obtained from the analytical method alone. This, however, shows that the stress pattern in the tank will be completely different for the two deflection patterns, and the qualification by the FE analysis alone may not be completely reliable.

10.4.4 Updated FE Model Method (Sinha et al., 2006)

This is the most reliable approach, as suggested by Sinha et al. (2006). It exploited the strengths of both the experimental and analytical methods for enhancing the confidence level in the seismic qualification. Here, the experimentally obtained modal data by in situ experiments are used to tune the initial FE model by the model updating method (Chapter 8), and then the updated FE model perfectly reflecting the dynamic behavior of structures is used for the seismic qualification. The advantage of this approach has been demonstrated on a typical in-core component of a power plant (Sinha et al., 2006). Although this approach is much more involved than others, it possesses a distinct advantage of having more numbers of degrees of freedom than the mathematical model derived alone from the modal experiments. Hence, the study by Sinha et al. (2006) is further discussed in the following paragraphs.

10.4.4.1 Experimental Setup and Modal Tests

For the design qualification of the tube, a full-length tube was erected in a tank representing a quarter of the reactor vessel geometry for a typical pressurized heavy water reactor (PHWR). The schematic of the setup is shown in the Figure 10.6. The tank in the setup does not exactly simulate the reactor vessel. However, the setup was useful in estimating the influence of fluid mass on the tube dynamics and the modal damping of the tube by conducting the modal tests. These two parameters are expected to closely agree with the as-installed values. The modal tests were conducted on the tube when the tank was empty (nonsubmerged condition) as well as when the tank was filled with water (submerged condition). This is discussed in Section 7.6.1. The identified natural frequencies for the first three flexural modes are listed in Table 10.1. The measured modal dampings at the first three modes under submerged conditions are 3.66, 1.96, and 1.20%, respectively.

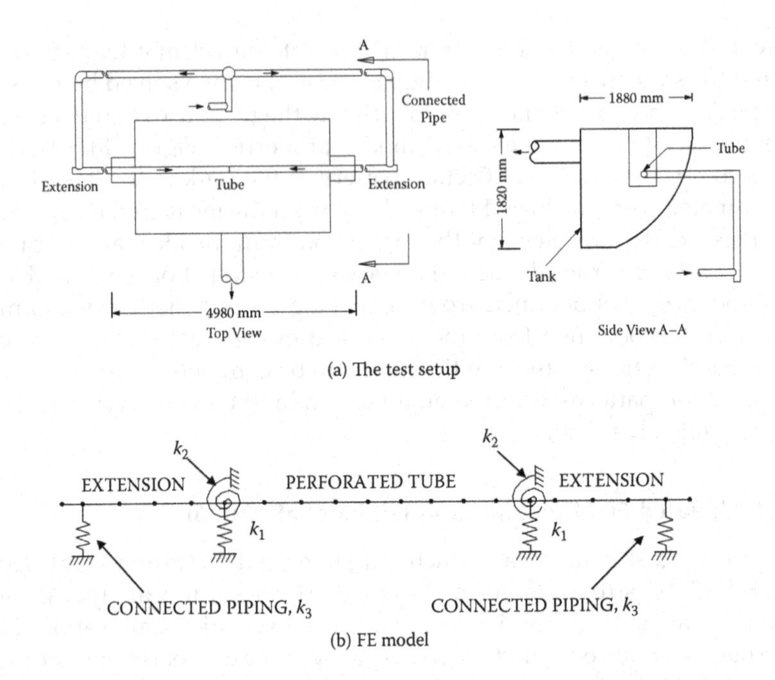

FIGURE 10.6
The test setup of the perforated tube and its FE model. (a) The test setup. (b) FE model.

TABLE 10.1

Natural Frequencies of the Tube and the Estimated Boundary Stiffnesses of the Test Setup

Updating Parameters		Nonsubmerged (in air) Modal Test Data (a)	Nonsubmerged (in air) Updated FE Model (b)	% Error (a and b)	Submerged (in water) Modal Test Data (c)	Submerged (in water) Updated FE Model (d)	% Error (c and d)	Reactor Condition
k_1, MN/m			21.63	—		21.63	—	
k_2, kNm/rad		Not known	0.00	—	Not known	0.00	—	Clamped–clamped condition
k_3, MN/m			22.08	—		22.08	—	22.08
Modes (Hz)	f_1	17.24	18.24	+5.800	12.65	12.23	−3.32	13.90
	f_2	46.56	49.95	+7.281	37.12	32.76	−11.74	37.92
	f_3	103.9	103.8	−0.096	64.06	65.120	+1.65	72.72

10.4.4.2 Updated FE Model

An FE model of the tube in the setup was also developed. Two-node Euler-Bernoulli beam elements (each node with two DOFs—one translational and the other rotational) were used to model the perforated tube and the extended portion of the pipe on either side of the tube. Since the object for the study was the tube, the tank and its supporting structure were not included in the FE model. However, their effects on the dynamics of the tube were considered by providing the translational and rotational springs of stiffnesses k_1 and k_2 at the location of the tube support at the side plate of the tank. A translational spring element of stiffness (k_3) has also been used on either side of the extended pipe to simulate the stiffness effect of the connected piping. The FE model is shown in Figure 10.6(b). To account for the inertial and stiffness effects due to the large number of holes distributed all over, thickness equivalent to the volume of holes was removed uniformly from the inner diameter of the tube in each element of the FE model. Since the material and geometrical properties of the tube and the extended pipes were accurately known, the boundary stiffnesses, k_1, k_2, and k_3, in the FE model were then updated by the gradient-based sensitivity approach (Chapter 8), using the experimental modal data of the tube in air.

The computed natural frequencies, listed in Table 10.1, for the updated FE model are close to the measured frequencies for the nonsubmerged tube. The maximum error is well within 7.3%, which is very much acceptable in practice, which confirms that the simulation with three springs truly represents the tube dynamics in the setup. Hence, the updated model is a true representation of the tube's behavior in the setup. On this updated model, the effect of water on the dynamics of the tube under submergence was included as the added mass. The added mass of the water was estimated according to the formula suggested by Sinha and Moorthy (1999). The computed natural frequencies are listed in Table 10.1 for the submerged tube, which are also close to the modal test frequencies, which further highlights the potential of the updated FE model.

10.4.4.3 The Reactor Conditions

The estimated values of the boundary stiffnesses during the process of model updating are also listed in Table 10.1. The estimated rotational boundary stiffness (k_2) close to zero indicates that the tube supported on the side plates of the tank is not thick enough to provide any rotational constraint. The small value of the translational stiffness (k_1) indicates that the tank supports are flexible and allow some deflection, which is also evident from the experimental mode shapes shown in Figure 7.33. However, the tube boundary conditions would be much different in the reactor. The side plates in the reactor are thick enough (50 mm) to restrict the rotation of the tube, and the tube is expected to have zero deflection at rolled joints. Hence, to simulate the reactor conditions, the values of boundary stiffnesses—k_1 and k_2—are increased to higher values to realize the clamped–clamped conditions of the tube as in the reactor. The computed natural frequencies for the tube in the reactor are listed in Table 10.1. The stiffness k_3 would not influence the dynamics of the perforated tube in the reactor; hence, the value of k_3 was assumed to be the same as in the setup.

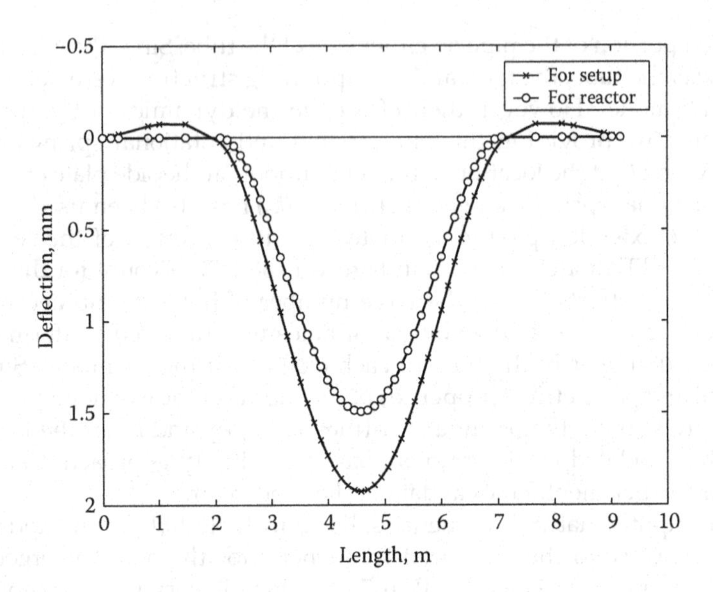

FIGURE 10.7
The maximum deflection of the tube.

10.4.4.4 Seismic Response Estimation

The updated FE model was used for the seismic response estimation for the following two conditions: (1) the perforated tube in the setup and (2) the perforated tube in the reactor. The estimated deflections of the tube in both conditions are shown in Figure 10.7 for the design seismic loads. The maximum stresses due to these deflections were estimated to be 9.51 and 9.90 MN/sq.m at the support end for the setup and the reactor conditions, respectively. It is important to note that the tube deflection for the reactor condition is smaller but of higher stress than the setup condition, which is expected due to stiff boundary conditions in the reactor. The stress, including conservative stress concentration factor (SCF), is found to be well within the allowable stress limit of zircaloy-2 material. Hence, the designed in-core component is considered to be seismically qualified.

10.5 Machine Installation and Commissioning

The satisfactory operation of any machine is always important for plant safety, increased productivity, reduced downtime, and low maintenance overhead costs. Such satisfactory requirements can be achieved through proper design and installation of machines on site. The design aspect is perhaps more mature in most cases. However, installation may play a significant role in the dynamic behavior even for properly designed machines. Vibration-based condition monitoring and codes are well known and widely respected for most of the rotating machines, like pumps, motors, turbines, etc. However, it has often been observed that many newly installed machines show high vibration, and quite a few fail frequently. In fact, experience shows that the vibration measurements and analysis during the commissioning stage are important to resolve the problems related to the machine installation, if any, and enhance the availability of the machines. A few typical examples are briefly discussed here, to bring out the usefulness of the vibration-based diagnosis for machine installation and commissioning (Sinha, 2003).

10.5.1 High Vibration

The moderator system of a nuclear power plant consists of five vertical pumps, with four in operation and one available as a standby (Rao et al., 1997). Figure 10.8 shows the schematic of the layout of the pumps' locations and piping. The pump assembly is shown in Figure 10.9(a). A vibration survey during the commissioning showed normal vibration on top of the motors (i.e., farthest location from the pump base) as per the Hydraulic

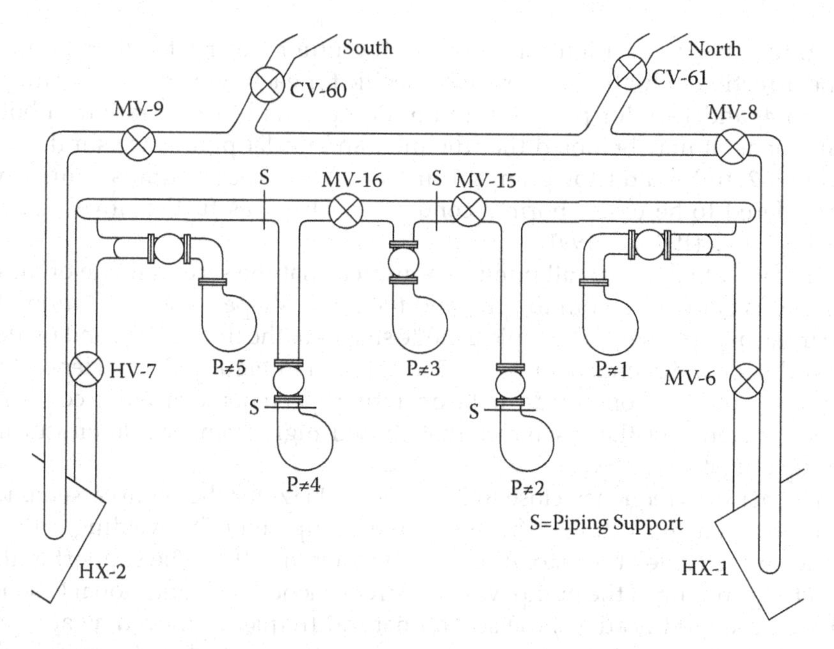

FIGURE 10.8
Schematic layout of pumps and piping.

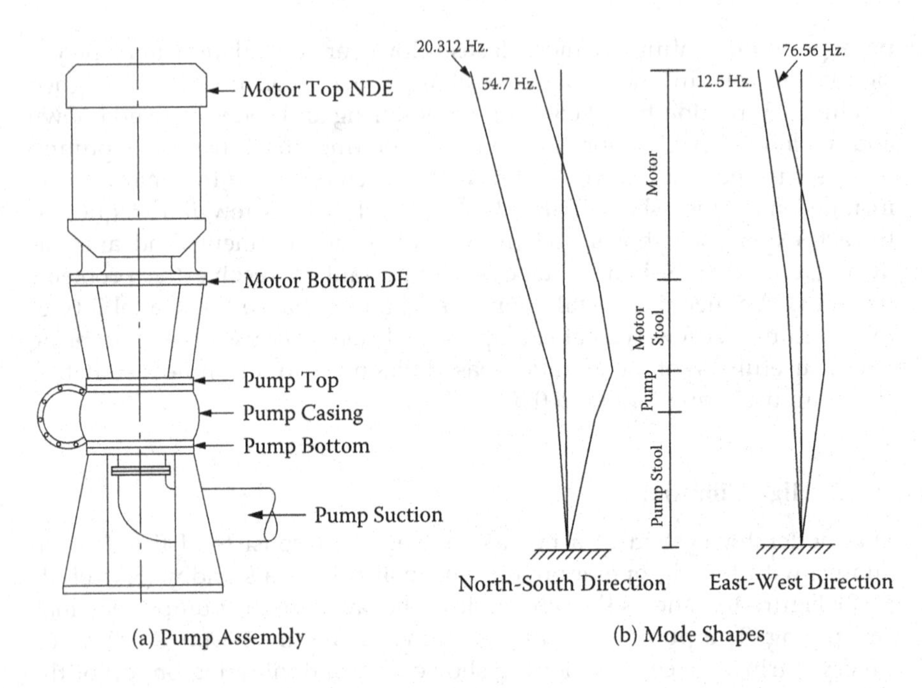

FIGURE 10.9
Schematic of the pump assembly and its mode shapes. (a) Pump assembly. (b) Mode shapes.

Institute (1983), but high vibration on the pump casing for four pumps. The direction of high vibration was specific for each pump: N-S for pumps 2 and 4, and E-W for pumps 1 and 5. Pump 3 had low vibration in both directions. It may be noted that the inlet and outlet piping are similar for pumps 2 and 4, and for pumps 1 and 5. Hence, these pumps cannot be considered to be under normal conditions. This was further investigated by Rao et al. (1997) in 1991.

The modal testing on all pumps confirmed that the second mode of these pumps is close to the pump rpm, and the mode shapes displayed antinode near the pump casing. The typical mode shapes at the first and second modes in orthogonal directions for pump 2 are shown in Figure 10.9(b). Hence, the closeness of the second mode to the operating speed of the machine confirms the occurrence of the resonance that caused high casing vibration during pump operation.

The natural frequency close to 3000 rpm (50 Hz) for the pump assemblies was changed by adding stiffeners to the pump stool (by welding a thick plate on each side of the stool), as shown by Figure 10.10. The support to the discharge piping of the pump was also strengthened with additional U bolts. With these modifications, the second natural frequency moved away from the pump rpm, and hence the casing vibration was reduced significantly (Rao et al., 1997).

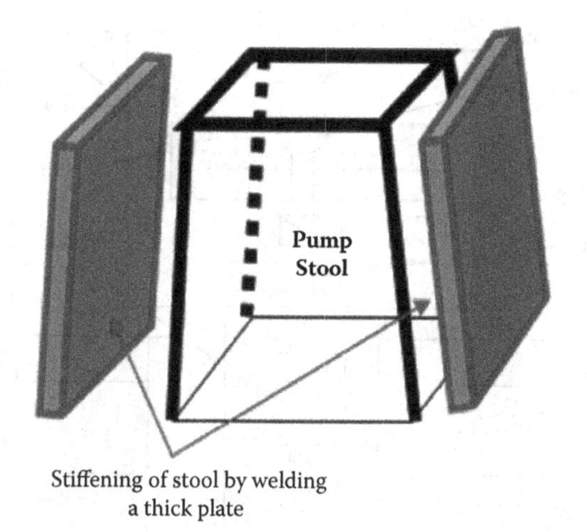

FIGURE 10.10
Stiffening of pump stool.

10.5.2 Different Dynamic Behavior of Two Identical Pumps

Figure 10.11 shows the schematic of the assembly of a motor-pump and foundation structure (Balla et al., 2005). The motor-pump assembly rests on top of three steel blocks (attached together by bolts) on either side. The pump is situated at the top of the motor. As shown in Figure 10.9, the motor and pump units are hung on their support. Total support height is 1.81 m and the length of the pump-motor unit is 1.75 m, which is anchored to the foundation about 1 m from the bottom of the motor. The nominal diameter of the shaft is 100 mm and supported through three bush bearings. The motor-pump shaft has a six-bladed impeller at the top. Figure 10.12 shows the piping layout of the two pumps for both suction and discharge. The pump has an axial suction and a radial discharge pipe of 200 mm OD. The suction and discharge lines for pump 2 are three times lengthier than those of pump 1.

The pump can be operated at 1500 or 3000 rpm depending upon the requirements. Vibration amplitudes for both pumps were found to be well within the severity limits as per the Hydraulic Institute (1983) when operating at 1500 rpm, and hence both pumps can be considered to be safe. However, pump 2 was showing relatively high overall vibration, with sub- and higher harmonic peaks of the rotating speed (25 Hz) in the spectrum when compared to pump 1. Typical waterfall spectra for both pumps on pump casing are shown in Figure 10.13. The waterfall spectrum consists of two start-up events of the pumps. These observations confirm that the dynamic behavior of these two identical pumps is completely different.

To understand this difference in dynamic behaviors, modal tests were conducted on both units. The first few natural frequencies for pump 1 were 23.76, 31.81, 37.76, 45.58, and 59.69 Hz. Similarly, the first few natural

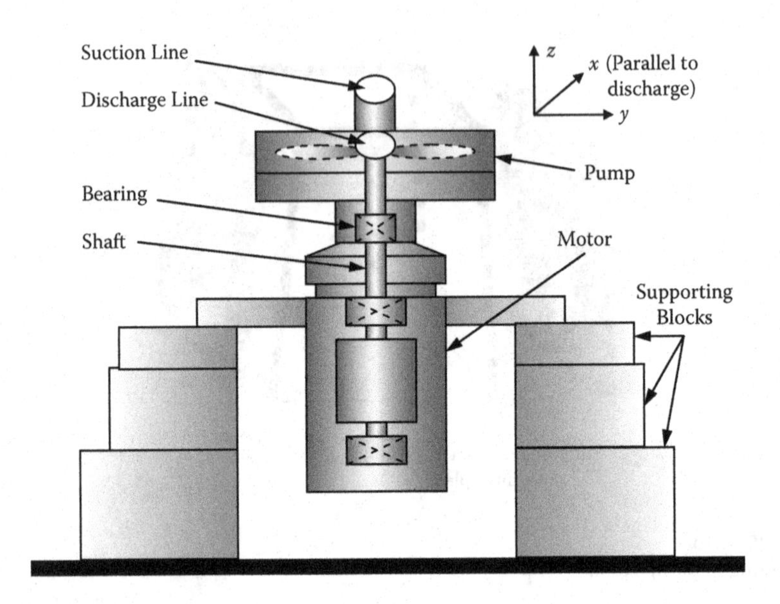

FIGURE 10.11
A schematic of the pump assembly.

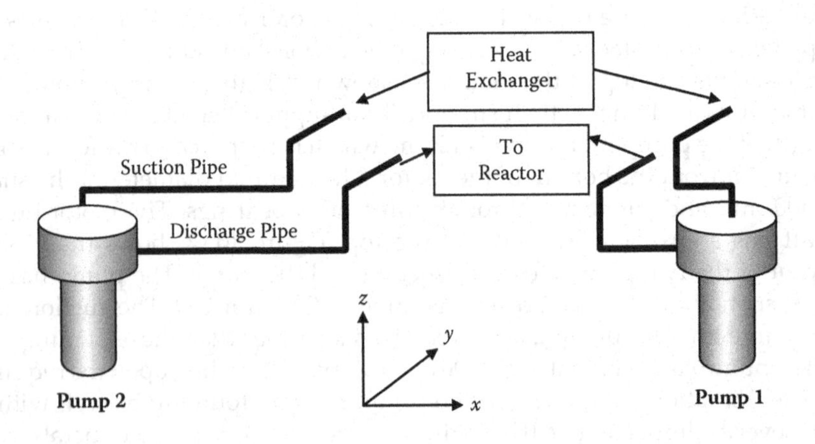

FIGURE 10.12
Piping layout for pumps 1 and 2.

frequencies for pump 2 were identified as 21.92, 25.80, 32.77, 34.99, 45.14, and 49.65 Hz. These experimentally identified natural frequencies were then used to understand the dynamics of both pumps during their operations. Since both pumps are supported on similar foundations, the differences in the identified modes for pumps 1 and 2 were suspected to be due to differences in suction and discharge lines. Hence, the study by Balla et al. (2005)

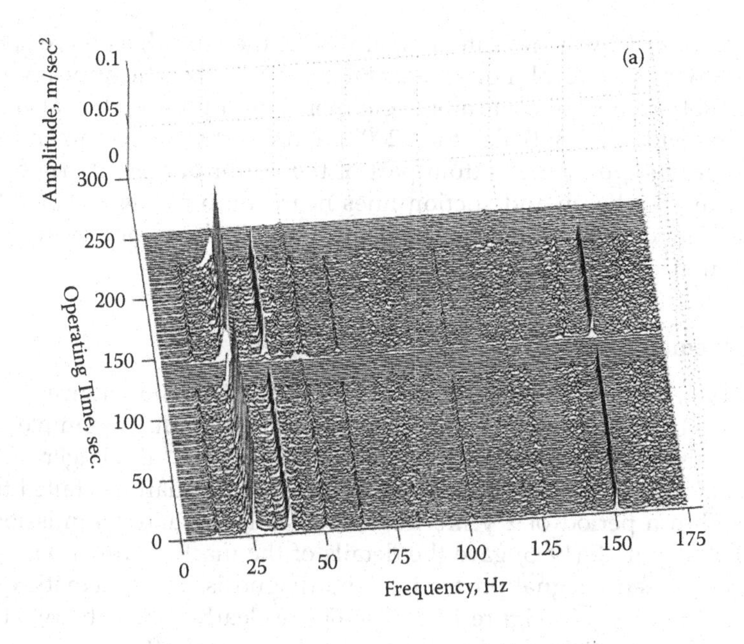

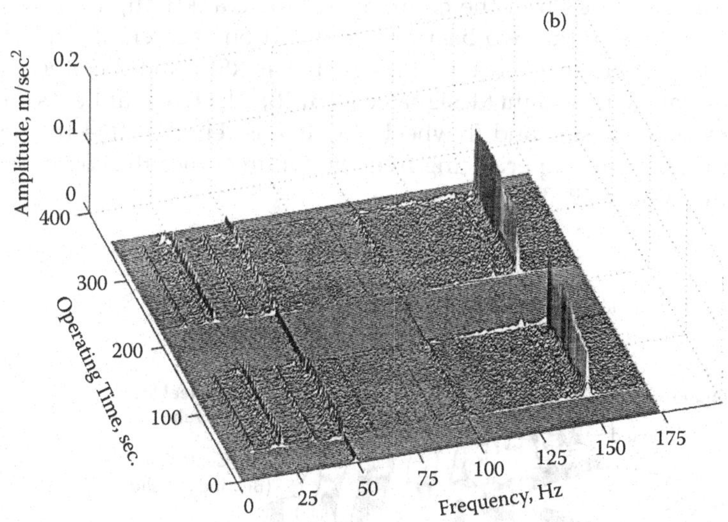

FIGURE 10.13
The waterfall spectrum in the x-direction at pump casing. (a) Pump 1. (b) Pump 2.

gave the details of the vibration measurements, modal tests, and their results, including mode shapes.

Based on the vibration behavior during the pump operation and the experimental modal data, it was concluded that pump 1 can be operated at 1500 and 3000 rpm, as there were no structural modes close to these speeds.

However, pump 2 was not safe to operate at either speed, as these speeds are close to its structural modes at 25.80 and 49.65 Hz. The appearance of several subharmonic and harmonic components in pump 2 has also been explained (Sinha, 2003; Balla et al., 2005). Finally, it was recommended to either rigidly support the bottom part of the motor or provide rigid supports to the discharge and suction lines near pump casing to shift these structural modes for pump 2. However, pump 1 does not require any modification.

10.5.3 Frequent Failure

Sinha (2003), Rao et al. (1997), and Sinha et al. (1996) have shown several cases of frequent failures of machines just after their installation. The simplest one is briefly discussed here (Sinha et al., 1996). It is a case of the Pilger mill, in which one of the five split roller bearings on the crankshaft has failed three times within a period of 2 years from its installation and commissioning in 1990. Sinha et al. (1996) gave the details of the machine function, vibration measurements, signal processing, and diagnosis. The schematic of the Pilger mill is shown in Figure 10.14. The figure clearly shows the two bearings mounted on each of the crankpins of the crankshaft, and one on the center pin. One of the two bearings mounted on each crankpin is for the connecting arm of the roll-stand (designated as RSB), and the other bearing is for the balancing weight (designated as BWB). The two crankpins are identified as operator side and flywheel side. It was reported that the operator side BWB has been experiencing frequent failures since the commissioning of the machine in 1990.

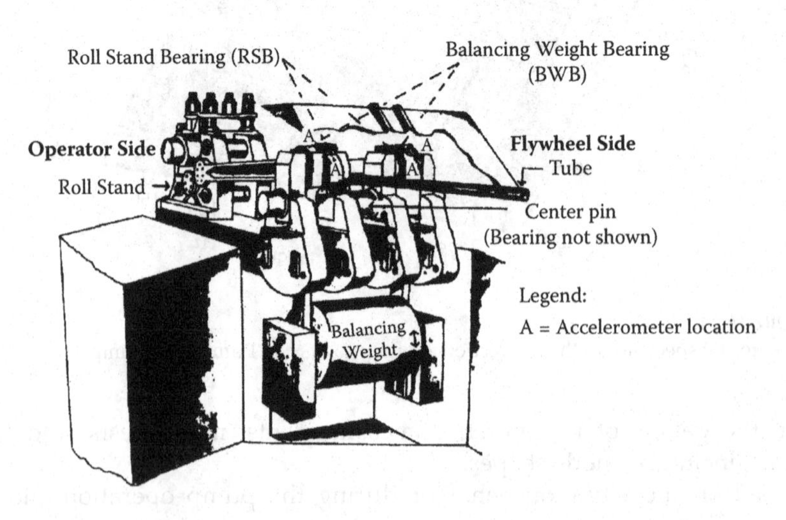

FIGURE 10.14
Pilger mill.

A simple diagram showing the movement mechanism of the crankshaft and the balance weight is shown in Figure 10.15. The measured acceleration responses of the balancing weight on the operator and flywheel sides using the piezoelectric accelerometers during the machine normal operation are also shown in Figure 10.15 for easy comparison. From the time waveforms and Figure 10.15, it was concluded that the balancing weight and its connecting arm on the operator side was not moving freely. A bend in the connecting arm was suspected as one of the reasons for such a time waveform, which was subsequently confirmed on site.

During maintenance, the bent connecting arm was replaced by a new arm, and the machine was put back into service in 1992. Since then, the machine has been working satisfactorily. Following this finding, several similar machines were acquired, installed, and commissioned based on vibration survey and diagnosis certification (Sinha, 2003, 2006).

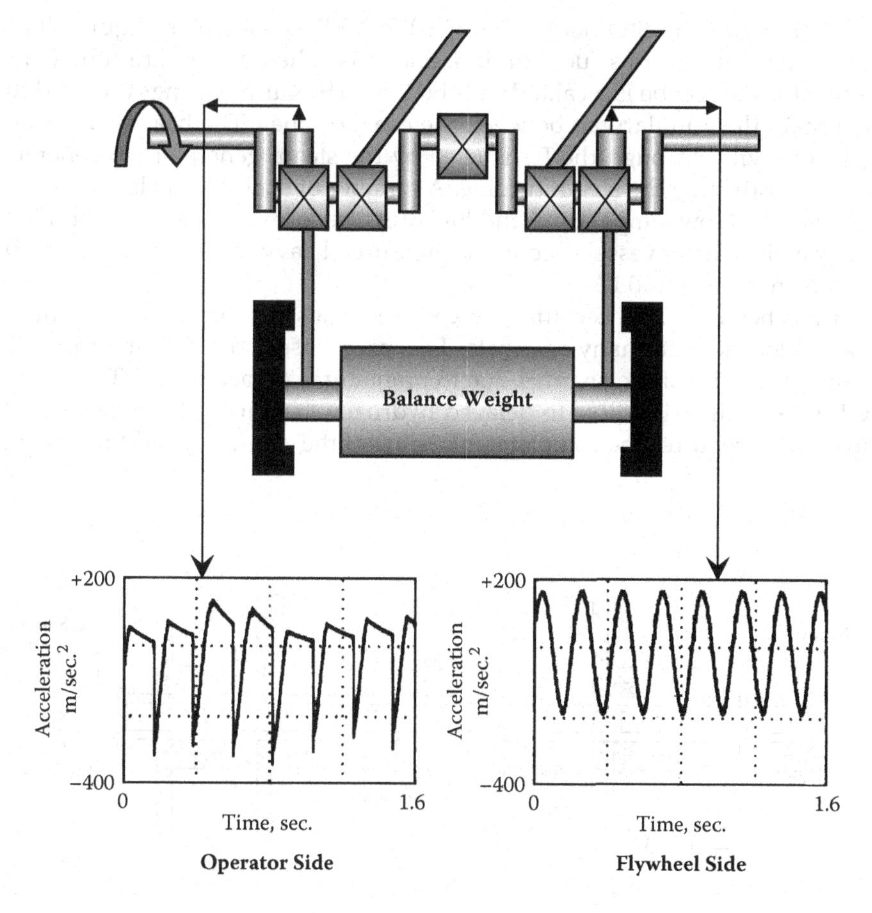

FIGURE 10.15
The vibration response of the balancing weight.

10.6 Aging Management for Machines and Structural Components

Vibration-based condition monitoring and diagnosis also play a vital role in the assessment of the health of aging structures and machines in a much quicker and reliable manner, so that the safety of the plant is not compromised at any stage of operation. In fact, vibration-based diagnosis has been found to be a reliable tool in extending the life of aged structures and machines in many cases. A few typical examples related to the structures and machines are discussed here.

10.6.1 Structural Components: The Coolant Channels (Moorthy et al., 1995)

A typical coolant channel of a 235 MWe NPP is shown in Figure 10.16. The inner tube carries fuel bundles, and it is called a pressure tube (PT), while the outer tube is a calandria tube (CT). The garter springs are used to maintain the annular gap between the coaxial tubes. The heat of the coolant conveying through the PT is used by the steam generators to generate steam for driving the steam turbine. A number of such channels immersed in a vessel at low temperature and low pressure form a reactor. A simplified view of the reactor vessel used in the pressured heavy water reactor (PHWR) is shown in Figure 10.17.

It has been experienced that the garter springs get displaced from their design locations in many channels. The large unsupported span of the PT restricts the life of the channel due to premature contact of the PT with the CT, making it susceptible to delayed hydrogen cracking. The conventional inspection requires the complete unloading of the channel, in addition to an

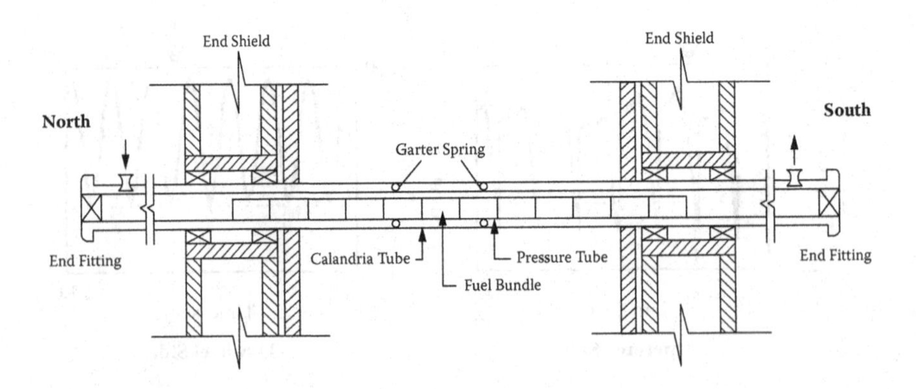

FIGURE 10.16
A typical coolant channel.

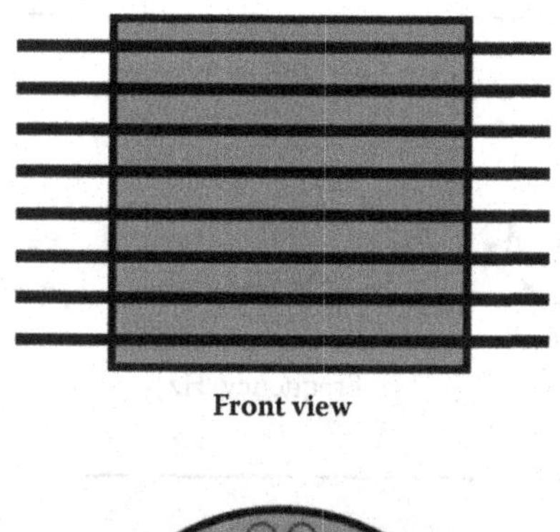

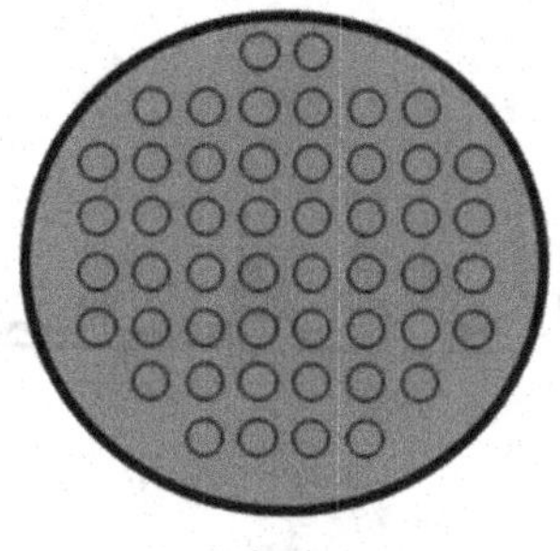

FIGURE 10.17
A simplified view of a reactor vessel with a number of coolant channels.

extended shutdown period for inspecting all coolant channels of the reactor. Vibration-based diagnosis of the channel was found to be accurate enough to detect the channels with direct CT-PT contact in a quicker manner. The measured vibration responses at either ends of the coolant channels due to flow-induced vibration were used to identify the CT-PT contacting channel. Typical measured vibration spectra for a contacting and a noncontacting channel are shown in Figure 10.18. It was observed that the contacting channels showed narrowband responses in comparison to the noncontacting channels. Hence, this became the tool for identifying the channels with CT-PT contact, which was also confirmed by physical intrusive inspection on several channels to enhance the confidence level of the vibration-based diagnosis.

Further FE analysis has also been done to understand the narrowband response of the contacting channels. Figure 10.19 shows the FE model where the PT and end fittings were modeled using beam elements, and the CT was modeled with a single mass and spring corresponding to its first mode. Since the contact generally takes place around the center of the span, and the first mode of the CT is the most significant contributor to the response, such a simplification would approximate the interaction between the PT and the CT adequately enough to bring out the discriminating feature. The response of

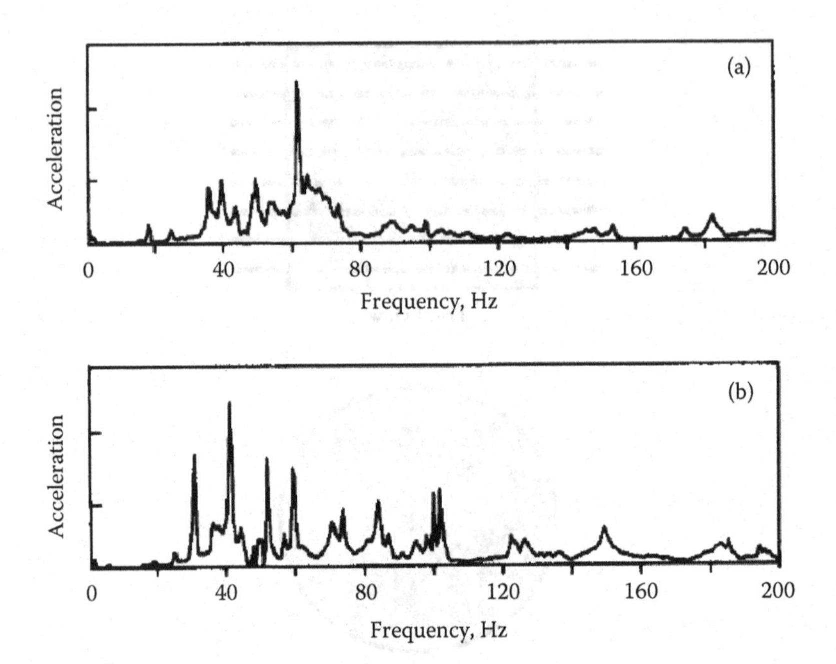

FIGURE 10.18
Typical measured spectra of coolant channel. (a) Contact. (b) No contact.

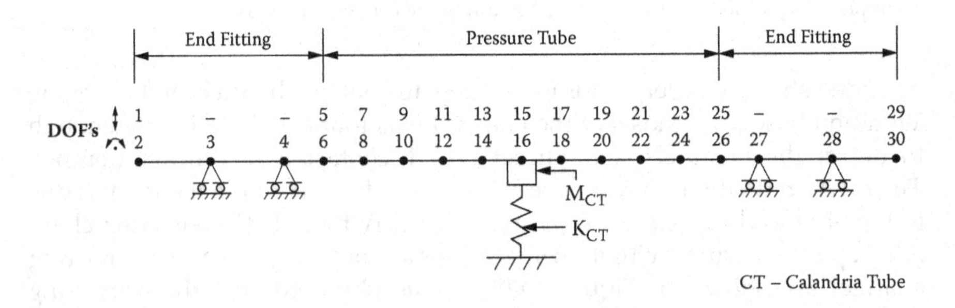

FIGURE 10.19
A FE model of the coolant channel.

the model to an impulse given at one end was computed. The response was computed using the Newmark-β method (Chapter 4). Figure 10.20 shows the response of the center of the channel to the impulse at one end. The figure shows the nonlinear interaction between the PT and CT. In the actual reactor, the measurement can only be taken at the ends of end fittings. Figure 10.21 shows the spectrum of the response of the end of the channel opposite the one where the impulse was given. The figure also shows the spectrum when such a nonlinear CT-PT interaction is absent. It can be seen that some of the modes are attenuated in case of direct CT-PT contact.

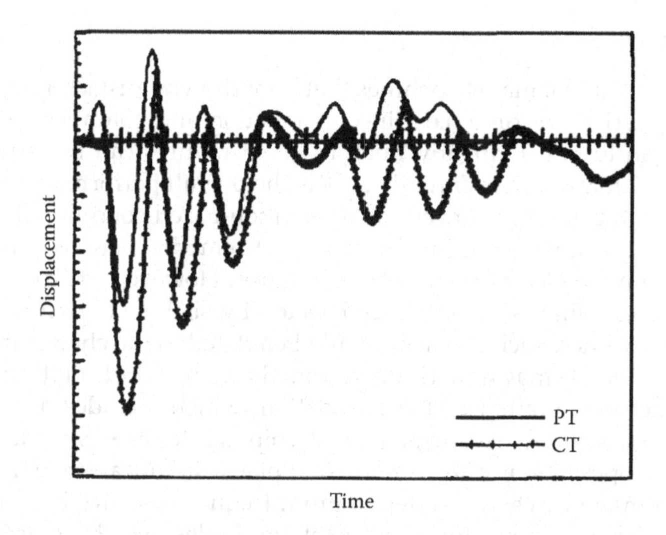

FIGURE 10.20
Typical nonlinear interaction between the CT and PT.

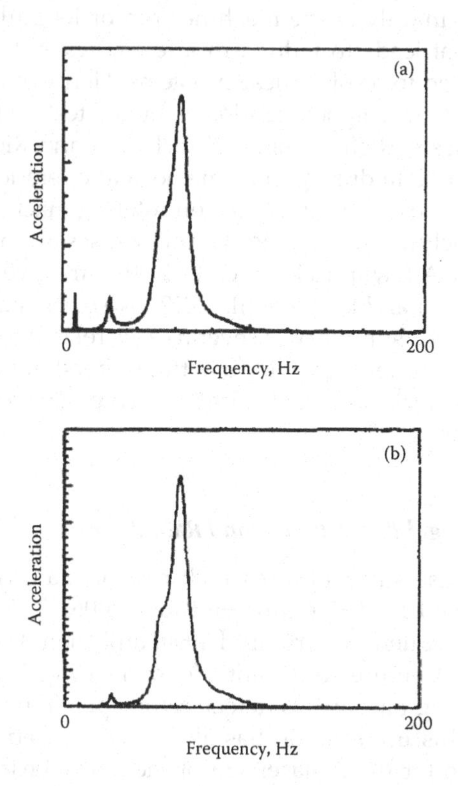

FIGURE 10.21
Simulated vibration spectra of coolant channel. (a) Contact. (b) No contact.

10.6.2 Machines

It was observed that many machines that have shown satisfactory operations over long periods of time (i.e., since installation and commissioning) suddenly began to fail frequently. Such frequent failures can be attributed to the aging factor, which eventually affects the overall performance of plants. Even for such cases, the conventional vibration-based condition monitoring is essential for identifying the faults so that remedial actions can be completed before they lead to catastrophic failures. However, the root causes of such frequent failures are often undetected by such condition monitoring techniques. Hence, such problems could be related to the change in the system dynamics that may have resulted from the aging effects of the machines. In fact, the deterioration in the foundation (which includes all auxiliary-connected structures, drive arrangements, piping, etc.) over extended periods of machine operation is often observed to play a significant role in frequent failure. In many cases, the system natural frequencies have been observed to shift closer to the machine rpm or its multiples, which often results in resonance. These resonance effects consequently lead to malfunctioning and fatigue failure. In many other cases, the operation deflection shape (ODS) of the foundation, mainly at the machine rpm or its multiples, reveals the malfunctioning that leads to failure. In a few cases, both tests (modal test and ODS) are needed to resolve these problems. Hence, it has been observed that the in situ dynamic characterization by modal testing (Chapter 7; Ewins, 2000) or ODS analysis (Richardson, 1997; Schwarz and Richardson, 1999) of measured vibration data during machine normal operations, or a combination of both tests (modal tests and ODS) on rotating machinery, can prove to be a viable approach for identifying the root causes and their rectifications for aged machines. A few individual case studies are given in the papers by Sinha and Rao (2006) and Sinha et al. (2012). Also, the paper by Sinha and Balla (2006) provides a summary of several cases related to the application of these techniques. In order to provide a better understanding of the practical application of these techniques, a typical case study (Sinha and Rao, 2006) is used for illustration.

10.6.2.1 A Centrifugal Pump (Sinha and Rao, 2006)

This is a typical case study of a research reactor main coolant centrifugal pump commissioned in 1985 (Sinha and Rao, 2006). This pump had never experienced any frequent failure until 1990, and then it suddenly began to experience frequent failures of its antifriction bearing. The vibration-based condition monitoring has always detected the appearance of bearing faults in advance, and this information has always been used to initiate a well-planned shutdown for the replacement of the faulty bearing. However, the condition monitoring could not identify the root cause for this frequent failure of the bearings.

Figure 10.22 shows the schematic of the pump assembly. It is a horizontally mounted centrifugal pump with an axial inlet and a radial outlet. The pump and the motor shafts are rigidly coupled to a shaft carrying a flywheel (FW). The FW is supported by a grease-lubricated radial bearing on the pump side and an oil-lubricated taper roller thrust bearing on the other side. The pump is driven by a 540 kW electric motor operating at 1492 rpm. The pump is mounted directly on the base plates embedded into a rigid concrete floor.

The modal test conducted on the complete pump assembly by Sinha and Rao (2006) revealed that the mode at 57.96 Hz had significant deflection only at both bearing pedestals (Figures 10.23 and 10.24). This can be clearly visualized from Figure 10.24. This frequency in the frequency response function

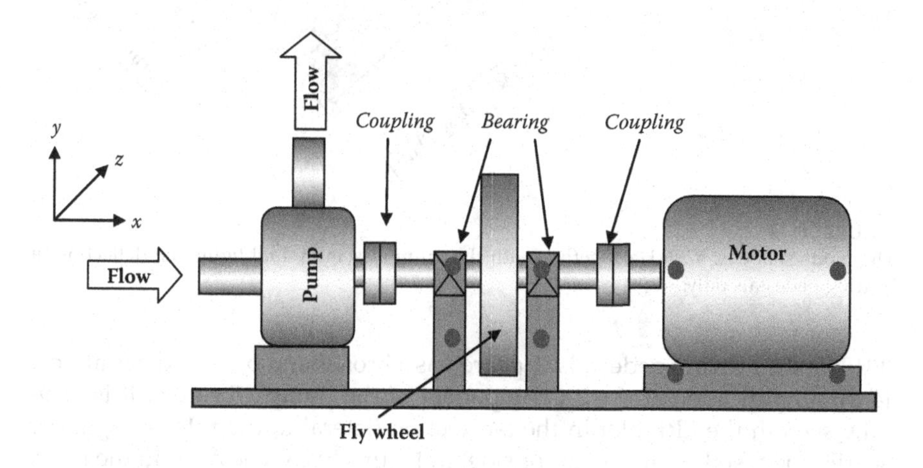

FIGURE 10.22
Schematic of a pump assembly and the measurement (dot) locations.

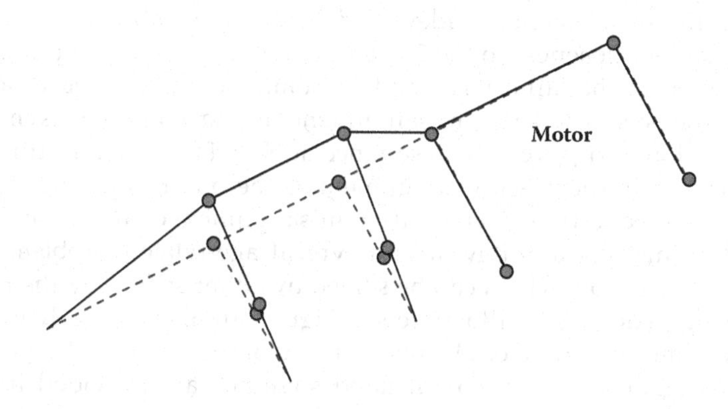

FIGURE 10.23
The mode shape at 57.96 Hz of the pump assembly.

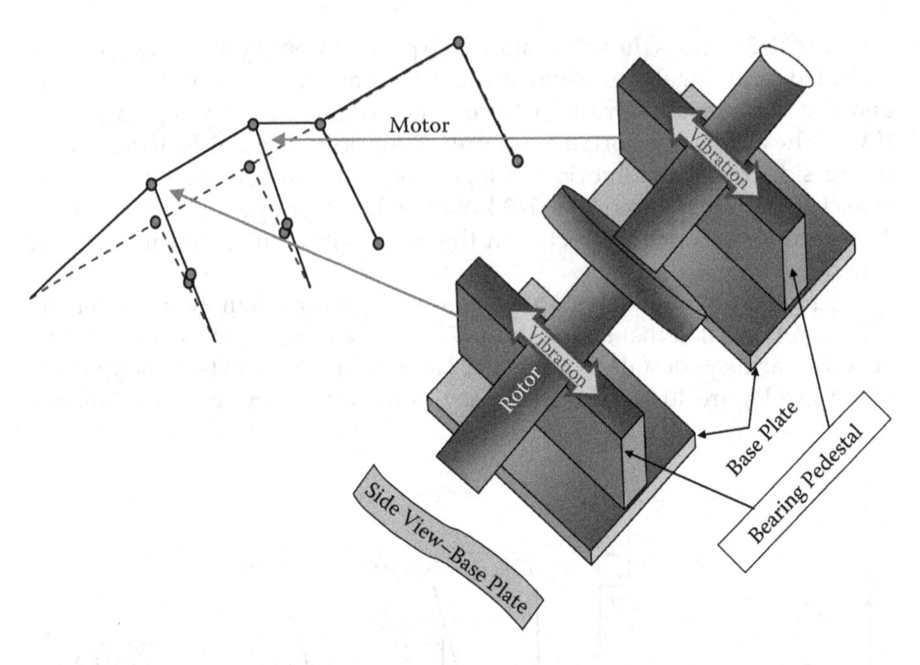

FIGURE 10.24
The mode shape at 57.96 Hz together with the pump assembly highlights the deflection of bearing pedestals only.

(FRF) at the bearing pedestals appeared as a broadband peak and has almost 14 dB amplification at the 2X component of the pump vibration. It is typically seen in the FRF plot in the *z*-direction (lateral to the rotor axis) at the bearing pedestal (near the motor side) in Figure 10.25. The drop in the natural frequency close to the 2X, and its broadband nature, must be due to the looseness between the base plate and the concrete, resulting in the nonlinear interaction between the base plate and the concrete surface.

Hence, the experimentally identified broadband natural frequency at 57.96 Hz and its closeness to the 2X component have been identified as the main reason for the failure. A small 2X component generated as a result of even the smallest shaft misalignment at the coupling/asymmetric shaft must have triggered the resonance at 57.96 Hz, which in turn leads to an increase in the shaft misalignment as the pump continues to operate under this condition. Such induced misalignment could cause damage to the bearings prematurely. It is a typical age-related problem of the machine foundation, which can be solved by either stiffening the roots of the bearing pedestals, as illustrated in Figure 10.26, or properly grouting the base plate in concrete. Details of the vibration measurements, data analysis, diagnosis, and recommended solutions are provided in Sinha and Rao (2006).

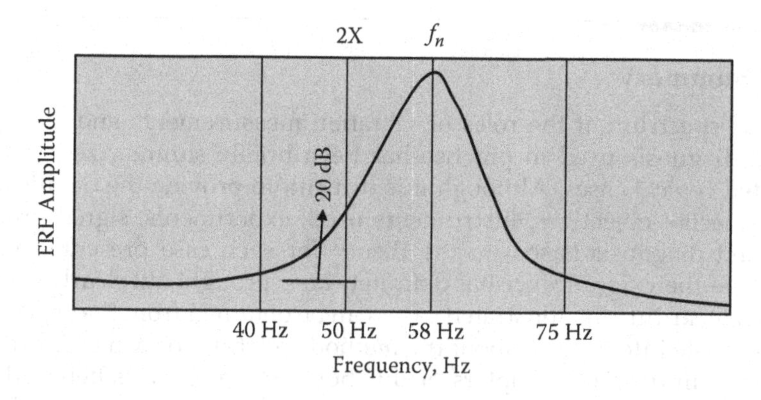

FIGURE 10.25
A typical FRF plot at bearing pedestal (near the motor) in the *z*-direction.

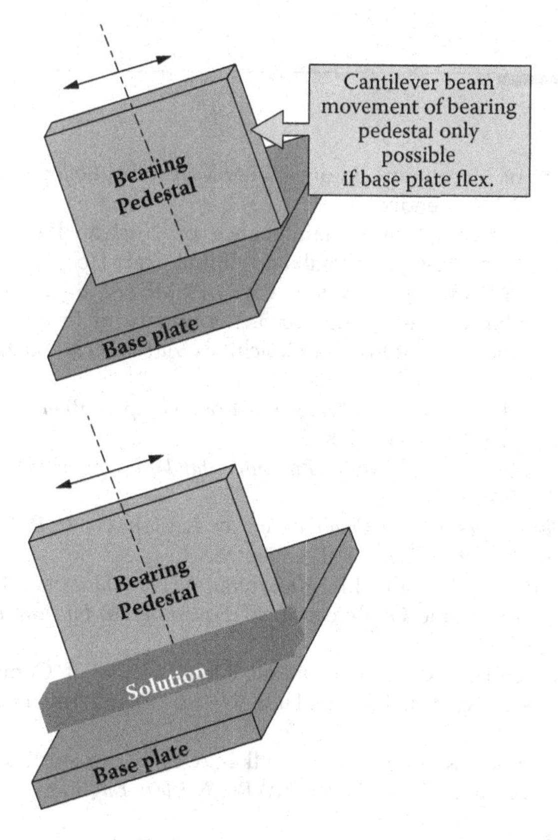

FIGURE 10.26
Typical pump pedestal with suggested solution.

10.7 Summary

A wide spectrum of the roles of vibration measurements and vibration-based diagnosis used in practice has been briefly summarized through selected typical cases. Although it is difficult to provide the details of all five aspects—objectives, instruments used, experiments, signal processing, and diagnosis based on the theory for each case presented in the chapter—their significance has definitely been brought out clearly. Further details regarding the illustrated cases can be obtained from the cited references, while information about the methods relating to each case can also be found in different chapters of this book. However, it is believed that this summary would be a very useful piece of information for engineers and researchers involved in vibration-based condition monitoring and diagnosis.

References

American Society of Mechanical Engineers (ASME). 1981. *Boiler and Pressure Vessel Code*, Section III, Appendix N.

ASME Code on Operation and Maintenance of Nuclear Power Plants. 1990. Requirement for Pre-Operational and Initial Start-Up Vibration Testing of Nuclear Power Plant Piping System, Part 3. ASME/ANSI-O&M.

Balla, C.B.N.S., Sinha, J.K., Rao, A.R. 2005. Importance of Proper Installation for Satisfactory Operation of Rotating Machines. *Advances in Vibration Engineering* 4(2), 137–142.

Ewins, D.J. 2000. *Modal Testing—Theory, Practice and Application*, 2nd ed. Research Studies Press, Hertfordshire, UK.

Hydraulic Institute. 1983. *Hydraulic Institute Standard for Centrifugal, Rotary and Reciprocating Pumps*, 14th ed.

IAEA. 1992. *Seismic Design and Qualification for Nuclear Power Plants Safety*, IAEA Safety Series No. 50-SG-D15. IAEA, Vienna.

Moorthy, R.I.K., Rao, A.R., Sinha, J.K., Kakodkar, A. 1996. Use of an Unconventional Technique for Seismic Qualification of Equipments. *Nuclear Engineering and Design* 165, 15–23.

Moorthy, R.I.K, Sinha, J.K. 1998. Dynamic Qualification of Complex Structural Components of Nuclear Power Plants. *Nuclear Engineering and Design* 180(2), 147–154.

Moorthy, R.I.K., Sinha, J.K., Rao, A.R., Kakodkar, A. 1995. Diagnostics of Direct CT-PT Contact of Coolant Channels of PHWRs. *Nuclear Engineering and Design* 155, 591–596.

Rao, A.R., Sinha, J.K., Moorthy, R.I.K. 1997. Vibration Problems in Vertical Pumps—Need for Integrated Approach in Design and Testing. *Shock and Vibration Digest* 29(2), 8–15.

Richardson, M.H. 1997. Is It a Mode Shape, or an Operating Deflection Shape? *Sound and Vibration Magazine*, 30th anniversary issue.

Schwarz, B.J., Richardson, M.H. 1999. Introduction to Operating Deflection Shapes. Presented at CSI Reliability Week, Orlando, FL, October.

Sinha, J.K. 2003. A Keynote Lecture: Significance of Vibration Diagnosis of Rotating Machines during Commissioning: Few Case Studies. Presented at National Symposium on Rotor Dynamics, IIT, Guwahati, India, December 15–17.

Sinha, J.K. 2006. Significance of Vibration Diagnosis of Rotating Machines during Installation and Commissioning: A Summary of Few Cases. *Noise and Vibration Worldwide* 37(5), 17–27.

Sinha, J.K. 2008. Vibration Based Diagnosis Techniques Used in Nuclear Power Plants: An Overview of Experiences. *Nuclear Engineering and Design* 238(9), 2439–2452.

Sinha, J.K., Balla, C.B.N.S. 2006. Vibration-Based Diagnosis for Ageing Management of Rotating Machinery: A Summary of Cases. *Insight* 48(8), 481–485.

Sinha, J.K., Hahn, H., Elbhbah, K., Tasker, G., Ullah, I. 2012. Vibration Investigation for Low Pressure Turbine Last Stage Blade Failure in Steam Turbines of a Power Plant. Presented at Proceedings of the ASME TURBO EXPO Conference, Copenhagen, Denmark, June 11–15.

Sinha, J.K., Moorthy, R.I.K. 1999. Added Mass of Submerged Perforated Tubes. *Nuclear Engineering and Design* 193, 23–31.

Sinha, J.K., Moorthy, R.I.K. 2000. Combined Experimental and Analytical Method for a Realistic Seismic Qualification of Equipment. *Nuclear Engineering and Design* 195, 331–338.

Sinha, J.K., Rao, A.R. 2006. Vibration Based Diagnosis of a Centrifugal Pump. *Structural Health Monitoring: An International Journal* 5(4), 325–332.

Sinha, J.K., Rao, A.R., Moorthy, R.I.K. 2003. Significance of Analytical Modelling for Interpretation of Experimental Modal Data: A Case Study. *Nuclear Engineering and Design* 220, 91–97.

Sinha, J.K., Rao, A.R., Sinha, R.K. 2006. Realistic Seismic Qualification Using the Updated Finite Element Model for In-Core Components of Reactors. *Nuclear Engineering and Design* 236(3), 232–237.

Sinha, J.K., Sinha, S.K., Moorthy, R.I.K. 1996. Diagnosis of the Bearing Failure in a Pilger Mill. *Shock and Vibration Digest* 28(2), 11–14.

Index

F

J

K

L

M

N

W

X

Y

Z

Printed in the United States
by Baker & Taylor Publisher Services